PLANNING AND CONT
THE BEST LAID PLAN

PLANNING AND CONTROLLING CONSTRUCTION PROJECTS: THE BEST LAID PLANS . . .

MICHAEL MAWDESLEY
WILLIAM ASKEW
University of Nottingham

MICHAEL O'REILLY
Kingston University

LONGMAN

The CHARTERED
INSTITUTE OF
BUILDING

© Addison Wesley Longman 1997

Co-published with The Chartered Institute of Building through
Englemere Services Limited
Englemere, Kings Ride, Ascot
Berkshire SL5 8BJ, England

Addison Wesley Longman Limited
Edinburgh Gate
Harlow
Essex CM20 2JE
England

and Associated Companies throughout the World.

Cover designed by Designers and Partners, Reading, Berks
Illustrations by Margaret Macknelly Design, Tadley, UK
Typeset by Dobbie Typesetting Limited, Tavistock, UK
Transferred to digital print on demand, 2002
Printed & Bound by Antony Rowe Ltd, Eastbourne

First printed 1996

ISBN 0-582-23409-3

British Library Cataloguing-in-Publication Data

A catalogue record for this book is available from the British Library

Library of Congress Cataloging-in-Publication Data is available

1006234093

Misce stultitiam consiliis brevem:
  Dulce est desipere in loco.

HORACE
(ODES)

Mix a little foolishness with your serious plans:
  it's lovely to be silly at the right moment.

CONTENTS

ABOUT THE AUTHORS

Mick Mawdesley BSc, PhD is a senior lecturer in construction management at the University of Nottingham. A Civil Engineering graduate, he obtained his PhD with a thesis entitled 'Resource Scheduling', and then spent some time in industry as a financial controller for a contracting company. During his academic career he has acted as a planning and control consultant for a range of companies including clients, designers, contractors, management consultants and claims consultants. His work in the UK, Europe, the Middle East, the Far East and South America has encompassed all stages from initial concept to final settlement, on projects including roads, pipelines, water supply and treatment works, and industrial, domestic and office buildings. He has specified and developed software for a number of small and large companies, and has taught planning and control to undergraduates, postgraduates and practising engineers.

Bill Askew MA, CEng, MICE, MASCE is a lecturer in construction management at the University of Nottingham. Prior to an academic career, he worked for a major construction company between 1972 and 1983 as an engineer involved in the management and planning of construction projects in the UK, as a designer of temporary works, and as a developer of computer software for planners and designers in head office. His experiences in project planning and control come from this background of site and office work, and from subsequent dealings with industry in research and consultancy projects as an academic. He has helped organise events concerned with construction automation and robotics, has presented planning research at conferences and has taught construction management on undergraduate and postgraduate degree courses. He has also delivered short courses for practising engineers, managers and planners.

Michael O'Reilly BEng, LLB, PhD, CEng, MICE, FCIArb, Barrister is Professor and Head of the School of Civil Engineering at Kingston University. He has worked for contractors and as a consulting engineer, and is a director of Thrustbore Contractors Limited, a specialist tunnelling company.

FOREWORD

Anyone blessed with the challenging disposition to take high-cost risks with the guaranteed promise of hard-earned returns could do no better than undertake a piece of construction, assuming firstly that five or six other well-meaning optimists have been beaten off by your attractive offer. All that is then required is to meet the client's wishes on time, for the price you first envisaged.

Clients (or their representatives) do not have an easy time either, putting forward high-risk proposals which may cost millions to bring to fruition. They run the considerable risk of not getting what they want and losing control of cost and time escalations on hugely complex operations.

The means by which a project is completed demand either hope, guesswork and good fortune, or regular planning and control. The former method will almost always lead to bad fortune. The latter is a step in the right direction, although the avoidance of failure cannot be assured due to the variance in the quality of the planning. The modern age has given us sophisticated computerised techniques as important aides to control our construction projects, but has simultaneously presented the opportunity for concealing unreliable plans when in the wrong hands. A little knowledge is a dangerous thing.

The authors of this book have a passion for ensuring that the users of planning techniques and the recipients of the results understand the assumptions and limitations inherent within the methods chosen. This is particularly important when relying on computers for appropriate solutions. The authors have all done their fair share of time in the construction workplace prior to meeting each other in an academic environment. Hence, practical advice based on theoretical knowledge is the strong theme throughout the course of the text. The range of advice is extensive, appreciating that both top management and junior employees have the need to plan and control their work. The section on short-term planning is not to be missed by anyone who aspires to controlling resources and finances on a day-to-day basis. This goes right to the heart of obtaining the best from the resources available.

Furthermore, it is stressed that planning and control does not just lie at the door of the people responsible for construction. Design and procurement establishments are called upon to ensure that they have adequate control over their time-scale and budgets for proposed schemes before introducing them to contractors or design-build consortia as tenders. In recent times, far too many schemes have failed to reach reality or if they have, proved to be too difficult to construct in the required time period. One has to ask how good the conceptual planning is. This book can certainly help to reduce the frequency and associated cost of failure.

For those seeking planning as a career discipline, or students wanting to test their knowledge practically, there are some very thorough worked examples, each lending themselves to different planning techniques, thus demonstrating the need to choose the most appropriate tool for the particular job at hand. The vital subject of how much detail to go into when selecting activities is carefully demonstrated by the examples. In my experience, plans can be frustratingly uninformative if the detail chosen is unfit for the purpose.

As is to be expected from supervisors of our industry's research, up-to-date topics such as risk management and uncertainty are given a full airing in order to help us make better decisions. The flavour given is worth a taste.

Over the years, I have shared and exchanged many views with the authors on planning and its pitfalls and am surprised when I look back at what I used to take for granted and now view with care or suspicion. There is a lot to learn and more to be aware of as being unreliable or misleading. This extremely thorough book of great practical value will enhance knowledge and help to protect practitioners from making the wrong choice. The perfect choice will probably remain unknown because even the best laid plans...can go astray, especially if they are not followed.

Colin Stevens, Chief Planner, Tarmac Construction

PREFACE

Project planning is not easy. It takes a long time and a lot of effort to produce a good plan. Project control is not easy. It is very difficult to control a project to a poor plan or to base control action on largely qualitative information. To help with project planning there are many techniques available. These need to be introduced and understood, their implications discussed, and project planning techniques need to be extended to tie in with project control.

With increasing reliance on computers in project planning, it is important that planners and managers are fully aware of the implications of what they are doing when they use software planning packages. Seldom is this the case and the outputs from planning software are often not understood because the methods and assumptions used in the software are not appreciated. This usually becomes apparent either when changes are made to plans and the computer output changes (perhaps not as expected), or when the 'same' plan yields different results when analysed by different software. With so much information now available from company and project databases for the management team on a project to use, there is great scope for improving the planning and control of projects. There is also great scope to generate excessively complex plans which conceal errors or do not model what the planner intended. Thus we get the familiar story of 'rubbish in – rubbish out', which is a shame.

Through our experiences in the management and planning of construction, we have produced a text which introduces techniques for planning *and* control, and which concentrates on recommending appropriate practice for the practical problems which readers may come across when using both computerised and non-computerised planning methods. We discuss the whole of the planning and control cycle as a function of management, and aim to address the interests of the many different levels of personnel and types of organisation involved in construction.

This book is intended for anyone interested in *real* project planning, from inception to completion. It is for professionals in all sectors of the construction industry and in organisations involved in some way with construction projects. Clients, consultant engineers, contracting organisations, architects, subcontractors and suppliers should all benefit from the practical instruction and advice on the application of the techniques available.

While the book is directed towards construction professionals who already have some knowledge of planning and control, the techniques that readers find new to them are hopefully covered in enough detail to give understanding, particularly concerning when and where a technique is appropriate. Cautionary notes based on experience and research will aid those seeking to understand and interpret the plans and planning methods they use. This book is appropriate reading, too, for students and junior managers who wish to know more than just what planning techniques are available and how to use them for classroom exercises.

This book has been written to be general so that it can be of use to people from many different backgrounds. No particular contract system is addressed by the book, although occasional reference is made to different payment systems when examples are being used. The bill of quantities forms the basis for payment on many contracts and is used occasionally in this book to illustrate points being made in the text. As the bill of quantities is not used universally in industry, the general discussion acknowledges this.

This book examines the often neglected area of activities, resources and finances; brings together techniques for planning and control; shows how to approach uncertainty and risk; handles short- and long-term planning; analyses the management and organisation of planning and control; discusses the choice of technique and evaluation of plans; and presents ways of using plans in the evaluation of entitlements. Many simple examples are used throughout to illustrate the points being made. Some of the points will be surprising to many who read them, and may change attitudes to planning.

It should be remembered that planning is not an end in itself – it should be an aid to completing a project. To demonstrate how different techniques can interact on a project, two long examples (for a civil engineering project and a building refurbishment project) are included. They are set in a typical town, Tingham, which could be anywhere – except of course *Not*tingham, England, the city where this book was conceived.

ACKNOWLEDGEMENTS

We would like to thank all those people in academia and industry who have willingly or unwittingly shared with us their own knowledge and experiences of construction and have given encouragement to write this book. These people include students, researchers and staff at the

University of Nottingham and elsewhere; and practising engineers, planners and managers working in the construction industry. Numerous staff at Tarmac Construction are particularly to be thanked, including Nick Ellis who originally produced the Tingham Bypass example.

Mick Mawdesley, Bill Askew, Michael O'Reilly
August 1996

PUBLISHER'S ACKNOWLEDGEMENTS

Translations of quotations given in the book epigraph and in the epigraphs to chapters are taken from *The Concise Oxford Dictionary of Quotations* (2nd edition, 1981) and are reproduced by kind permission of Oxford University Press.

We are grateful also to J M Dent & Sons for permission to reproduce the epigraph taken from Jerome on page 68; and to A P Watt Ltd for permission to reproduce the lines from Kipling on page 214.

xxv

LIST OF ABBREVIATIONS

aod	above ordnance datum (a level in metres)
CPA	critical path analysis
CPD	critical path duration (in network analysis)
DUR	duration (of activity or project)
EET	earliest event time (usually in activity-on-arrow networks)
EFT	earliest finish time (of an activity)
EGL	existing ground level
EST	earliest start time (of an activity)
FF	finish to finish (connection in precedence networks)
FS	finish to start (connection in activity-on-node or precedence networks)
GERT	graphical evaluation and review technique
HO	head office
HVAC	heating, ventilation and air-conditioning
LET	latest event time (usually in activity-on-arrow networks)
LFT	latest finish time (of an activity)
LST	latest start time (of an activity)
M&E	mechanical and electrical (services)
NPV	net present value (in financial evaluation)
PERT	program evaluation and review technique
SF	start to finish (connection in precedence networks)
SS	start to start (connection in precedence networks)
TF	total float (of an activity)

1

INTRODUCTION

'The best laid schemes o' mice an' men
Gang aft a-gley.'

Robert Burns
(To a Mouse)

This famous line states that the best plans often go wrong. But why? And how often? Has the planning process taken into consideration all that it should? Is the reality too complex to model, however painstaking the approach? How much detail is required in plans anyway? Does the use of 'often' suggest that things go wrong 'too often' despite our best endeavours? Hopefully this book will reduce the frequency with which plans go wrong, and will help planners understand why and how they go wrong when they do.

1.1 MANAGEMENT

Management is a broad subject which is practised, often without training, by almost everyone. This book looks at an area of management, the planning and control of construction, with the broad aim of providing an insight into and analysis of some of the techniques and practices encountered. As such the book presents discussions which are primarily intended for a readership of practising planners (or other interested professionals and managers), while presenting techniques in a way that students or others unfamiliar with the material can understand also. Before explaining the specific objectives and layout of this book, the broad area of management is introduced to emphasise the background against which the book is set.

There are many functions of management and many people have attempted to define, refine and extend them. Essentially, the main functions of all managers can be summarised as:

- planning
- organising
- staffing
- leading
- controlling.

This list is good irrespective of the level at which the manager operates although, at different levels in a company, managers have a different balance of these functions. At low levels of management, there is more 'leading' and less emphasis on other functions, whereas at high levels of management there is less 'leading' and more emphasis on the other aspects. To perform these functions well implies the need for good communication. Indeed most criticisms of managers usually, when analysed, relate to poor communication. Chapter 2 is dedicated to this important topic, and discusses the major 'hard copy' methods for communicating information in planning and control.

From the list above, planning and control are significant functions of management at all levels of an organisation. Their success is dependent, to an extent, on the techniques used and also on the company's organisational structure which must allow these functions to be carried out properly with good information and to be integrated with the other functions of management. Another key function of management can now be identified, namely the monitoring of work in progress in order to make necessary information available. Later chapters cover the topics of managing the planning process, and monitoring and controlling construction.

The following sections of this chapter introduce the management context of planning, monitoring and controlling construction. Against this background, the aims of this book and its layout are then discussed.

1.2 PLANNING, MONITORING AND CONTROL

Before looking in detail at any aspect of planning, or indeed at the contents of this book, it is worth defining what 'planning' actually is, since the term is used widely throughout the construction industry and means different things to different people.

A good dictionary (here *Collins English Dictionary*) might offer two definitions of a plan:

> *'a detailed scheme, method, etc., for attaining an objective'*
> *'a proposed, usually tentative idea for doing something'*.

Planning is thus not an end in itself but an aid to the completion of a project.

Interestingly, a related term is a programme, which is more specific in definition:

'a diagram or list showing work to be done, with associated times'.

By this definition a plan is a detailed scheme or method for attaining an objective (which in this case is the completion of a project), and this 'detailed scheme' includes the method, the activities to be carried out, the timing of the activities, the resources to be used and the finances required.

Planning is thus a general term which is used to encompass the ideas which are commonly referred to as programming, scheduling and organising. Its aim can be defined to be:

Making sure that all the work required to complete a project gets done:
- *in the correct order*
- *in the right place*
- *at the right time*
- *by the right people and equipment*
- *to the right quality*
- *in the most economical, safe and environmentally acceptable manner.*

There is thus much that this book should cover in order to address satisfactorily the topic of planning and controlling construction work.

This chapter continues with an introduction to the *who, what, where, when* and *why* of planning in its broadest sense (as introduced above). The *how* is left to the later chapters which introduce and discuss the techniques available.

1.2.1 The who, what, where, when and why of planning

When considering the questions:

- Who should plan?
- What should they plan?
- When should they plan?

the correct answers are very easy, namely:

- Everybody should plan.
- They should plan whatever work they are either responsible for or actually carry out themselves.
- They should plan continuously.

Everybody involved in a project, from the chairperson of the client organisation to the labourer of the subcontractor, should plan (and hopefully does). Problems arise if they do not plan or if they spend their time planning the wrong things. The following are some of the major parties to construction and personnel involved, together with suggested planning considerations for them.

CLIENT

The client should plan a construction project because the construction would usually be part of a larger project or the rest of the client's work. It must therefore be integrated with these other aspects. The expenditure on the construction will be recouped from the completed project either directly or indirectly; the client must be aware of the expenditure pattern and the required dates for the integration of the construction within the overall project.

PRINCIPAL CONSULTANT (ARCHITECT OR CONSULTANT ENGINEER)

The principal consultant needs to plan for three distinct reasons. First, a financial and physical plan of a project is required before the construction work starts in order to brief the client on the potential commitments. Second, a plan is required in order to produce the drawings and other necessary information at times which would not adversely affect the running of the project. This may be based on the construction plan produced by another organisation but would include the timing and resource demand for the consultant's own work. Finally, the consultant often requires a plan in order to ensure that the demands being placed on the contractor at tender time are reasonable and would not result in excessive costs being incurred by the client.

CONTRACTOR

It is the contractor whom most people associate with project planning and indeed it is usual for the contractor to carry out the majority of planning for the construction phase of a project. The contractor needs a programme of work for many reasons including being able to:

- report to the client,
- monitor progress,
- control the work,
- muster the required resources ahead of their use,
- ensure that the resources are utilised effectively,
- ensure that information is available on time,
- ensure that the cash flow is satisfactory,
- ensure that the work is performed as economically as possible,
- ensure a safe system of work,
- ensure environmental matters are properly considered.

In this list the term 'resource' is used as the general term for all types of labour, plant, materials, subcontractors and overheads (such as engineers).

With so many actual and potential uses, it is not surprising that many programmes are produced for a single project within a contractor's organisation. While on smaller projects programmes may be combined, reflecting people's combined roles, on a large project the following people would have different programmes:

Estimator The estimator requires an outline programme of work in order to determine the cost of the work properly. This may be checked and changed by senior management at the finalisation meeting before tender.

Project manager The project manager needs to have programmes of work in order to control the finances of the project, to make sure the correct resources will be available and to report to the client, engineer or architect.

Agent The project agent, here defined as the person directly responsible for the construction of the works or any major section of them, needs to have programmes of work in order to make sure the correct resources are available and to ensure that they are used efficiently on the project.

Site engineer The engineer needs to have programmes of work in order to ensure that work such as setting out and checking is done at the correct time. The programme is also used to make sure that the resources are used efficiently on the project.

Ganger The ganger, or other person in charge of a team of operatives, needs to have programmes of work in order to make sure that the resources are used efficiently on the work. The ganger must make sure that all the resources are available to carry out the task in hand and that they work together to carry out the task.

Operative Although it may be thought that individual operatives do not need a programme, it is important that they think ahead (plan) in order to ensure that the system of work is safe. If this is not done, problems arise and control becomes re-active.

Subcontractor Subcontractors have similar requirements to the main contractor as far as project planning is concerned.

For all people on a project, programmes are needed to make sure the work is done safely, to the required quality and with due consideration for the environment.

All these plans and programmes are not produced just once. Planning is a continuous process which carries on throughout the project as described in the following section.

1.2.2 Planning and control cycle

The management of planning is an important part of the overall management process. To manage planning it must be recognised that the planning process incorporates not only the planning function but also the functions of monitoring and control. The whole process can be represented in the planning and control cycle shown in Figure 1.1. From the cyclic nature of the process it can be seen that planning is not something that happens once in the life of a project and is then forgotten until the next similar period. A discussion on the elements of the figure now follows.

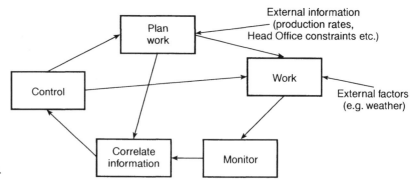

Figure 1.1 The planning and control cycle.

PLAN WORK

It is impossible to do any work, and hence achieve any progress, without work planning of some kind, even if this is just in the very broad sense of deciding what is to be done and how it is to be performed. It should be obvious that the better the work planning the fewer the problems that can be expected to be encountered in the execution of the work. This self-evident truth is often obscured by the skill of the site operatives who are able to perform most construction tasks, given time. Indeed they often ignore the proposed method of working in order to 'get the job done'. This should not give the impression that work planning is useless; rather, it should indicate that planning and control are essential to ensure that work progresses in a timely, economical and orderly way.

It is very important to realise that planning means not just the physical planning of the work to be done. It also includes all financial aspects. In particular, a plan should provide information on the amount of money flowing into and out of the project and when this happens. Since money is generally earned before it is received and committed before it is spent (and a project will generally incur real or notional cost for being in debt and accrue benefits for being in surplus), there is a need for detailed consideration of the timing of the work. From this it is a simple matter to calculate the profit and loss over time.

WORK

Following the work planning, the work can be carried out. The work should be done in a way which achieves the objectives set in the planning phase. Various objectives may have been set and the detail will depend on the level at which the planning and control is being exercised.

MONITOR

The third phase of the process is the collection of information regarding the doing of the work. The monitoring should provide both physical and financial information, which should be in a form suitable for comparing with the plan. Only when this has been done can the fourth phase, correlation of information, be carried out.

CORRELATE INFORMATION

The information here is the progress information (collected in the monitoring phase) which is to be correlated with the planning information (from the planning phase). In this phase the achievement (from the work) is compared with the targets (from the plan). Several techniques are available to help with this and most texts on 'control' concentrate almost exclusively on the phases up to and including the correlation phase. Control, however, cannot exist without some action and this is the purpose of the next phase.

CONTROL

Control action should be based on the results of the correlation of information, and may be of the traditional *re-active* type in which action is taken to affect the work output based on recent information and past experience. Increasingly, however, people are realising that the time-scale of construction activities often renders this type of control of little use, and more emphasis is being placed on *pro-active* control. Pro-active control attempts to change the plans to achieve the long-term targets or, if necessary, to use the plans to prove that the long-term targets are unrealistic and thus should be changed.

The basic planning and control cycle of Figure 1.1 can be combined with the concept of hierarchical planning introduced in Section 1.2.1 to give a hierarchical planning and control system, as shown in Figure 1.2. This is discussed further in Chapter 11.

1.2.3 Types of plan

The previous section dealt with planning in its general sense, for all types of personnel in all types of organisation on all types of project and at all times. Several specific types of plan can be identified in construction. These can be classified at several levels within a management hierarchy, for example:

- corporate strategic plans
- pre-tender plans
- master plans or pre-contract plans
- short-term construction plans.

These classifications serve as a reminder that construction is based around contracts which often impose specific requirements on programmes for construction.

Pre-tender plans usually aim to identify, for the benefit of costing and/or as an element of a bid the:

- methods of working
- organisational structure for the site
- subcontracting arrangements.

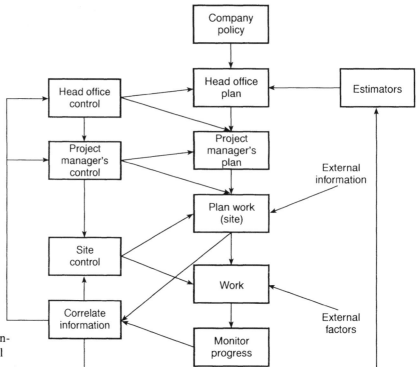

Figure 1.2 Hierarchical planning and control system.

Preconstruction plans consider in greater detail the:

- breakdown of major work items
- timing of the work
- sequencing and phasing of the work
- safety, quality and environmental aspects
- risks and opportunities
- resource requirements
- dates
- information-required-by dates
- selection of the subcontractors and suppliers
- work schedules.

Short-term construction plans consider in even greater detail:

- the utilisation of the available resources
- mustering the resources required
- doing the work safely, economically and to the right quality.

Each type of plan has different objectives, as indicated above, but the overall aim of planning remains as stated at the start of Section 1.2.

The types of plan given above would be different if produced by different organisations because their particular objectives would be different. It is evident in summary that the nature of a plan will depend on the:

- type of organisation (for example client or contractor)
- level in the organisation (for example managing director or project manager)
- time of production (for example tender stage or post-contract stage)
- objectives.

1.2.4 Benefits of planning

If undertaken properly, planning is a time-consuming and expensive process. It requires time from experienced personnel who could be employed on other tasks and increasingly it requires the use of computers. The justification for incurring this expenditure lies in the benefits which can accrue.

The following list of possible benefits is by no means exhaustive but indicates the areas in which advantages can be gained from good planning. These benefits were produced by a senior planning engineer with a large international contractor and might therefore be considered biased towards that branch of the construction industry.

- Planning predicts the timing of activities and their sequence and hence the total construction period.
- Planning enables the safety, quality and environmental impact of the work to be properly considered.
- Planning allows the risks and opportunities to be evaluated.
- Cost and therefore price are reflected by the working methods and the time taken, thus providing a basis for the estimate.
- Planning provides a basis for monitoring.
- Planning predicts flow of cash, i.e. cost against return.
- Planning provides a basis for claims evaluation, in particular extension of time entitlement calculations.
- Planning pinpoints when materials are required, thus optimising storage space and minimising breakage and loss.
- Using histograms for major bulk materials, such as concrete and quarry products, provides suppliers with average and peak levels of demand.
- Planning predicts resource levels of labour, staff and plant.

Additional benefits to the clients (or their consultants) are:

- Planning provides deadlines for latest dates for release of information.
- Planning yields the client's expenditure forecast, thereby predicting the flow of payments to the contractor through a monthly valuation process.
- Planning predicts the staffing levels required for adequate super-vision.
- Planning assists in giving an information service to the public and in the organisation of site visits for outside parties.
- Planning enables the coordination of other schemes in adjoining areas, for example in highway schemes it allows the integration of major traffic diversions and road closures.

Planning will only be able to provide these benefits if, of course, the quality of the planning is high. This can be achieved through a planning cycle where plans are evaluated before being used and where plans are monitored in use. Feedback from these two processes (evaluation and monitoring) can be used to provide information for replanning or future planning.

1.2.5 Objectives of planning

Having set out the broad aim of planning and listed the benefits of producing a range of plans, it is clearly difficult to declare the detailed objectives of all forms of plan. The detailed objectives are clearly related to the specific function of the plan for the particular organisation producing it. A useful clarification of the objectives of planning can, however, be given in relation to the time-scale covered by a plan. The main objectives of long-, medium- and short-term planning are as follows (using the example of a project of at least one year's duration).

Long-term plans (produced 3 monthly for the remainder of the project)

To ensure that the project will be carried out within the specified constraints and in the most cost-effective manner, with due consideration to safety and the environment.

This requires that the timing of the work and the resources to carry it out are considered. There will usually be considerable uncertainty in long-term plans for construction projects because of the nature of the industry.

Medium-term plans (produced monthly for next 3 months)

To ensure that the targets set in the long-term plans are going to be achieved.

These provide the basis on which the required resources can be mustered, ready for work. There is less uncertainty in these plans than in the long-term plans but it is still considerable.

Short-term plans (produced weekly for next two weeks)

To ensure that the resources provided by consideration of the medium-term plans are utilised in the most efficient manner to achieve the project objectives stated in the higher level plans. To ensure that work proceeds with due regard to safety and quality.

There is little uncertainty in these plans.

1.3 LAYOUT OF THE BOOK

The preceding sections have introduced project planning in the broad context of management and the particular context of the planning and control cycle. The general aims of planning have been identified and the benefits of good planning have been explained. The difficulties in defining the specific objectives of a particular plan have been introduced by examining some of the many types of plan that are prepared by people at different levels in different types of organisation.

This book addresses the subject of planning construction work in the context set out above. It covers in some detail the use of some of the more common planning techniques used to produce project models, and the models produced are analysed and potential problems identified. Practical tips for planners are included and summarised at the end of each chapter. Two major case studies, one for civil engineering and one for building, are used to illustrate the application of various techniques. The management of the planning process and techniques for monitoring and control are also addressed, to complete a practical book on project planning.

It is hoped that this book will be useful to all parties involved in construction work, not just contractors, sub-contractors or builders. Although many of the examples relate to planning the construction phase, the roles of the major parties are discussed and some examples appropriate to clients and their representatives are included. Students should be able to see some of the benefits and, importantly, the limitations and possible consequences of using the techniques that they learn about on real projects.

1.3.1   Outline of contents

The importance of effective communication in management has already been mentioned. Chapter 2 uses communication as the theme for presenting the outputs from some of the planning techniques commonly encountered in construction. The advantages and disadvantages of the techniques behind these forms of output are discussed and some explanation of, and guidance in, the use of the techniques is included.

Important in any plan are the resources to be deployed in the planned work and the financial implications of their deployment. Chapter 3 introduces the various types of resources and finances that might be encountered in construction and explains how they can be classified and how they should be considered in planning.

Project plans are usually built up from components referred to as 'activities', which are elements of work or operations required to be done to complete a project. An activity is almost certain to require resources and finances in order to carry it out. Getting the selection of activities right is thus very important to the success of the plan, but is rarely given enough consideration. Chapter 4 introduces the formulation of activity lists and discusses what planners need to consider in order to get this selection stage right.

A plan has to go further than dividing the project into under-standable activities. It has to re-combine the activities into a model of the project, as the planner sees it, if it is to meet the objectives of the project.

Many problems can occur when sequencing or linking activities in a plan. When the logic links remain in the planner's head and are not shown explicitly in the plan, the plan can be misinterpreted and important links not appreciated. When logic links are shown explicitly, such as on a network, the correct representation of a link can be difficult for the planner to convey (and even at times understand), and the correct interpretation of the link can be difficult for the plan user to infer. Chapter 5 examines the connections between activities in some detail and demonstrates, through examples with networks, that correct representation and interpretation can be very difficult.

Further complications in planning, such as working with more than one calendar for different activities and the use of fixed dates, are covered in Chapter 6.

The subject of planning with resources and finances is considered in Chapter 7. Various techniques for aggregating, smoothing, levelling and scheduling resources are covered.

Having by this stage introduced and discussed numerous techniques for planning projects it would be wrong to assume that plans will now be correct in all that they represent. There is considerable uncertainty in construction which makes it difficult to assign durations to activities and to be sure about the sequencing of the work. Techniques to help deal with various forms of uncertainty are introduced and discussed in Chapter 8.

Making decisions in an uncertain world raises the topic of risk management, which is covered in Chapter 9.

The particular topic of short-term planning also rarely gets enough consideration and so useful techniques for this are examined in Chapter 10.

Coverage of planning then concludes with a look at the management of planning and its organisation at all levels. This is covered in Chapter 11 with reference to the planning and control cycle (Section 1.2.2).

More detailed examination of the related functions of monitoring and control follows in Chapter 12, including an introduction to some useful techniques which can be applied to short- and long-term plans.

Having examined a range of techniques for the planning and control of construction, Chapter 13 discusses which technique to choose in particular situations and offers ways of evaluating plans that have been produced.

Plans are often referred to in contract documents such as the legal contract between client and builder. Contractual matters associated with planning are discussed in a general sense in Chapter 14, together with the use of planning techniques in establishing contractual entitlements, or in claims' evaluations.

All the important areas of construction planning have now been discussed and the book concludes with two case studies in Chapters 15 and 16. These are based around a civil engineering project (a new tank farm) and a building project (a hotel refurbishment). Appendix A introduces the background information for some other simple projects which are used as examples elsewhere in the text.

1.4 SUMMARY

- Management is a broad subject which is practised by almost everyone.

- The main functions of managers are planning, organising, staffing, leading and controlling.

- Good communication is essential for good management.

- The success of planning and control is dependent on the techniques used and on the company's organisational structure.

- The objective of planning is to ensure that all the work required to complete a project gets done in the correct order, at the right time, by the right people and equipment, to the right quality, in the most economical, safe and environmentally acceptable manner.

- Everybody should continuously plan the work they are responsible for or which they actually carry out.

- Different programmes for the same project are likely to exist for the client, consultant and contractor.

- Different programmes for the same project are likely to exist for different people within an organisation, such as the contractor's estimator, project manager, agent, site engineer, ganger and operative.

- The planning and control cycle of management includes planning the work, doing the work, monitoring the work, correlating plan and progress information, and controlling the work.

- Planning the work should cover financial planning as well as just scheduling the physical work.

- Project monitoring should provide information in a form suitable for comparing with the plan for the work.

- Project control may be re-active or pro-active.

- Plans exist in an organisation's management hierarchy, such as corporate strategic plans, pretender plans, (pre)contract plans and short-term construction plans.

- Good planning takes time and is expensive.

- There are many benefits of good planning, including cost savings when the work is carried out.

- The main objectives of long-, medium- and short-term planning are different but related.

2

PRESENTATION OF PLANNING INFORMATION

‘ “What is the use of a book,” thought Alice, “without pictures or conversations?” ’

LEWIS CARROLL
(Alice's Adventures In Wonderland)

For 'book' read 'plan'. The most common criticism of management at all levels is its apparent inability to communicate with the people who matter, be they more senior, subordinate or simply performing a different function within the organisation. The format of communication is important for a clear understanding of the information conveyed. 'Pictures' are particularly useful in planning and 'conversations' can help explain what plans mean and how to use them.

2.1 IMPORTANCE OF COMMUNICATION IN PLANNING

Construction planning is achieved using a variety of techniques. The techniques may vary considerably, encompassing both the specific, mathematical techniques available for particular tasks or types of work, and the less well defined, intuitive approaches based on experience (and used by so many). It is important to appreciate that no matter how realistic, sophisticated or appropriate the various techniques are, they have little practical value unless the results can be communicated to those interested in them, such as the people required to implement or evaluate the plan. Different planning techniques lend themselves to different methods for representing results. Further, in different circumstances a given plan might best be communicated through a different medium. In some situations, such as a discussion between planners in the office, a plan might best be presented using a medium which enabled all the internal thinking, or logic, to be expressed. If the same plan were to be shown to site foremen to explain how a section of work interfaces (in time and space) with adjacent sections, a much simpler medium would probably be better.

The best medium for communication depends on:

- the type of project
- the type of plan required
- which model (mathematical or otherwise) has been used to produce the plan
- the type of person to whom the plan is to be shown
- the purpose of communicating the plan.

Planners have developed various media for representing (or visualising) the results of their analyses. Amongst the most common are:

- bar charts
- space–time diagrams
- target-completion-time diagrams
- networks
- resource profiles
- financial graphs
- reports
- method statements.

These and other forms of representation, or output, from the planning process are discussed in this chapter.

Discussion in the following sections centres around the strengths and weaknesses of the various forms of output when used in real situations. For some of the lesser known planning methods the underlying techniques are briefly outlined. For networks, resources and finances, detailed discussion is deferred until important issues on modelling projects are discussed in subsequent chapters. Some knowledge of what is meant by the terms *resources* and *activities* is expected for this chapter, but they too receive more detailed coverage in later chapters. There are many texts which describe the main techniques used to produce some of the outputs shown in this chapter. In most cases such texts use simple examples which conceal some of the problems with using the techniques on real projects.

In discussing the forms of output generated from the planning process it should be appreciated that in some cases the method of representation may itself be the planning tool or working medium, which in its final form becomes the representation of the plan. The prime example of this is the bar chart where, in many cases, the planner sits down with a squared sheet of paper and emerges some time later with a bar chart and a pile of waste paper containing earlier attempts. There would be few 'formal' calculations involved but much thinking, jotting and sketching.

2.2 BAR CHARTS

2.2.1 Gantt chart

The Gantt chart is by far the most common form of bar chart and is indeed what most people refer to as a bar chart. Developed by Henry Gantt (1861–1919) the bar chart is still the best known method for representing plans. A bar chart shows the activities of a project and how they are scheduled in time, by using lines or bars proportional in length to the scheduled duration for the activities. The activity names are often grouped as a block down the left (and sometimes right) of the chart, or written above the activity bar. A time scale extends across the chart from left to right. Figure 2.1 shows a typical bar chart for the Tingham Bridge project (Appendix A).

Bar charts are popular because they do not get excessively complicated as the number of activities and number of time periods increase, and they can be drawn for various levels of detail e.g. long-term, monthly, weekly, daily or hourly time-scales.

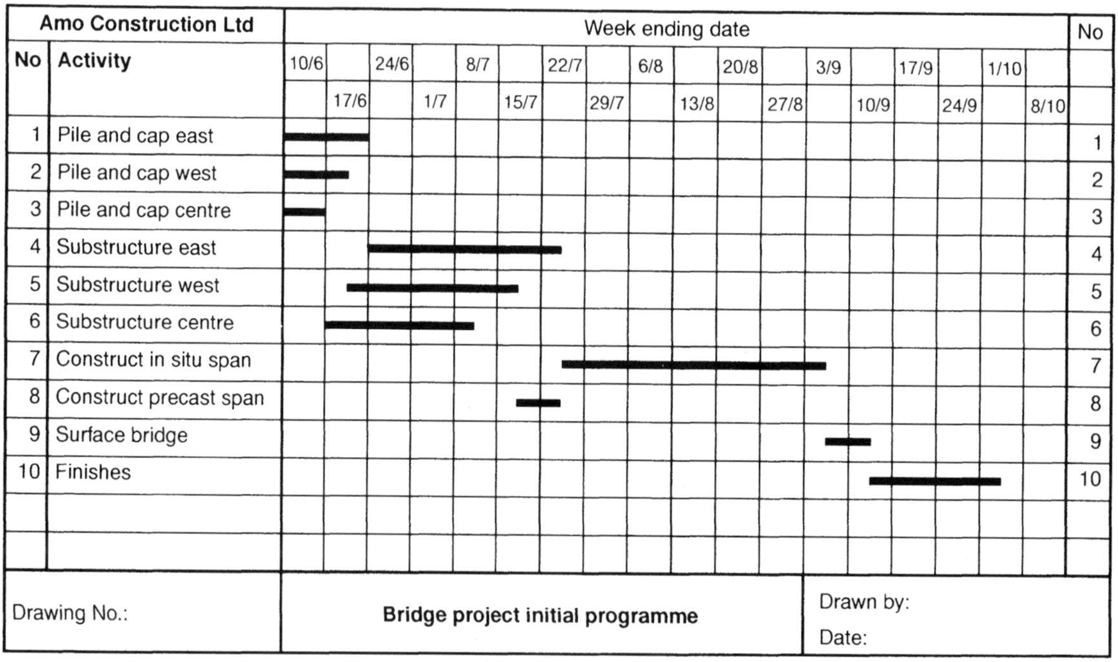

Amo Construction Ltd		Week ending date												No
No	Activity	10/6	24/6	8/7	22/7	6/8	20/8	3/9	17/9	1/10				No
			17/6	1/7	15/7	29/7	13/8	27/8	10/9	24/9	8/10			
1	Pile and cap east	▬												1
2	Pile and cap west	▬												2
3	Pile and cap centre	▬												3
4	Substructure east		▬▬▬											4
5	Substructure west		▬▬▬											5
6	Substructure centre		▬▬▬											6
7	Construct in situ span				▬▬▬▬▬									7
8	Construct precast span			▬										8
9	Surface bridge								▬					9
10	Finishes								▬▬▬					10
Drawing No.:		Bridge project initial programme						Drawn by: Date:						

Figure 2.1 A simple bar chart or Gantt chart.

Bar charts are also useful for monitoring progress but, if monitoring is extended to record keeping, careful thought is needed as to what is actually recorded and how. Project monitoring is covered in more detail in Chapter 12. For now it is worth considering what the bars on a bar chart mean.

Essentially a bar on a bar chart means the activity is going on within the time period indicated by the ends of the bar. The following assumptions might also be made (perhaps without realising so) either in the production or subsequent interpretation of the bar chart:

- the rate of progress is constant throughout the length of the bar;
- the resource use is constant throughout the length of the bar;
- the start times shown are the times at which the activities *will* start rather than when they *can* start.

Rarely are these planning assumptions stated, so it is interesting to consider how bar charts are, or may be, interpreted when used for other purposes such as monitoring, record keeping or deriving plans for the shorter term. Later chapters help to appreciate these issues.

Because of possible complications and misinterpretations, bar charts are particularly appropriate for simple projects or for *outlining* complex projects. If this is seen as a limitation it should be remembered that the limitation is usually far outweighed by the fact that bar charts are excellent as a medium for communicating important planning information; and in fact their use is far more widespread than just simple or outline programmes. The reason why bar charts can communicate particularly well is *because of* their simplicity and adaptability.

Bar charts are also versatile in that they can be produced to show the work for a single resource, a single gang, a particular subcontractor or a specific area of a project. For this application they can be drawn showing only the relevant information. Often a Gantt (or activity) bar chart (Figure 2.2) is used but a useful alternative for short-term planning is the resource bar chart (Section 2.2.4). If a Gantt chart is used for a single work type undertaken by several gangs, the bars can be linked to show resource movement, although care must be taken not to reduce the clarity of the presentation (Sections 2.2.2 and 2.2.3).

A bar on a bar chart is identified by its name, usually an activity name. For some types of work (for example, work on roads) a useful chart can be achieved only if large activities are broken down into smaller ones (for example, drainage 0–100 m and drainage 100–200 m) or if small activities are combined into larger ones. This can affect the spatial feel for a project. On linear projects in general, such as pipelines or highway projects, where distance is a particularly important parameter, an alternative to a bar chart is to plot time against distance (chainage) rather than time against activity name. The result is a useful tool in planning (Section 2.3).

A consequence of the simplicity of bar charts is that it is often difficult to see immediately the effect of changes to the project (for example, variations or change orders) because the information used to

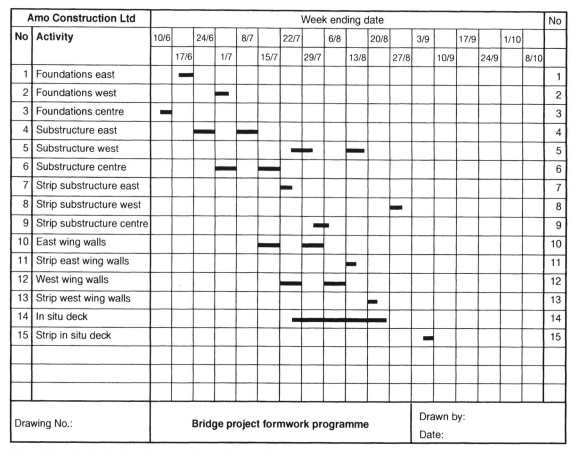

Figure 2.2   A Gantt chart for a single work type (formwork).

derive the plan is not shown on the bar chart. Evaluating or re-evaluating the project duration can thus be a problem unless more planning information is available, such as activity interdependencies and resource requirements. This brings us back to the importance of communication, not just of the final output but also of how it was reached.

2.2.2   Linked bar chart

A problem of simple bar charts mentioned in Section 2.2.1 was that they do not show activity interdependencies or logic sequences. In a linked bar chart for the Tingham Bridge project (Figure 2.3), arrows are used to provide links showing some (usually the important) interdependencies of activities. Whilst this appears to be sensible and helpful it can soon become confusing. As the size and complexity of projects increase it becomes difficult to avoid lines intersecting, overlapping, crossing bars or being lost against the vertical and horizontal background grid.

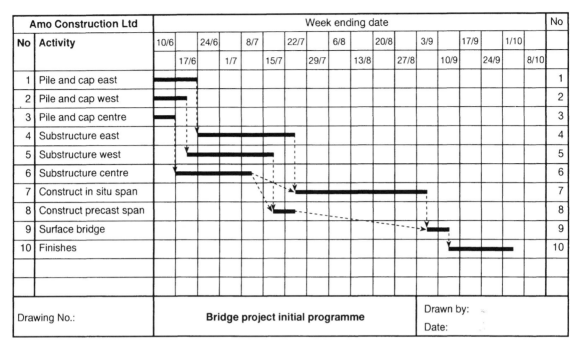

Amo Construction Ltd		Week ending date																	No
No	Activity	10/6		24/6		8/7		22/7		6/8		20/8		3/9		17/9		1/10		
			17/6		1/7		15/7		29/7		13/8		27/8		10/9		24/9		8/10	
1	Pile and cap east																			1
2	Pile and cap west																			2
3	Pile and cap centre																			3
4	Substructure east																			4
5	Substructure west																			5
6	Substructure centre																			6
7	Construct in situ span																			7
8	Construct precast span																			8
9	Surface bridge																			9
10	Finishes																			10
Drawing No.:		**Bridge project initial programme**										Drawn by:							
												Date:							

Figure 2.3 A linked bar chart.

The arrows sometimes show construction logic but are also often used to represent resource connections to show the proposed progress of key resources through the project. This is good only until something changes, which it invariably does, and then these links have the same problems as resource constraints in networks (Section 5.8). Without explanation of what they mean, the logic conveyed by the arrows on a linked bar chart can be misinterpreted.

Sometimes the linking arrows are drawn from or to points other than the ends of activity bars. As discussed above (Section 2.2.1), when trying to understand exactly what a bar means (in terms of the work required to be done in the activity), it is difficult with linked bar charts to know what exactly must be done to comply with the arrows representing the links. It can be helpful to draw a bar chart by listing the activities in the order in which they start. This leads to a cascade effect on looking down the bar chart and the result is often called a *cascade chart*. This representation can make the chart easier to read and understand, so that links between activities become less necessary, enabling some to be left out. It can also be argued that arranging the activities in this way can destroy other, perhaps more sensible, groupings of activities.

In practice, the use of arrows or cascades to represent links on a bar chart should be seen as a balancing act between conveying extra planning information and losing the clarity of the bar chart as a medium for communication. In reality the linked bar chart is most useful on simple projects or for outline programmes where the scope for confusion is less.

2.2.3   Other information on bar charts

Bar charts can show information other than activity names, durations and connections. Examples are 'information required by' dates, 'access required by' dates, 'hand over' dates or just general notes written on the chart.

Figure 2.4 shows some of these features on a bar chart for the Tingham Bridge project. In particular the following features are illustrated:

- An imposed date on the start of Activity 8, 'Construct precast span', is brought about by the unavailability of the precast units until that date.
- The activities have been moved from their earliest possible times to the times at which they are programmed to be done (as determined in Section 2.6).
- The total float (Section 5.2) for the activities is shown, with the float (or slack) before the activity represented in a different way to the float after the activity.
- A note describes the piling requirements.
- The required project completion date is given.

There are no standard symbols for these features although companies might find internal standardisation useful to increase the usability of the charts for communication purposes.

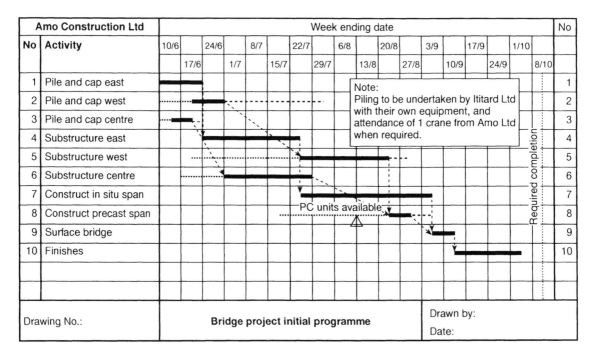

Figure 2.4   An annotated bar chart.

As can be seen, the extra information which can be included on a bar chart does not always improve the legibility and hence usefulness. Even in this very simple project, confusion can arise (Figure 2.4) because:

- The links between the activities, drawn here between the programmed end and start of activities, could appear to be between arbitrary points of activities crossed by the links. For example, the line linking the starts of Activities 3 and 6 could be misinterpreted as a link from somewhere near the start of Activity 4 to the start of Activity 6.
- The links are sometimes coincident and often cross each other.
- The quantity of information on the chart is becoming excessive.
- A key becomes necessary to understand the symbols.

As the bar chart gets bigger in both time and number of activities, the ability of the user to follow connections, either explicitly included or implied, reduces to a point where only the timing of the activities can be obtained from the chart. A colour scheme might help clarify matters but this can be a problem when multiple copies of the bar chart are required.

Another common variation of a standard Gantt chart is illustrated in Figure 2.2 where a single activity is shown as two bars (for example, Activity 4 'Substructure east'). This type of multiple-bar activity could be a simplified representation of what might otherwise be several activities. Without further information it may be impossible to determine what the two bars mean in terms of the work included in each, so the apparent simplification in fact becomes a complication.

A further variation, which is sometimes used when the positioning (in time) of the activities is not critical, is to join multiple bars with dotted lines or indeed to draw the whole of the activity as a dotted line. Since this does not show either the duration or timing of the activity but only the allowable time window, this representation should be avoided for planning purposes.

2.2.4   Resource bar chart

A different form of bar chart which can be useful in certain situations is the resource bar chart. An example of this is given in Figure 2.5 for the Tingham Bridge project. Basically the resource bar chart lists each resource under suitable groupings and indicates the activities that each is to do throughout the period of the plan. The time-scale shown is very short – days rather than weeks, months or years – as it is very difficult to plan specific resources or gangs in detail over a longer period.

If used to cover a long time period, the resource bar chart is not as good as a Gantt chart (Section 2.2.1) because of what can happen on a construction project. Unforeseen occurrences such as resources not being known, different rates of progress for resources or the announcement of changes to a design all introduce uncertainties which could be expected to affect the future resourcing of the project. If long-term plans show resources specifically allocated to activities, this can be a constraint on the

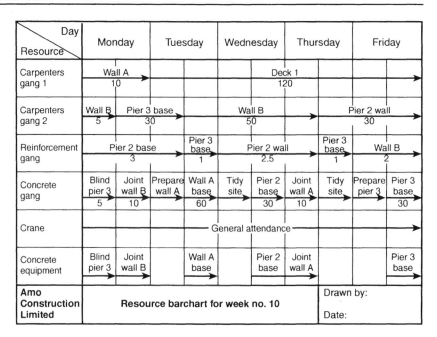

Figure 2.5 A resource bar chart.

flexibility of the plan to accommodate unforeseen changes. This suggests that short-, medium- or long-term considerations in planning might be different – and indeed they are because the fundamental objectives are different (Section 1.2.5). Long-term planning allows decisions to be made on what resources will be required and gives time for them to be mustered. Short-term planning should be directed to making the best possible use of resources which are already available. These issues are discussed further in Chapter 10.

Figure 2.5 shows a week's allocation of the resource gangs to specific activities. As well as including the activities, other useful information can be given such as the major quantities (for example, concrete volumes or steel reinforcement tonnages) calculated for the week or period, as shown below the bars. This is useful when evaluating a plan and communicating information to the operatives. A fuller description of this, including why some activities may not always have quantities associated with them, is contained in Chapter 13.

The resource bar chart is thus a powerful display tool as it focuses attention on the resources and not just the work identified as activities. It is, after all, the resources that actually do the work.

2.3   SPACE–TIME DIAGRAMS

On linear projects, such as pipelines or highway projects, chainage (distance) is an identifiably important parameter. On repetitive projects, such as housing estates or multi-storey buildings, where there are a lot of similar activities, it is reasonable to consider that the repetition number (that is, how many floors or how many rooms) is important too. In such circumstances, the second important parameter (distance or repetition number) can be better incorporated into a plan by using alternative methods to bar charts.

Space–time diagrams exist in a variety of formats. If consistency with bar charts is a high priority, time should be the horizontal axis and space or distance (or unit no.) the vertical axis. In practice many planners prefer, for roads and pipeline projects, the horizontal axis to be a section along the site chainage with time plotted vertically downwards. For buildings, the conventional representation of time plotted horizontally is preferred and this is closely associated with the line-of-balance method (Section 2.4). Whatever the convention for the axes, the work represented on the diagram is modelled by a shape drawn between times and locations.

Many people are unfamiliar with the use of space–time diagrams so some explanation of their make-up is now given in order that the advantages and disadvantages can be appreciated.

2.3.1 Linear projects

Using the convention of plotting time downwards, the five most common shapes used on a space–time diagram are shown in Figure 2.6 and are interpreted below for possible activities on a road project:

Horizontal line This identifies an operation that happens instantaneously over a significant length of the project, such as traffic switches, start of holiday and fixed dates.

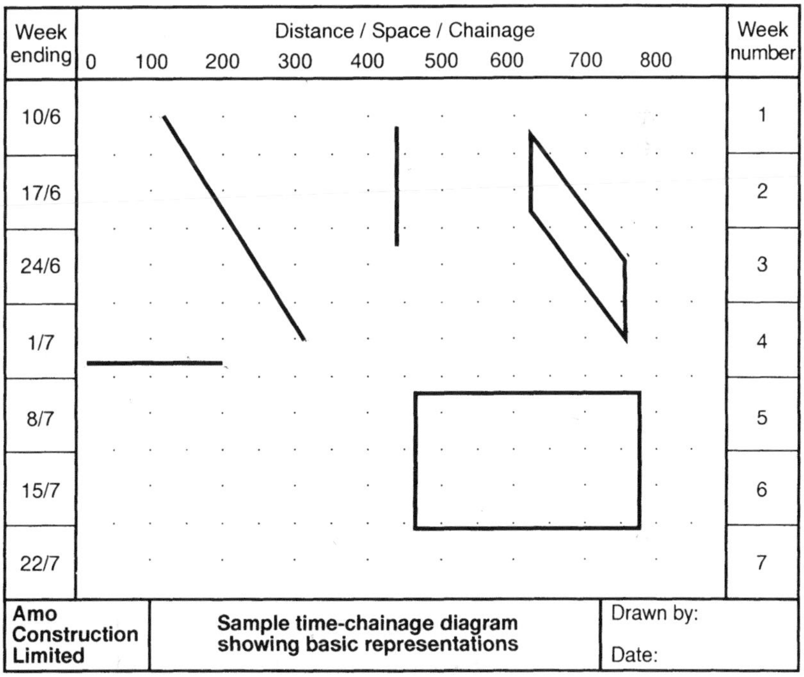

Figure 2.6 A general space–time diagram with time advancing downwards.

Vertical line   This identifies an operation that is fixed at a particular chainage or occupies a relatively short chainage along the main longitudinal section; examples are bridges, service diversions, side roads and drainage outfalls.

Sloping line   This is used if a linear operation has a relatively negligible duration at any particular location and is scheduled to progress along the project, such as for drainage, fencing and road markings.

Sloping box   This is a parallelogram which is used when an operation has a significant duration at each chainage and is scheduled to progress along the project, for example surfacing and topsoiling.

Rectangular box   This is used for an operation that occupies a significant chainage of the project. The box indicates that the operation may be happening at any location along the indicated chainage at any time during the period of occupation of the area, for example earthworks and retaining walls.

The use of the above symbols for the Tingham Bypass project (Appendix A) is shown in Figure 2.7. This shows a section of new road with earthworks and other features at various locations, and is drawn with a longitudinal section across the top of the chart and with time plotted down the chart. On a road project the space–time diagram is often referred to as a time-chainage chart.

As well as being a means for communicating a plan, the time-chainage chart can, like a bar chart, be used as a planning tool. For example, the time-chainage chart can be used to help prevent clashes between resources by isolating work areas for the resources for certain lengths of time. Figure 2.8 shows an area of cut in the region from chainage 200 to chainage 500 which takes all the time (15 days) between the ends of days 4 and 19. The drainage extends beyond the chainage where cut is required and is shown as an inclined line to represent a constant rate of working between two chainages. The diagram shows how it is relatively easy to programme the drainage to be continuous without interfering with the cut at a particular location. The drainage is programmed to commence at chainage 870 m on day 15 (before the cut has finished) yet leave a clear day between the cut and drainage at chainage 500 m.

The basic shapes for the symbols (Figure 2.6) are simplifications that are usually adequate for planning purposes but they involve several assumptions which are, at best, approximations to reality. The most important of these are that:

- work progresses at the same rate at all locations;
- there is the same amount of a specific type of work to do at each location;
- work of a specific type takes the same time to carry out at each location.

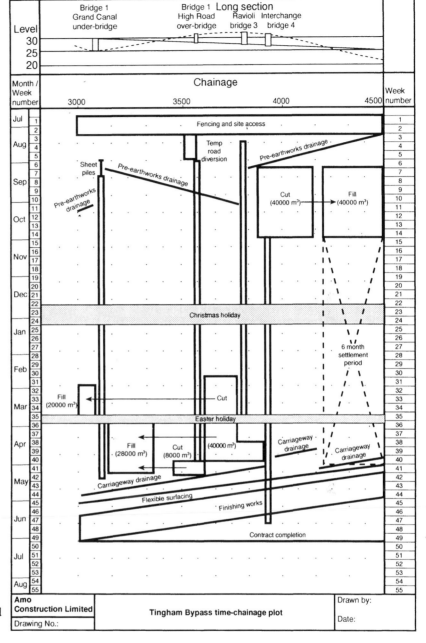

Figure 2.7 Time-chainage chart for a road project.

These assumptions are all closely linked and may collectively be considered unrealistic, as the quantities of work and the degree of difficulty in carrying out the work are likely to be different at different locations. This means that shapes other than those identified above would need to be used – such as distorted parallelograms and curved shapes – to represent the plan for the works. If progress was likely to

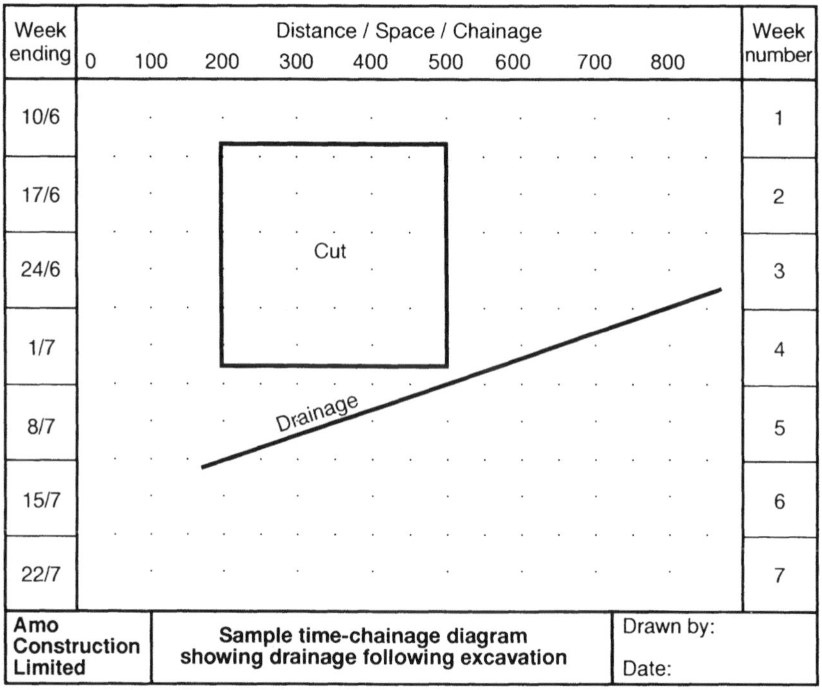

Figure 2.8   A time-chainage chart used as a planning tool.

vary at different locations or times, the need for such shapes could generally be overcome by considering tasks occupying shorter chainages, for which the linear representations would once again become reasonable approximations. Chapter 4 covers this point in greater detail for work activities in general.

A particular simplification encountered in practice is the use of a straight inclined line instead of a parallelogram or box. It is a simplification because its use implies that an activity progresses past a particular spot in a very short space of time. In reality all activities occupy a finite length of the site for more than an instant. The general case is shown by the parallelogram in Figure 2.9. Note that approximating the parallelogram by the straight line CD might be taken to imply that the activity would normally start at point C and would normally end at point D. This distinction may be important when the system is tied to a network-based computer planning package.

2.3.2   General comments

Some contractors have traditionally used time-chainage charts for their planning and control of major roads contracts. These have been hand-drawn and would not, in general, integrate with network-based computer planning packages. The use of time-chainage charts in the planning and control function has thus been limited because of the difficulty of updating them manually; allowing for alterations has been difficult and time-consuming. Some computer packages are able to produce time-chainage charts from networks. With these advances, it is expected that

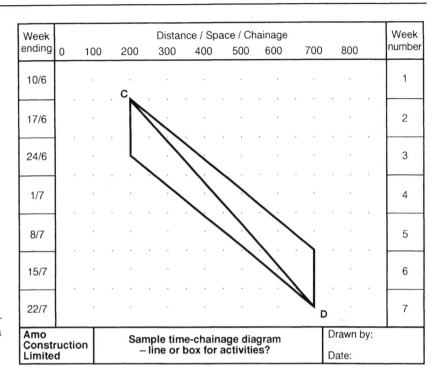

Week ending	Distance / Space / Chainage									Week number
	0	100	200	300	400	500	600	700	800	
10/6			C							1
17/6										2
24/6										3
1/7										4
8/7										5
15/7										6
22/7								D		7
Amo Construction Limited	**Sample time-chainage diagram – line or box for activities?**							Drawn by: Date:		

Figure 2.9 Line or parallelogram on a space–time diagram.

time-chainage charts and other space–time diagrams will become more popular.

Time-chainage charts are very good for improving the visualisation of a plan provided there are not many overlapping activities or much work scheduled to be going on at the same chainage at the same time. A conflict may arise, for example, where a retaining wall on one side of a road is being constructed at the same time as drainage on the other side of the road. The time-chainage chart would allow planners to check this apparent overlap for safety or feasibility; this is in fact a strength of this planning method. Overlapping activities do, however, represent a weakness in the method as a tool for communicating with other people. The use of colour can help overcome this weakness and indeed can be used to clarify the chart for all users. Colour coding is often standardised in companies using this method of planning. Irrespective of how clearly they are drawn, the general complexity of time-chainage charts makes them suitable for communication with certain personnel only, for example planners and senior site management, and not with operatives.

Time-chainage charts are really only suitable for projects which exist principally in one dimension. Examples of this are pipelines or roads without significant junctions. If, for example, a project consisted of two roads which intersected, the time-chainage representation of this is complex. It is suggested that two charts be produced, one for each road. On each chart, the other road would appear as a box between specific chainages (for example the interchange in Figure 2.7). With this method it is possible to show many but not all of the interactions. Both charts require careful checking for resource utilisation.

It is reasonable to expect that the charts could also be useful for monitoring (Chapter 12) and record keeping, because rates of working as well as work in progress could be gleaned from an as-built chart. In practice this is rarely achieved due to the complexity and scales of the charts.

It was implied above (start of Section 2.3) that linear projects could be considered as special cases of repetitive projects. It will be shown now that the space–time diagram is closely related to the line-of-balance method.

2.4   TARGET COMPLETION TIME AND LINE OF BALANCE

A similar method to the space–time diagram involves the use of a graph of target completion times for repetitive activities. This is covered in a little more detail here. The technique is illustrated using the construction programme for a 10 floor building, Tingham Offices (Appendix A).

2.4.1   Target-completion-time graph

The rate of completion of the project phases is often determined from the project requirements, stated here to be that Tingham Offices must be finished in 300 days and floors must be handed over at a rate of 4 per 100 days, once construction is up and running. These requirements can be represented graphically (Figure 2.10) as a target-completion-time graph.

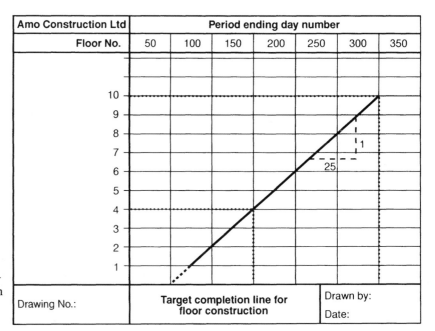

Figure 2.10 Target-completion-time graph for building floors.

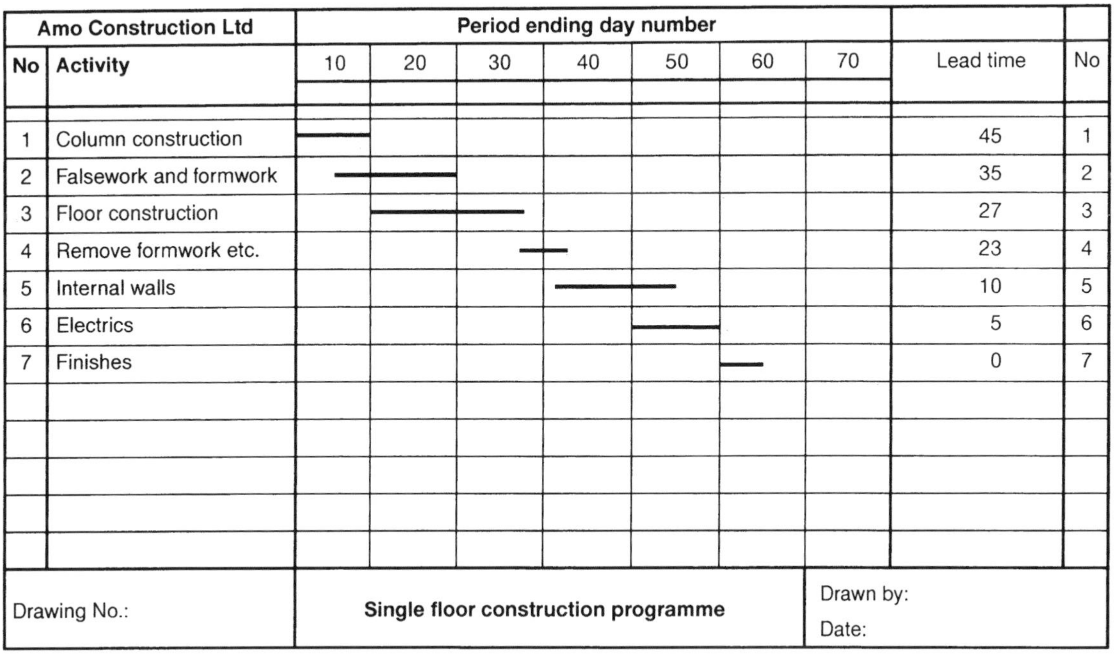

No	Amo Construction Ltd Activity	Period ending day number							Lead time	No
		10	20	30	40	50	60	70		
1	Column construction								45	1
2	Falsework and formwork								35	2
3	Floor construction								27	3
4	Remove formwork etc.								23	4
5	Internal walls								10	5
6	Electrics								5	6
7	Finishes								0	7
Drawing No.:	Single floor construction programme						Drawn by: Date:			

Figure 2.11 Programme for construction of one floor of Tingham Offices.

This shows when particular floors are required to be completed to just meet the project requirement.

It is useful to have a plan for the work that must be completed on each floor as well as knowing when the floors must be completed. Each identical floor could be completed in accordance with the same schedule, often shown as a bar chart for a cycle of construction. Figure 2.11 shows a bar chart for one cycle (that is, one floor of the office building). This bar chart, representing work which would be repeated 10 times to complete the office block, is now analysed relative to the required completion time for the cycle by working back from completion to establish *lead times* for the activities comprising the work. The lead time is the amount of time that a particular *control point* is planned to precede the end of a cycle by. A control point is typically defined as the finish of an activity which is part of the cycle (for example finish column construction, finish falsework and formwork).

In this example the construction cycle is 55 days per floor and the lead times on 'construct columns', and 'falsework and formwork' are 45 days and 35 days, respectively, being the difference between the finish of an activity and the finish of the cycle.

Progress against the target completion line can be checked by observing how much work has been done and comparing it with what should have been done to stay on schedule. Consider time 150 in Figure 2.10. In order to be on schedule, four floors should be finished.

This information can be extended to all of the control points by drawing lines parallel to this target completion line for each of the other control points (Figure 2.12). (As an example the lines for 'column

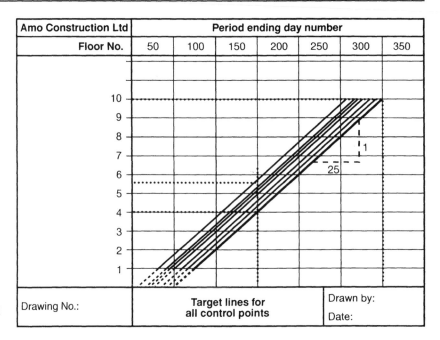

Amo Construction Ltd	Period ending day number						
Floor No.	50	100	150	200	250	300	350

Drawing No.:	Target lines for all control points	Drawn by:
		Date:

Figure 2.12 Target lines for all control points.

construction' and 'falsework and formwork' are drawn 45 days and 35 days, respectively, to the left of the line for the finish of the cycle.) From this graph the number of units (or part-units) to be completed can be read off for the activity associated with each control point. Here the column construction should be completed for almost six floors and the falsework and formwork for slightly less.

Instead of producing a potentially confusing target-completion-time diagram with many target lines (one for each control point), it is possible to use just one line, typically that shown in Figure 2.10. The progress required against a particular control point can now be determined by reading off the number of units against the designated time *plus* the lead time. For example, to establish the columns required to be completed by day 150 the graph in Figure 2.10 should be read off at day 150 + 45 or day 195, to reveal that the columns on almost six floors should be completed by day 150.

The target-completion-time graph is thus a very powerful planning tool in its own right, as it enables what should be done by what time to be established easily. As a plan it should really be used in conjunction with the cycle bar chart and other information, to see such things as the interdependencies of the work and the activity durations.

2.4.2 Line of balance

From the information about all the control points, a line-of-balance chart for time 150 can be drawn (Figure 2.13) in the form of a histogram. It is what this histogram represents that is the line of balance referred to in the

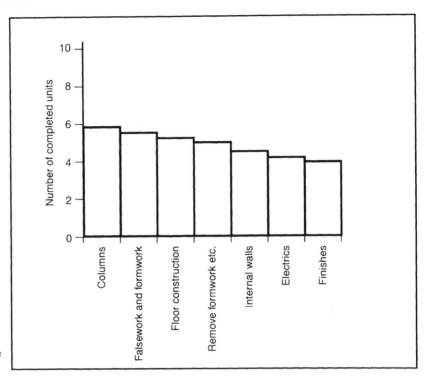

Figure 2.13 Line of balance at day 150.

method; the target-completion-time lines are often erroneously referred to as the line(s) of balance. Because the graph of target-completion-time lines for the various control points can cover the whole duration of a project on a single sheet, it is this graph which is usually the representation of the plan, derived using what is called the general line-of-balance method.

The line-of-balance histogram is useful for monitoring progress; progress can easily be checked against the histogram bars. However, the usefulness for monitoring is limited by the fact that information on it is constrained to the instant of time for which it is drawn. The information shown is clear and this makes the method a good one for communication, but in reality it is useful only for identifying short-term targets.

The target-completion-time lines are rather more useful for monitoring progress and for indicating the need for control action, as shown in Chapter 12. By plotting the planned and actual lines on the same axes, the trends in work completion compared to the plan can be seen for each activity in addition to the position at any instant in time. Any divergence between planned and actual achievement can be detected and early, sensible control action can be taken.

This introduction to target-completion-time graphs and the line-of-balance method has considered only what must fundamentally be achieved to meet the overall progress requirements for cycle completions. As a plan in its own right, the technique so far lacks any information about resources.

2.4.3   Resource considerations

Resource considerations are obviously important and, if incorporated in the line-of-balance method, may lead to target-completion-time graphs similar to Figure 2.12. In practice, each activity type may have a different gradient due to different or variable rates of working for the resources. Examples of why this could happen would be when:

- optimising gang sizes for productivity, quality and safety;
- planning for continuity of resource utilisation;
- varying gang sizes during a project;
- keeping gang sizes constant if the amount of work or degree of difficulty changes between units.

A resultant target-completion-time graph for the Tingham Offices project may look like that in Figure 2.14. This graph shows different rates of working for the different gangs involved in the project.

Figure 2.14 appears to show continuity in resource utilisation, because the use of continuous lines implies continuity in each activity throughout the cycles of a project. However, planners should be aware that this may be a false impression, depending on the number of cycles in the plan, the time to carry out the activity on a single cycle, and the speed of completion required. This is illustrated in Figures 2.15 and 2.16. Figure 2.15 shows a resource completing work on Cycle 1 then moving immediately to Cycle 2, and so on. Figure 2.16 shows the resource

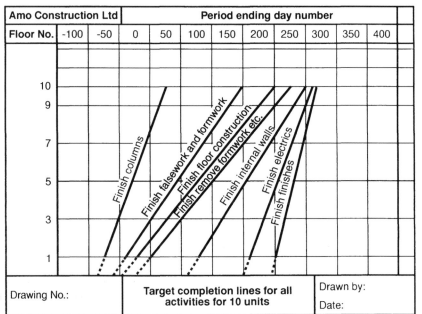

Figure 2.14 Resource-based target-completion-time graph.

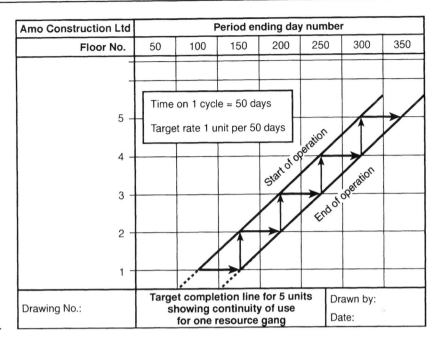

Figure 2.15 Progress line showing continuity of working for a single resource gang.

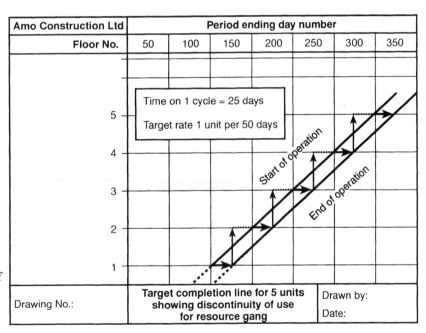

Figure 2.16 Progress line showing discontinuity of working for a single resource gang.

finishing Cycle 1 but not able to move to Cycle 2 until some time later. Without further scrutiny (or calculation), the sloping lines on Figures 2.15 and 2.16 appear to show continuous progress through all the units.

2.4.4 General comments

In general terms the target-completion-time graph can be a powerful planning tool (as well as a medium for representing the plan), in the same way that a bar chart or a space–time diagram can. This benefit comes through developing the plan as the graph is put together on the drawing board or computer by considering and allowing for such things as buffer periods between activities (e.g. for safety and drying out), the time that resources need to spend in one location, and special relationships between activities.

A special relationship in a building project can be seen in Figure 2.14 where the line denoting the finish of 'Remove formwork' is drawn parallel to the line denoting the finish of 'floor construction'. In this case it is not possible to delay curing, which is what defines the formwork removal rather than resource-use considerations.

As methods of representing planning information, both the target-completion-time graph and the line-of-balance diagram are useful in that they have the potential to convey relevant information to construction teams and project managers. In complex projects the target completion graph for the whole project may be too complex – with densely packed and possibly overlapping lines – for effective use as a communication tool.

Compared with the space–time diagram, the target-completion-time graph and line-of-balance diagram possess no direct representation of the spatial aspects of the project. This can be a disadvantage or an advantage, depending on the type of project. On a housing estate construction project, for example, it is reasonable to expect an available resource to be able to move to any property on the estate where it can find work. The line-of-balance method can therefore be used to good effect as a tool to keep track of how many units should be completed by a given time, rather than which specific units should be finished.

On some projects, such as housing estates, some of the work is repetitive and some is not. This suggests that a combination of planning techniques is appropriate, and in such cases it is possible to find a bar chart and a target-completion-time graph on the same sheet depicting the plan of work.

The mathematical aspects of the line-of-balance method are covered more fully in other books on project planning techniques.

2.5 NETWORKS

A network is a diagram showing all the activities which are needed to complete a project and the order in which they *must* be done. It therefore comprises two parts:

- the activities
- some method of displaying the 'technological' logic of construction.

NETWORKS **35**

The word technological is used here to mean the basic construction logic which must be followed to complete the project (for example, foundations coming before the roof). The inclusion of other forms of logic (for example, resource logic) might be unnecessarily restrictive on the plan and its execution (Section 5.8). An understanding of activities *and* logic is crucial to the use of networks. A brief outline of networks is given here to indicate how they operate as conveyors of information. Standard network analysis theory, in particular the mathematical aspects, is included briefly later (Section 5.2). Practical considerations and extensions to standard theory, necessary to model some construction situations, are discussed in great detail in Chapters 4 and 5.

There are two basic ways of representing networks, activity-on-arrow networks and activity-on-node networks.

2.5.1 Activity-on-arrow networks

An activity-on-arrow network for the Tingham Bridge project (Appendix A) is shown in Figure 2.17.

Each arrow represents an activity between two event nodes. Each different activity must be uniquely represented by two nodes as it is the two nodes which identify the activity joining them. This fact may necessitate the use of 'dummies', which are logical links requiring neither time nor resources. Dummies can sometimes be included to assist in the production of a neat network diagram or, more usually, to show logical relationships which cannot be shown any other way. The dummies used in Figure 2.17 (the dotted lines) are essential to show the logical relationships.

Using arrows between events as the basic building block of the network, all the identified activities in the project are combined to produce a representation of the construction showing all the logic connections.

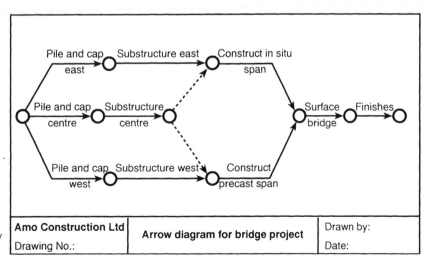

Figure 2.17 An activity-on-arrow network (arrow diagram).

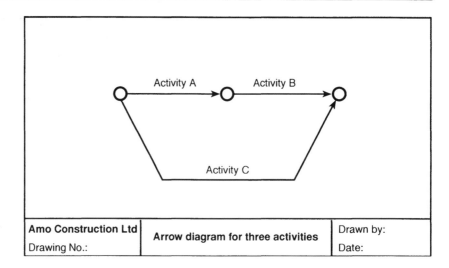

Amo Construction Ltd	Arrow diagram for three activities	Drawn by:
Drawing No.:		Date:

Figure 2.18 Can arrow lengths be proportional to activity durations?

One of the advantages generally claimed for activity-on-arrow networks is that the duration of the activity can be represented by the length of the arrow for that activity. In practice this claim is a fallacy. It is not possible in the general case to construct a network in which the arrow lengths are scaled to the activity durations. For instance, considering the three-activity network shown in Figure 2.18, it can be drawn to scale only if the duration of A + the duration of B equals the duration of C, which is rarely going to be the case.

Figure 2.19 shows another attempt at drawing the network in Figure 2.18 to a time-scale, when all the activities are of equal duration. It can be seen that the line representing Activity C must be split into at least two and perhaps three separate parts. If the activity were to be carried out at its earliest time then a 'following dummy', representing the free float of the activity, would be required; if it were to be done as late as possible a 'leading dummy', representing the float before the activity, would be required. In Figure 2.19, the activity is planned to take place somewhere between its earliest and latest times and two dummies become necessary as shown. New events are not inserted to mark the dummies; the activity cannot be located exactly (in time) between the existing events. This would have knock-on effects as the network was built up.

Further, if an activity has a very short duration, the practicality of drawing it against a time-scale is further decreased. Eventually, if all the activities are drawn on a time-scale and the connections added, the effect is very similar to a linked bar chart, the problems of which have been discussed in Section 2.2.2.

Because of this supposed advantage, the use of activity-on-arrow networks has the disadvantage that many observers believe that they can gather more information from the network than is in fact there; it is in any case questionable whether it is an advantage to show arrow lengths scaled to activity durations. Since activity durations are bound to be subject to change and uncertainty, such a network would need to be very

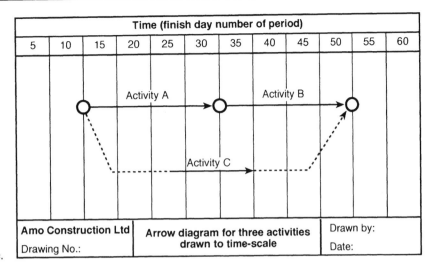

Figure 2.19 Arrows drawn to a time-scale.

flexible. Yet there is rigidity inherent in fixing the arrow length–activity duration relationship. As the duration of most activities is dependent on resource loading, the duration is not a fundamental property of the activity and so should be kept independent of the length of the activity arrow on the network. In summary, networks should be accepted as representations of the logic between activities, and other tools such as bar charts should be left to show the durations, limited logic and a time-scale.

The value of activity-on-arrow networks for communication could be improved if the need for dummies could be eliminated, as they can confuse in trying to clarify. The use of a different type of network, the activity-on-node network, enables dummies to be avoided altogether.

2.5.2 Activity-on-node networks

An activity-on-node network for the Tingham Bridge project (Appendix A), showing the same logic as in Figure 2.17, is shown in Figure 2.20. Here the activities are represented by boxes and the arrows denote the logical sequences of the activities.

There is no need for dummy connections in these networks, as all logical relationships can be represented using the basic building blocks of activity boxes and linking arrows. Because there is no pretence of showing activity durations in the geometry, this format is more robust. It shows only interactions, thereby enabling uncertainty, changing duration and other such factors to be readily accommodated.

As a communication tool the activity-on-node network can be both clear and unclear. The boxes enable activity names to be readily identified but long names can soon become illegible. A temptation is to draw networks with the nodes identified only by numbers to be read against an activity list. This destroys the ability of the network to be a one-stage communication tool. A further problem arises when the logic links become lines without direction (that is, without arrow heads), which can

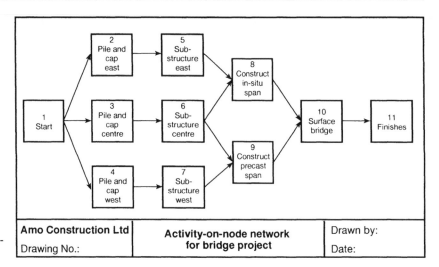

| Amo Construction Ltd | Activity-on-node network | Drawn by: |
| Drawing No.: | for bridge project | Date: |

Figure 2.20 An activity-on-node network.

confuse, even on simple projects. It is also difficult to understand the timing implications from the network itself as there is no time axis. Although progression through the network is generally from left to right, it is difficult to follow this convention all the time in complex networks.

2.5.3   More complex networks

On complex projects, networks are often unusable because of problems with legibility of text and the ability to follow the logic lines. Hundreds of activities may be involved with many more logic links. This is one form of complexity. Another form of complexity is in exactly what symbols are used to model a project. Precedence networks (activity-on-node format) and ladder diagrams (activity-on-arrow format) are complex forms of network that contain advanced symbols requiring trained personnel to interpret them.

Many countries produce standards documents which specify or define terms for use in networks. It is not usually compulsory to use the recommended terminology, although a knowledge of it is useful (Chapter 5). In many organisations it is usual to produce networks which do not correspond wholly with standard guidelines, particularly when showing activity information. In some cases, such as when trying to model the links represented on a time-chainage chart, it is usually necessary to go further than is provided for in the standard documentation. An interpretation of standard connections between activities and related matters, including an appreciation of network analysis, is given in Chapters 5 and 6.

Under some standard systems, the information provided about an activity as a result of a completed network analysis – for example, an activity's earliest and latest times – can be shown on the network (as if it were a fundamental property of the activity). Discussion in Chapter 5 shows that this is not the case in general. Indeed putting the times on the network diagram gives them an authority which they should not have, and is the source of many practical problems in using network planning techniques. Again, it is suggested that networks should show *only* activities and logic links. The other information which must be part of a plan, such as durations and timing, should be kept elsewhere and represented in a different form, not on the network. A network in itself is not a plan as it does not indicate when activities should take place, and no attempt should be made to make it a plan. It is a planning tool.

A precedence diagram for the Tingham Bypass (Appendix A) is shown in Figure 2.21. This can be compared with the time-chainage chart for the same project (Figure 2.7).

A standard notation has been used which allows the timings to be shown on the network. The warnings about doing this are reiterated.

- The timings may change, while the connections remain the same, thus making the overall diagram misleading.
- When using the network to illustrate a situation where some work has been completed, the timing information may be confusing.
- The times displayed relate to the critical path analysis (Chapter 5) and may bear little relationship to the scheduled times for the activities.
- The times are calculated using a particular set of assumptions, yet an alternative, equally valid, set of assumptions may give different answers (Chapter 5).

When considering networks as a communication tool, there are several areas of potential confusion if lots of connections cross or are close together, which occurs frequently, whether networks are computer generated or not. Figure 2.21 places the activities on a regular grid – a common method of positioning. For precedence networks this results in a lot of kinked lines, especially for start–start (SS) and finish–finish (FF) connections (Section 5.4). It may be better to use a staggered layout (Figure 5.12) in such cases. The two activities 'Excavate acceptable 3880–4150' and 'Acceptable fill 4250–4500' are linked with SS and FF connections to show there is a progress constraint between the activities, not just that they are linked at the start and the finish. On close scrutiny these connections are not adequate to model the constraint that the excavated material would be used for the fill material, so the activities *must* go on together. These activities should be linked (tied) in some way but no standard method of illustrating this is available (and few computer packages handle this situation). The use and handling of tied activities are discussed in Section 5.7.

In summary, networks are generally useless as a medium for communicating with the operatives as it is difficult to see what needs to be done and when. As a medium for communication between planners or organisations, networks are potentially useful because much information can be gleaned from the analyses that can be carried out on them. Sadly,

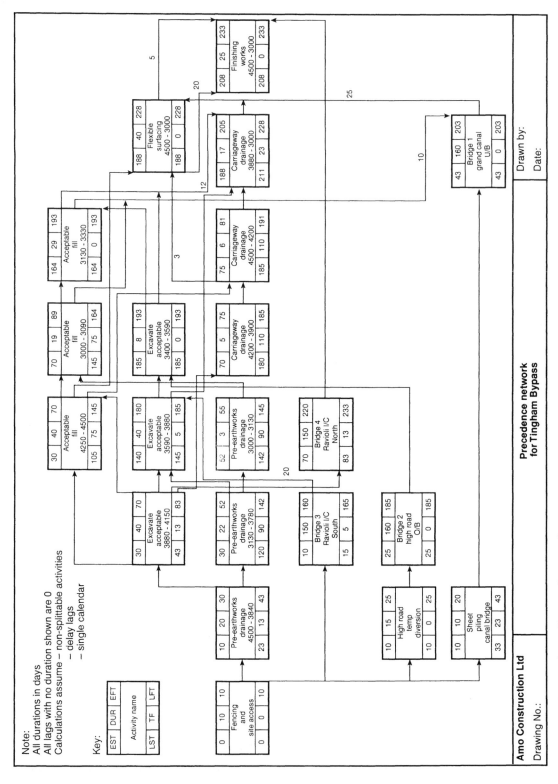

Figure 2.21 A precedence diagram for a project.

too few people know how best to model projects with networks and how to interpret the results of such powerful planning tools. It is hoped that the following chapters will help to increase that number.

2.6 RESOURCE PROFILES

As work is actually carried out using labour and plant resources, it is useful to both planner and construction manager to have some indication of the planned resource utilisation. The multiple resource requirements of the activities constituting a project can be determined as part of the planning process, using a variety of techniques (Chapter 7). The result of this part of planning would be a schedule of resources which could be succinctly presented as a set of graphical resource profiles covering the duration of the project. One such resource profile for one of the resources on the Tingham Bridge project is shown in Figure 2.22. It is derived more fully in Chapter 7.

This is a resource histogram which can be used to help identify problems of peak or intermittent demand. Such histograms are useful ways of representing the deployment of resources on a project. Taken in isolation they do not show the interactions between resources, but they can usefully highlight problems with plans when plans are being produced. The histograms are useful for communicating planning information to certain people but would not on their own be considered the main representation of the project plan. An alternative representation of the information in Figure 2.22 is cumulative, rather than incremental, as shown in Figure 2.23. It can be derived simply by aggregating the incremental quantities in Figure 2.22.

Resource information is useful in the planning process, but it is difficult to compare the histogram for one resource with that for another.

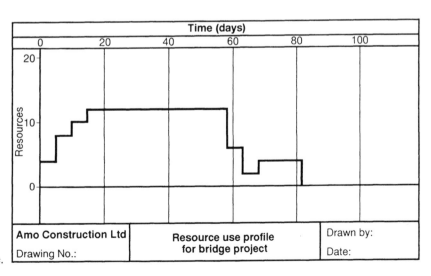

Figure 2.22 Incremental resource profile.

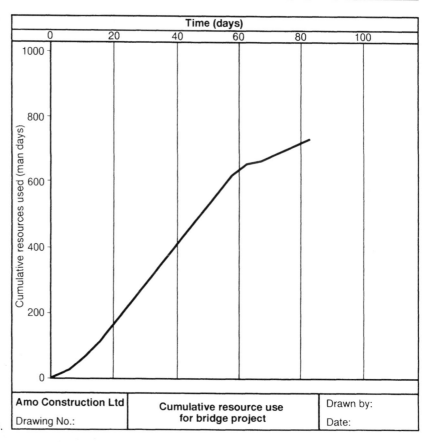

Figure 2.23 Cumulative resource profile.

It would be useful to combine the resource (and other) information in a way which could present a better overall view of the project. This can be done by considering the project finances, which are closely associated with the resource costs of labour, plant and materials.

2.7   FINANCIAL PLANS

Financial plans should be produced at all levels of focus of a project (short-, medium- and long-term) and are particularly useful for the higher managerial teams of a project or organisation. This means they sometimes get ignored at the short-term level of planning, which is a mistake. They would normally represent the planned financial position throughout a project and as such would be concerned with the following:

- expenditure or liability
- income or earnings
- cash flow or surplus.

Information could be represented periodically or cumulatively, and could be broken down for sections of a project or for the different resources on a project. The graphical portrayal of this information would be in a form similar to resource profiles and is shown in Figures 2.24–2.27. It is difficult to draw exact financial curves since they depend on the particular way in which work is carried out on the project and the methods of payment defined in the contract. The terms income, earning, expenditure and liability are discussed in Section 3.3.

These four diagrams make the following assumptions:

- The work will be carried out at a constant rate throughout any given period (month).
- The resources on the project will be constant throughout any given period (month).
- The client will pay the contractor 90% of the money earned 3 weeks after the end of each period and the remainder in two equal parts, 3 and 6 months after agreed completion of the contract.
- The contractor will pay bills (to labour, plant companies, subcontractors, suppliers etc.) throughout the contract in such a way that the expenditure throughout any period is even.

In the graphs the period figure might be given as a single point (or rectangular block) for the interval rather than the saw-tooth effect which represents the cumulative position within the period increments. This would effectively mean that the time-scale was no longer continuous but instead divided into discrete monthly blocks.

It is often useful to see the financial plan against a background of the work activities. This is achieved by superimposing, with appropriate scales on the axes, the financial plan on to a bar chart of the project. The time axis would be common to both the bar chart and the financial information. This representation can be accomplished by some computer planning packages.

The subject of financial planning is covered in more detail in Chapter 7.

2.8 MASS-HAUL DIAGRAMS

Earthmoving is often a key element in construction projects. Efficient earthmoving can improve the profitability of a project, particularly on road projects where the earthworks form a significant part of the work (in terms of time and money), and also affect much of the other work on the project.

On projects of any size it is desirable to match the volumes of excavation (cut) and fill such that all the cut material can be used and no imported fill is required. Factors which complicate matters are when the volume of fill is greater than the cut volume and when some cut material

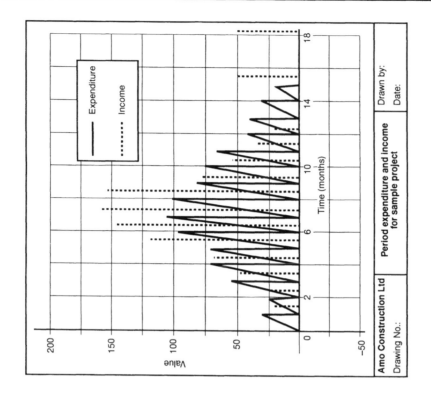

Figure 2.25 Financial plan for expenditure and income in period increments.

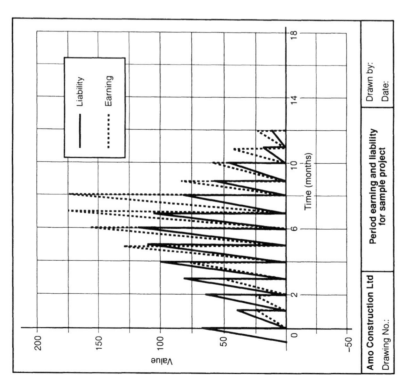

Figure 2.24 Financial plan for liability and earning in period increments.

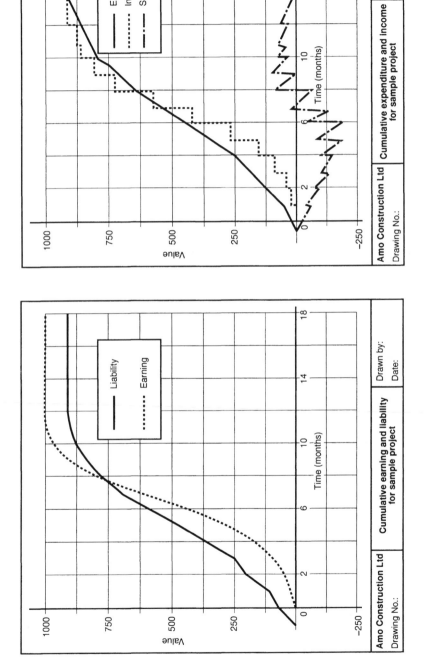

Figure 2.27 Cumulative expenditure and income graph.

Figure 2.26 Cumulative liability and earning graph.

is unsuitable as fill, possibly because it is contaminated. The task of balancing the cut and fill operations by determining the finished earthworks profile is usually undertaken by the designers, with the builders having little or no input into the process.

An important factor to be considered in earthmoving is ensuring that the volumes of earth are transported the shortest possible distance. In some cases, usually on roads projects, the distances may be excessive, making it cheaper to dispose of cut material and to borrow locally in the vicinity of fill locations. Contractors have considerable influence over this process as large costs can be incurred or large savings made.

A technique commonly used to assist with the planning of earthmoving operations is the mass-haul method, shown in Figure 2.28 in its various stages of production. Although a mathematical technique, it does result in a useful diagram which can communicate concisely the necessary information about areas of cut and fill and the movement of earth. Clearly this is a planning tool rather than a plan.

To show the simplicity of the technique, consider a road where the planned formation and the lie of the existing ground are as shown in Figure 2.28 (Step 1). From this, it is a simple matter to develop the cut-chainage diagram shown in Step 2. This is a diagram showing the depth of earthworks along the road. In this example, as is normal, the area to be cut is shown above the base line and the area to be filled is shown below it.

The next diagram, Step 3 in Figure 2.28, shows the cumulative cut volume along the section of road starting at the left end and moving across to the right. Cut is shown as positive and fill as negative. Over the whole length there is a net cut, indicated by the right hand end of the diagram being above the zero line, and so some material must be tipped. *Any* horizontal line drawn across the diagram is a balance line (one is shown dashed) and converts the diagram into the mass-haul diagram. One particular horizontal line can be calculated mathematically to minimise the earth haul distances but for various reasons this may not be the best line to use. For the balance line shown in Step 3, the cut volume between B and C balances the fill volume between C and D, cut EF balances fill DE, cut AB would be tipped off site, and so on.

The mass-haul diagram is used to help the planner to decide how the excavated material should be moved on site. It is rarely, if ever, a straightforward choice. Proponents of the mass-haul method suggest that the technique enables the minimum mass-haul, the product of volume of material and the distance which it must be transported, to be computed. While this may be true for an idealised project, the question remains as to whether minimising the mass-haul necessarily provides the optimum construction solution.

It should be noted that there is no need for the balance line to be continuous, but if it is not then additional tipping or borrowing will be required at intermediate points in the project. This in fact often produces the optimum solution, because of other factors such as the type and

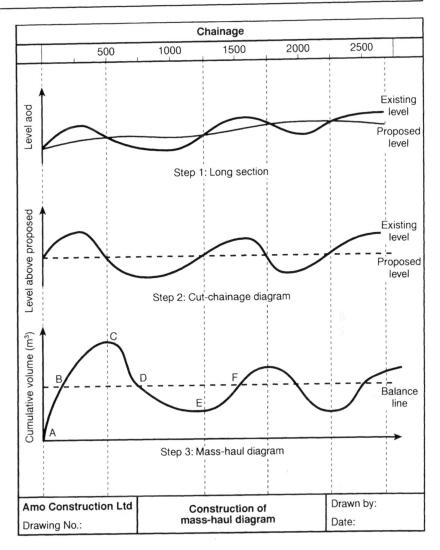

Figure 2.28 Stages in the production of a mass-haul diagram.

suitability of the material at all locations, the presence of rivers, or other obstacles to the provision of haul roads. Bulking and shrinkage of materials should also be considered.

In summary very simplistic assumptions form the basis of the whole process, and in particular the choice of the position of the balancing line to provide a minimum mass-haul and consequently minimum cost solution. Some of these assumptions, with appropriate comments, are:

- There are no access or structural works constraints.
- All the material is usable.

- The material will be moved the shortest distance. (In practice the newly placed fill often cannot be trafficked, so that unless a specific haul road is constructed the contractor will have to start filling at the far end of the section of fill.)
- The principal determinant of cost is the product (volume × distance). (In practice the loading and tipping operations are very expensive and occur whatever the haul distance.)
- The changes in volume which the earth undergoes from *in situ* to loose and then to compacted are often ignored. (This can have severe consequences.)
- The cost of travelling in both directions is the same. (This is unlikely even when only small hills are involved.)
- It is possible to travel in the required direction along the line of the road. (If the excavation is being undertaken along a face, this may not be the case.)

Despite these very serious drawbacks, the mass-haul method encourages thought about the project and consideration of how the special features of the project might affect the earthmoving plan. It may be that the process of criticising the applicability of the mass-haul method is more important than the answers it gives.

2.9 TEXT-BASED PLANS

The preceding sections have covered ways of representing plans by schematic or graphical means. In many cases these will be annotated with notes or supplemented by drawings, reports, tables or other text-based approaches. Text-based ways of communicating planning information are discussed here.

2.9.1 Method statements

Method statements are documents describing how a project, or part of a project, is to be carried out. They are sometimes produced by all parties, but they reach a greater level of detail when it is a contractor producing them.

Method statements can be used for many things, including communicating with other parties and developing internal estimates. Detailed method statements are particularly useful for developing a safe system of working in order to achieve a specified quality.

Depending on their function, method statements may contain a few or many items from this list:

- a description and diagram of the project offices, stores and stock areas;
- a plan of work for the project;
- a list of plant and equipment to be used;

- phasing diagrams showing the proposed status of the project at various times and stages of construction;
- location diagrams showing placement and operational capabilities of major plant;
- location diagrams showing placement of temporary works;
- safety matters;
- environmental considerations;
- design of temporary works;
- budget calculations;
- descriptions of the construction process;
- descriptions and diagrams of access points for a site;
- information on the major subcontractors;
- details concerning major suppliers, the restraints on them and the phasing of the supply of materials from them.

The relative importance of each of these obviously depends on the project and the perceived use of the specific method statement.

In summary, method statements can provide a lot of information which would not be shown on other diagrams.

2.9.2 Lists and schedules

It was mentioned above that lists could be included in method statements. Lists could also appear on bar charts or any other diagrams. Lists are therefore simple and widely used ways of preparing, presenting and communicating planning information. Many people draw up lists or schedules of their daily work or prepare them for subordinates. Lists and schedules encapsulate, sometimes explicitly, the writer's thought processes and are thus simple and highly effective planning techniques as well as communication tools. They should not be underestimated or ignored.

Sequence should not be inferred from lists although they are often the result of thinking through the order of construction. Approximate sequencing on lists can be both an advantage (aiding legibility and understanding) and a liability (through error or omission).

Examples of lists which may be considered either as representations of plans or parts of plans include:

Critical path analysis outputs   These are derived from the results of calculations performed on networks (Chapter 5).

Logic reports   These show the relationships between activities, and are either the input for or the output from network-based planning analyses (Chapter 5).

Activity quantity schedules   These are a fundamental requirement for good planning and show the relationships between the activities and the quantities of work to be done in each.

Unfortunately many lists are destroyed when the work to which they refer is finished. This is a shame as these lists are potentially one of the best records of the project, supplementing diaries and other sources of information.

2.10 REPRESENTING TIME ON PLANS

The representation of time on plans is of utmost importance, especially when their main purpose is communication. As the reader will no doubt have noticed from the diagrams in this chapter, there are several ways of denoting time on plans.

The following is a list of the more common methods, followed by commentary on their advantages and disadvantages:

- calendar date
- project week numbers
- project day numbers
- project shift numbers
- calendar week numbers
- calendar day numbers.

The calendar date is an obvious method for putting a date in a programme and is well understood throughout much of the world. If there is more than one working shift in a day, the use of a date alone is insufficient and a shift number or reference must be added. If the programme is drawn up in terms of calendar dates and the start date of the project is changed, the whole of the programme should be redrawn because of the date changes and the different position of holidays within the project. Care should be taken, when referring to a date, that everyone understands whether the reference is to the start or the end of the day or to something happening within the day.

For long-term plans, the smallest time unit considered is often a week. In this case, project week numbers can be used. This overcomes one of the problems of calendar dates in that the programme is to some extent independent of the start date of the project. Care must be taken, however, to ensure that holidays are in the correct place and that the productivity of the various gangs corresponds to the time of year in which the work is to be carried out.

Much construction work is affected by the weather and consequently is likely to suffer more problems at certain times of year than other times. Using calendar-related dates or weeks highlights these areas, ensuring more accurate plans. Planning in units which are not date-related and then converting the plan to calendars can cause problems, as discussed in Chapter 6.

In conclusion, planners should spend some of their own time in deciding the best way to put time on their plans.

2.11   SUMMARY

- Plans must be communicated to the appropriate people.

- There are many methods available for representing plans.

- Some forms of representation are also the medium for developing the plan.

- Bar charts are well suited for:
  - simple projects
  - short-term programmes
  - outline programmes
  - communicating with people.

- Bar charts are not so well suited for:
  - detailed programmes of complex projects
  - displaying spatial information
  - projects on which there will be alterations
  - evaluating the durations of projects
  - displaying activity interdependencies.

- Linked bar charts can provide extra useful information regarding activity dependencies, but this should not be at the expense of clarity in communication.

- Resource bar charts are a useful planning tool for making sure that resources are properly allocated and deployed.

- Plans can be annotated with useful (clarifying or supplementary) information.

- For linear projects, time-chainage charts give a good appreciation of the location of the work in both time and space.

- Target-completion-time graphs give a good appreciation of planned progress on repetitive projects.

- The line-of-balance method is a useful mathematical approach especially for building works.

- Networks are useful at a senior level provided they are used properly.

- Planners should be aware of the advantages and disadvantages of the various types of network.

- Resource profiles and financial plans provide extra useful information.

- Financial plans can be compared with progress records and used for control purposes.

- The mass-haul method is a useful mathematical approach for earthworks planning, subject to considering all practical constraints.

- Lists and schedules are unglamorous but essential and should be kept as project records.

- Thought should be given to selecting an appropriate time measure for a plan.

3

RESOURCES AND FINANCES

'Which of us . . . is to do the hard and dirty work
for the rest – and for what pay? Who is to do the
pleasant and clean work, and for what pay?'

JOHN RUSKIN
(Sesame and Lilies)

Work in its many forms is carried out by the resource of labour, using
the resources of plant and materials. In construction, a particular type of
work in different environments could be hard, dirty, clean or even
pleasant, and the cost could vary accordingly. Seldom if ever do the
undertaking of the work and the exchange of money occur
simultaneously. Selecting the right resources for a job and understanding the complexities of finance are vital attributes for good
managers. Similarly, planners must understand the basics of resources
and finances to produce good plans.

3.1 RESOURCES AND FINANCES IN PROJECT PLANNING

Project planners often concentrate their efforts on the timing aspects of
planning and scheduling, and either ignore considerations of resources
and finances or consider them only superficially. The teaching of project
planning often concentrates on the theory and analysis of activity-based
models, with little discussion on how the activities are determined and
what they mean. Before tackling activities in Chapter 4, the important
topic of resources and finances is considered.

In reality the resources, which actually do the work, and the finances, which may be needed to build the works, should be planned as carefully as time is. Ironically it is usually constraints on the time, finances and resources available for the planning process itself that account for the lack of consideration of a project's resources and finances. Sometimes, planners believe that it is too difficult to plan the resources and finances in more detail and that the benefits of doing so would be marginal.

Plans are generally derived in terms of activities; activities are carried out by resources; the resources employed determine the duration of the work; and the way the activities are put together affects the financial picture, or model, of a project.

It is therefore essential to consider the influences on a project of the resources and finances, so that improved and justifiable project models can be developed and that better judgements might be made in future planning operations.

Resources and finances are being talked about here as if they were different. In fact finances are resources, although they are a particular type of resource. It is convenient to distinguish resources such as labour, plant, equipment and materials from the financial resources such as earning, liability, expenditure and income, and that is done where necessary in this chapter. Resources and finances are frequently considered separately because it is assumed that different people have responsibility for the planning and control of each. They are, however, very closely related to one another and a project manager should have control over significant parts of both and ensure that they integrate with the physical plan.

We know that resources are used to carry out a project and thereby have an effect on a project, but it is still difficult to define a resource. Definitions are either too restrictive or too general. To illustrate the difficulty consider the general definition that

a resource is anything which affects any element of a project in any way.

Yes, labour, plant, materials, subcontractors, managers and money are resources, but what of weather, space, luck or control? Definition is unimportant but the identification, characteristics and effects of resources are important.

Resources and finances are assigned to a project and allocated to the activities within it in order to complete the required work. Clearly the number of resources allocated to the work influences its duration, but there is a rather complex relationship between the resource levels and the duration especially when, as is usual, more than one type of resource is required for the activity.

The preparation of a financial plan and the cost of construction work can be predicted or ascertained by various means, one being to determine the resource costs of the work. It is important therefore that resources are appropriately classified so that they can easily be assigned to activities in the plan and allowed for in the financial plan or valuation, not omitted or duplicated as can happen if due care is not taken or if the resource classification is not understood.

This chapter deals with the classification of resources and finances. The allocation of resources and finances to activities and the consequences of the resulting activity models are covered in Chapter 4; the scheduling of activities to provide programmes of work which consider the importance of resources and finances are covered in Chapter 7; and the use of resources and finances to speed up a project also appear in Chapter 7.

3.2 RESOURCES

3.2.1 Classification

In order to estimate for, plan and control a project properly, it is necessary to maintain detailed records of the use of resources and their productivity. The extremely large number of resources involved on even the smallest construction project means that it is unrealistic to expect to consider them all individually without imposing some order or grouping. The most common form of grouping is a classification system which puts together the resources with similar attributes. This allows agglomeration of similar resources and facilitates communication between interested parties on the project. It also helps to ensure that all resources and finances are considered.

For most construction purposes, resources can be grouped into:

- labour
- plant
- materials
- subcontractors
- overheads.

This classification system can also be used by design, supervision and management companies, although the detailed resources within each group would obviously differ from those employed in the construction process. The separate subset, financial resources, is considered in Section 3.3.

The above groups of resources are different in many ways and cannot be treated in the same way. For example, if materials are available on a project one day, they can generally be used either on that day or at any later date (but see below) with no extra cost, but if a labour resource is on site and is not used on any particular day then the cost is incurred for the idle day. Resources can thus be considered storable or non-storable.

The groupings suggested above, although being useful for most projects, are not sufficient to include all the resources encompassed by the concept of a resource presented in Section 3.1. It may sometimes be desirable to consider a different classification:

- real
- abstract
- space.

The real resources are those which are definite entities such as labour and plant. They are considered in detail in Sections 3.2.2 to 3.2.6.

Space is something that can be defined, is required for the execution of almost all activities and may affect a project, so it is therefore a resource. Indeed, it is sometimes a lack of space which is the only restriction on the execution of an activity. Space cannot be positioned conveniently in any of the initial groupings, so it is recommended that, if the above classification system is to be used, an additional class be included to cover this.

The abstract resource is explained in Section 3.2.8.

There now follows a more detailed examination of the resource groupings to draw attention to other potential difficulties for planners and managers.

3.2.2 Labour

Labour resources are the human resources that carry out the required work. Typical examples on a construction site are:

- labourers
- joiners
- steelfixers
- bricklayers
- plasterers.

Typical examples in a design office are:

- structural engineers
- geotechnical engineers
- draughtspersons.

It is obvious that the first list does not contain all the people required to construct a project nor the second all the people for its design. Missing from the first list are people such as gangers, engineers, planners, quantity surveyors, quality managers and safety officers who have a supervisory and checking role over many activities as well as a direct input to particular activities. Missing from the second list are people such as computer operators and secretaries. In so far as these resources can be assigned to many activities at the same time they could be treated as overheads (Section 3.2.9), but a case could be made for treating them as labour.

A further possible confusion concerns the operators and drivers of plant or equipment (including computer operators). Although obviously human, the operators are often considered to be part of the plant (Section

3.2.3) and may be allocated to more than one grouping. This can cause errors in costing or scheduling resources, so care must be taken to ensure that these kinds of labour resource are included in a classification system for a project once and once only.

When planning, it should be remembered that the task of providing the labour resource can itself cause extra work to be done on the project. As an extreme example, if a large construction project were to be undertaken in a remote area, it would be common to provide accommodation for the labour force. This would require extra activities to be included in the programme of work and they would usually have to be carried out before any major work on the permanent works of the project. This could affect the financial resourcing of the project. A more common example would be the necessity to provide different facilities for the site staff, depending on the number employed. Different countries have different legal provisions but it is common to require greater facilities and facilities at a better standard for greater numbers of employees. The provision of the facilities affects the work to be done and its cost.

Resources often work in gangs and it can be useful to differentiate between resources which can easily operate in various gang sizes (for example carpenters and steelfixers) and those which cannot easily do so (for example piling and road surfacing gangs); the latter are considered as standard gangs. Resource gangs and gang sizes are discussed further in Sections 4.4 and 4.5.1.

3.2.3 Plant and equipment

This classification includes all the machines and tools which the labour uses to execute the work. For convenience, plant and equipment are often grouped into subdivisions within the main classification. Typical subdivisions are based on attributes such as:

- physical characteristics
- mode of operation
- task undertaken.

Further subdivisions based on *physical characteristics* might be:

- mechanical
- non-mechanical.

Further subdivisions based on *mode of operation* might be:

- driven
- not driven.

Further subdivisions based on *task undertaken* might be:

- transport
- lifting
- excavation
- piling.

Obviously these classifications are not mutually exclusive. A crane, for example, would be in the categories mechanical, driven and lifting. A back-actor/loader, which is a multipurpose machine, might be considered in transport, lifting and excavation, when classified by task undertaken. Thus although the task undertaken classification apparently offers the most detail, it also often produces the most problems, because plant is not uniquely defined and there is a danger of it being omitted or included more than once.

The plant and equipment must also be carefully defined to avoid other errors in scheduling or financing. For example, the drivers and other operatives associated with pieces of plant should not be included in both labour and plant classifications (Section 3.2.2). This distinction seems obvious, but it is not clear which is the best classification system. Consider the facts that it is necessary to pay drivers but not cranes, but a scheduled crane would be useless without the driver.

There may be several different types of plant required for a single activity. For example, a piling activity may require a piling rig (which could be recognised as piling equipment) plus a crane to lift the piles, a compressor, several shovels, perhaps an excavator and of course a sledge hammer for correcting minor mistakes. How many of these should be considered individually is a matter for the manager to decide. Sometimes the small items of non-mechanical plant (shovels and sledge hammer in this example) are grouped together and considered as a single monetary resource (if they are considered at all on individual activities). Sometimes they are considered only as overheads (Section 3.2.9).

Major items of plant often have additional costs and time associated with their operation because they have to be delivered to site, assembled (possibly), maintained and removed from the project. It is important to recognise these requirements when planning and to include them in the programme, either as extra activities, as increased durations of existing activities, or as increased costs.

The use of some types of plant can, in particular cases, necessitate the use of extra labour (and thus incur extra cost). Examples of this would be a crossing controller for earthworks plant crossing trafficked roads. These types of resource must be accounted for somewhere, either as a labour overhead or a labour resource or a plant resource.

Care should be taken when considering major items of plant to ensure that there is space on the site both for the operation of the machine and for access and egress. Such points should be considered in the method statement (Section 2.9.1). They may also require the inclusion of space resources (Section 3.2.7).

3.2.4 Materials

As with other resources, materials can be classified in several ways. The three most common sets of major subdivisions are:

- materials for permanent works
- materials for temporary works

or alternatively:

- storable materials
- non-storable materials

or:

- materials for a specific use
- materials for general use.

The first of these classifications seeks to distinguish between the materials built into the works and those that are only incorporated temporarily. For example, formwork would normally be classified as temporary and concrete as permanent, but sometimes formwork is built into the works, perhaps because the structure is designed to retain the formwork as a feature or because it would be too difficult or too costly to remove it. Similarly, concrete is sometimes used in a temporary manner, as for example in the foundations of a temporary bridge, and removed when it is no longer required.

Permanent works materials can be used only once, but *temporary works* materials may be capable of use several times over. With *temporary works* material, the quantity may change between uses because of irreparable damage to the material or because it proves impossible to remove some of the material from the work. This suggests very different cost implications for the project with the irreparable material usually incurring less cost than the abandoned material. For example, a piece of sheet piling can be driven and extracted several times although it may need cleaning or maintaining between the extraction and any subsequent use. Some pieces may be so badly distorted that they cannot be straightened enough for future use and have to be sold for scrap, and some pieces may have to be cut off and left in place because of the difficulty of extraction. Pieces sold for scrap generate some income. Care must be taken if the first classification system is used.

The second classification system considers storable and non-storable materials. Storable materials are ones which can be delivered to a project and used at some time in the future. Non-storable materials are those which must be used as soon as they are delivered. Examples of storable materials are reinforcing steel and aggregates for producing concrete. Examples of non-storable materials are wet concrete and macadam or asphalt for roads.

No material is completely storable and there is no definite rule that any given material is storable or not. Wet concrete has a short usable life and steel has a long one. Neither will last indefinitely if not incorporated into the works. Storable materials must be treated differently from non-storable ones in the planning process.

The third classification system considers the particular use of materials. Materials for a specific use are those which are to be built into a specific part of the works. An example is a reinforcing bar of a particular shape and size. Resources for general use are those which can be incorporated in many different places. Concrete and, perhaps, straight reinforcing bars are such resources.

Despite this distinction, construction workers are very inventive (!) and are likely to make a specific reinforcing bar fit exactly where it was not intended, in order to speed up progress. Unfortunately construction workers rarely consider the wider aspects of this type of action, and it invariably results in an increased duration and/or cost for the project.

Whatever classification system is adopted, one aspect which warrants special attention is wastage. Wastage means loss of material, which may be through:

- off-cutting
- damage
- theft
- incorrect application
- being lost.

Although wastage does not contribute to productivity or constrain the project in the normal sense, and hence does not appear to fit the standard definition of a resource, it nonetheless can hinder progress and represent a drain on project resources which could otherwise be applied more profitably.

In many ways material resources are very like the basic financial resources, liability and earning (Section 3.3). They could be treated identically if it were not for the different vocabulary that has grown around the two areas, and this should be borne in mind for the rest of this chapter.

3.2.5 Generated material resource

Resources are generally brought on to a project for use at a specific time and taken off when they are no longer required. Some resources (particularly materials) can be generated, as well as being used up, by the execution of an activity. Examples of such materials are earthworks fill material and formwork.

A plant resource which could be considered to be generated is a large crane or tower crane created on site by activities which construct it. The removal activities effectively destroy the resource. It is more realistic to consider generated material resources than generated plant resources in our planning considerations.

Fill material is usually generated by carrying out an excavation activity and the availability of fill material is governed by the progress of the work. Formwork is slightly different in that it is initially brought on to the project in the same way as any other material but, when it is used, it is erected and cannot be used by any other activity until it has been struck from its original position. In detailed programmes of work there could be an activity, *erect formwork,* which uses the material and another, *strike formwork*, which generates it for use on other activities. By programming in this way, an activity such as *pour concrete*, which requires the formwork to be there, is modelled without in fact using any formwork. All temporary materials can be considered in this way in the scheduling process (Chapter 7).

3.2.6   Subcontractors

Subcontractors are used very widely in the construction industry. The benefits claimed for their use include:

- they enable their employer (any organisation, not just the main construction contractor) to carry out specialised tasks without maintaining a skill base;
- they reduce the fluctuations in direct labour levels of their employer;
- they provide a degree of certainty with respect to cost and time.

The disadvantages include:

- they are less controllable than a company's own labour;
- they can be more expensive.

Planning for the use of subcontractors is often considered as time only planning or target setting and the requirement for resource considerations is ignored. This attitude can lead to many problems and should be avoided if possible.

When considering subcontractors in the planning process, it is suggested not that they are considered as a single resource but that their labour and plant should be considered as extra labour and plant resources for the main contractor. Any other method of treatment inevitably removes a great deal of detail and is likely to produce an unrealistic plan of work. The lack of information about the output rates and availability of subcontractor resources, which is used as a reason for not considering them in detail, must be overcome in order to ensure that proper targets are set and control can be exercised at the deployment stage (Chapter 12).

A key point here is the contractual agreement, which often specifies simply that the subcontract work must be completed by a particular date. In the contract, there is little if any opportunity afforded in the contract for the main contractor to exercise control over the subcontractor's resources to help meet the target date. If, however, the main contractor does direct the subcontractor to use resources in a specific manner, the subcontractor may experience disruption and be able to claim for this. The main contractor must therefore accept that having flexibility in the use of the subcontractors' resources may lead to higher financial costs for the project.

In addition to subcontractors' own resources, it is important to consider the supervision and attendance which would be necessary when subcontractors are operating. The supervision might be in terms of the staff required to:

- oversee the work,
- ensure that the work is not held up for lack of resources,
- ensure the smooth integration of various subcontractors.

These would all affect the timing and cost of the project as well as the durations of the individual activities. They would also affect the resource requirements of the project and potentially the resource requirements of other apparently unconnected activities.

Attendance on subcontractors would be the labour and plant which must be provided by the main contractor for subcontractors to carry out their work. Such attendance can be significant and, if not properly planned, can severely affect the performance of the project adversely. On construction sites, cranes are major items of plant which are often required for attendance on subcontractors. When carrying out such attendance, cranes are able to spend less time on the main contractor's own work, and the productivity of the contractor's own resources may decrease. If cranes are not available, the subcontractors' productivity may be affected, resulting in an increase in the cost to the main contractor and potential disputes.

Subcontractors offer much, but their integration into a project would be much enhanced if the main contractor planned for them properly.

3.2.7 Space

Space is often a limiting factor in construction projects; even if the works are on a green-field site the boundaries are likely to be close to the construction edge and access may be a problem. For various reasons the idea of a space resource could be useful for planners.

As an example, when constructing reinforced concrete it is often possible to erect the formwork before the reinforcement or vice versa, which might suggest that in some cases the two operations could take place together. If the structural element occupied a relatively small space the labour force required to fix the reinforcement and formwork might not all fit into the confined space (or if it did then safety would be compromised). In this case a space resource could be used to ensure that the two did not work together.

A space resource is represented, like any other resource, by a name and quantity available. When demanded by an activity about to be carried out, the space resource, like any other resource, would be assigned to the activity and thus would not be available for use by any other activity requiring the same space resource.

In a more extreme situation a spatial constraint may dictate which activity should be carried out first. This is illustrated by the construction of a reinforced concrete column where the reinforcement has to be fixed before the formwork can be erected, because the erection of the formwork produces such a small space in which the reinforcement has to be fixed that the steel fixers cannot fit into it. In this situation, it is suggested that a space resource need not be used but instead a technological constraint should be introduced between the two activities *fix column reinforcement* and *erect column formwork*. This could be in the form of a logic link (Chapter 5) indicating which activity must be done first.

It can be seen in the above example that the difference between a resource constraint and a technological constraint is not always as clear as might be imagined. In practice it is often a matter of judgement as to whether or not to include a constraint. Some computer packages allow users to define several sets of constraints and to decide which sets to apply for any particular run of the calculations. Resource and technological constraints are discussed more fully in Chapter 5.

3.2.8 Abstract resources

Abstract resources have been suggested as being useful for such things as ensuring that two activities do not occur at the same time. If it is really necessary to do this, there must be a practical reason and this, as such, should give rise to a non-abstract resource. An example of this might be the need to prevent two activities happening together for safety reasons, when a *safety* resource (or more likely, a space resource) could be used. It is usually possible to manage without abstract resources but they should be borne in mind in some situations.

Another use for abstract resources is in project control (Chapter 12). There are many resources on a project and it can be difficult to know which (if any) to select for monitoring progress against. For a project as a whole it may be appropriate to convert resources to money and use earnings and liabilities (Section 3.3) as control measures. In some cases it may be better to use an abstract resource such as *activity days* or *activity weeks*. Here each activity would have an activity day (or week etc.) value equal to the number of days the activity was planned to last. The progress of a project as a whole can usefully be monitored against this measure and is discussed further in Section 12.3.5.

3.2.9 Overheads

Overhead resources are those which are necessary for the completion of the project but which cannot be directly associated with any particular activity. Typical of such resources are the supervision provided, the offices, telephones and vehicles. For other resources, such as cranes, it is not always easy to decide whether it is an overhead or not. It is perhaps consistency of definition that is important, so if a crane is once considered to be an overhead then it should always be one. Overhead resources should be considered on a project but must be treated differently to ordinary resources.

3.3   FINANCIAL RESOURCES

It is possible to consider financial resources in a similar manner to other resources (Section 3.2), breaking down the types of finance to classify them. The most commonly used classification for finances is related to the direction of flow of the money and can be thought of as:

- income
- expenditure
- profit.

The provision of work, services or goods usually occurs at a different time to the exchange of money. In project planning and control, the actual flow of money is not usually as important as the scheduling of the work or the provision of the goods or services. It is, however, important to acknowledge and be able to allow for this time difference in many models of a project. This can be accommodated in planning by using a simplified model of finances which can be demonstrated as follows.

Consider a person doing some work for an employer at a particular time. In doing the work, the worker earns a reward but in most circumstances expects to get paid for it (income) sometime in the future. When the work is being done the employer is incurring a liability to pay the worker but it becomes an expenditure only when payment is made. In general a liability becomes an expenditure and earnings become income. An expenditure for one person or organisation becomes an income to another. The word cost has been avoided in this description as it is used loosely in many situations. It should be considered only as a generic term relating to finances and is used as such in this book.

Figures 2.24 to 2.27 clarify these terms. Figure 2.24 shows the monthly liabilities of a contractor on a project over a 13 month period. These liabilities are incurred continuously in carrying out the work. In return for doing the work the contractor expects payment related to the monthly earnings evaluated during each month over the 12 month period of the project. Figure 2.25 shows the monthly expenditure, idealised as a constant rate of expenditure per month, incurred over a 15 month period when the liabilities shown in Figure 2.24 are settled. Similarly the income received monthly as payment by the client in respect of the valuations of the contractor's earnings is also shown in Figure 2.25 as a one-off payment each month (less retention money). Figures 2.26 and 2.27 show the same information, plotted cumulatively rather than monthly, with the additional information, income minus expenditure, shown in Figure 2.27.

The simple model is complicated by various realities that make financial management a complex topic in its own right. Examples of reality are: where some work does not directly earn anything but must nonetheless be carried out; where inflation and interest rates are considered; where taxation and other dues are involved. In order to plan other resources it is important to understand the simple model and discuss its application in planning. The components of the model are discussed below.

3.3.1 Expenditure

When a project is carried out various payments are made. The owner's expenditure includes payment to:

- the main contractor(s) for the work.

The main contractor's expenditure includes payment to:

- various suppliers for the materials;
- subcontractors for their input;
- the plant hire company (or the contractor's plant department) for the use of the plant and equipment;
- the labour force for their work.

It is obviously possible for the main contractor to differentiate between the expenditure given to:

- suppliers of materials
- subcontractors
- labour
- plant.

It is reasonable to consider these as classifications for the expenditure. It should be noted that the proposed classification is identical to that used for other resources (Section 3.2).

In financial terms the classes are often called *cost heads*. Typical cost heads in a contractor's organisation are:

- labour
- plant
- permanent materials
- temporary materials
- subcontractors
- overheads.

Financial overheads in the above list are monies which are required for the completion of a project but which cannot be assigned to any individual piece of work. They can be lump sums, time-based or value-based. Overheads based on lump sums are those which occur only at a single time on a project. An example is the money payable as a bond at the start of a project.

Time-related overheads are those which depend on the duration of a project. An example of this is the salary of office managers on projects (since they would be paid for the duration of their project). Time-related overheads may operate over just a specific part of a project and might vary in when they occur and for how long. Hammock activities (Section 4.2) are useful for modelling such situations.

Value-based overheads are those which vary depending on the amount of work being done on the project. Engineers' salaries might be considered in this category because more engineers would be required if there was more work.

Many types of plant and equipment could be considered to be overheads. An example might be a crane on a bridge project where it might not be possible to allocate the crane specifically to any of the activities, such as *erect formwork* or *fix reinforcement*, but where the crane is an essential resource for the work and hence an essential expenditure. The cost of the work would increase as the duration of the construction of the bridge increased because the crane would have to be available for longer. This might be irrespective of the cost of the individual activities but could be because the crane could not be used as efficiently, there being an increase in its idle time. Such a piece of plant might be considered as an overhead. This would have the advantage of enabling true costs to be seen for the whole project but would not provide information for detailed control and future estimating (Chapters 11 and 12). If this were done, hammock activities (Section 4.2) would ease calculations for planning and control purposes.

It can be seen that overheads themselves can be split into labour, plant and materials, so the class could, conceptually, be eliminated. Because of the uncertain and arbitrary nature of some of the other overhead charges (for example the contribution that a project must make to the running of a Head Office complex) it is inadvisable to consider removing the overheads classification entirely.

3.3.2   Liability

The payments referred to in the expenditure classification above are rarely made as the work is done. The labour force on a project may be paid weekly in arrears, the site staff monthly in arrears, the subcontractors monthly in arrears and the material suppliers three-monthly in arrears. The longer the delay in payment the greater the advantage to the payer and the disadvantage to the payee.

When work is done by an employee (or subcontractor or supplier) a liability to pay for the work in accordance with the contract's payment system is incurred by the employer. The financial liability becomes an expenditure after a lapse of time.

The time delay between expenditure and liability varies from one contract to another, from one organisation to another and from one cost head (Section 3.3.1) to another. It may also be that there is a retention written into the contract which allows a percentage of the payment due to be withheld until completion of the project, or some time after it, in order to try to ensure the satisfactory execution of the work.

Projects measured on the comparison of expenditure with income may give false indications of profitability because of the large amount of money owed to creditors. Most site-based, and many quite senior, project managers have little or no control over when owed money is actually paid; this is a function of the administrative processes of organisations set up by their managers. For this reason site-based personnel should plan and control on liabilities rather than expenditures.

3.3.3 Income and earning

In the same way that money flows out of companies during a project, money also flows into them. Main contractors receive payment from owners; suppliers and subcontractors receive money from main contractors. The money flowing into a company is here called income or revenue. It is obviously very closely related to expenditure discussed in Section 3.3.1. Earning is the incoming equivalent of the liability outflow described in Section 3.3.2. For planning and control purposes the income is less controllable than the earnings and so is not covered specifically here.

Income is of utmost importance to a company and so attempts can be made to offset the effects of retention money and payment in arrears. This can be done by trying to generate higher margins (income minus expenditure) on the early work on a project. This and related financial matters are very important topics but are not discussed further in the context of this book.

3.4 SUMMARY

- Resources are as important in planning as time is.
- Resources and finances are largely interchangeable.
- Resources can usefully be grouped.
- The more common resource groups are labour, plant, materials, subcontract and overheads.
- Important resources should be identified and the others grouped appropriately.
- In any function all resources should be considered once and once only.
- The estimate and the plan should be cross-checked for each resource.
- Common cost heads (breakdowns of finances) are labour, plant, materials, subcontract and overheads.
- Subcontractors should be considered as one's own labour and plant for planning purposes.
- Space can sometimes be a useful resource.
- Abstract resources can be used to model such things as safety.

4

ACTIVITIES AND RESOURCES

'I like work: it fascinates me. I can sit and watch it for hours.'

JEROME K. JEROME
(*Three Men in a Boat*)

Doing work consumes resources, takes time and costs money. That should be enough to fascinate anybody involved in construction or management. Unfortunately, planners do not generally get the opportunity to spend time sitting and watching work and appreciating the intricacy of it all. If they did there might be a better understanding of how to use activities to model the work that must be done to complete a project.

4.1 INTRODUCTION

All construction work is carried out by resources and all project plans are based to a large extent on activities. The success or failure of a plan is often determined by the choice of the activities – the cornerstones of the plan. As this fact is seldom fully appreciated, this chapter examines some of the important factors to be taken into account when considering the choice of activities. It covers such things as how many activities to use, the level of detail in activities, the interaction of activities, and the consideration of activities and resources together.

In keeping with the spirit of the book, only practical aspects are considered and it is assumed that the reader is familiar with basic planning techniques. Where there is some debate as to the interpretation or use of techniques in practice, the issues are fully covered.

Resources and finances are treated together whenever possible throughout this chapter because they are very similar in behaviour and in any case, as stated in Chapter 3, finance is a type of resource. It is sometimes suggested that a single resource, *money*, is used and all other types of resource reduced to this. This is satisfactory in some instances but for detailed planning and scheduling it is suggested that this approach is insufficient and several types of resource should be considered.

A plan of work needs to allow for the conflicting resource demands of activities. The resources that exist in construction work need to be defined and assigned to activities which can then be scheduled to meet the objectives of the project. A construction activity normally represents a complex piece of work requiring several different types of resource to carry it out. A project needs an even greater number of types of resource. The greater the number of resources, the more realistic the model appears to be but, unfortunately, the more complex the process of planning becomes. Mistakes are also more common when large numbers of resources are used.

Project or site resources are assigned to activities to carry out the work. The progress and final duration of the activity are defined by, or at least closely related to, the number of each resource assigned to it. The process of assigning resources to activities can become very complex and have considerable influence on the plan and the cost of a project. This is covered in Sections 4.4 to 4.7.

Having decided on the resources to be used on a project and having assigned them to activities, some form of scheduling procedure is usually carried out (in a process sometimes rather confusingly known by the general name of *resource allocation*). The scheduling process determines the start and finish time of the activities, having considered the deployment of the planned resources. Descriptions of resource scheduling procedures and the process of planning with resources are contained in Chapters 7 and 11.

4.2 DEFINING ACTIVITIES – THE BASIS OF A PROJECT MODEL

As planning relies significantly on the use of activities, it should be possible to define and classify several types of activity. Various standard definitions of an activity exist, usually suggesting that it is a part of a project that uses resources and takes some time to complete. As some of the important parts of a project are events that happen virtually instantaneously, and consume little or no resources, it can be argued that an activity should simply be defined as a part of a project. Whatever the basic definition, several types of activity can be identified, as below. This chapter concentrates on normal activities although all types should be understood and their application appreciated.

Normal activity   One which operates over a set time using particular resources and for which it is possible to define its relationship with other activities clearly.

Hammock or summary activity   An activity, joining two specified events, that may be regarded as spanning two or more normal activities. Its duration is initially unspecified and is only determined by the difference between the times of the events concerned. For further information refer to Section 4.5.4 and, for clarification of events and activity interactions, refer to Chapter 5.

Process activity   This is similar to a normal activity except that it is not possible to define its relationship to other activities fully using just its start and end. In reality, process activities are normal activities which are geographically spread out (for example *surface road*). Most planners do not recognise the difference between these activities and normal ones, and use techniques developed for normal activities with process activities. This can lead to misunderstandings.

Chemical process activity   These are similar to a normal activity except that it is impossible to stop them after they have started. Often they require no standard resources and work 24 hours per day 7 days per week every day of the year. A construction example is *cure concrete*.

4.2.1   Choice of activities

Accepting that activities form the basic components of many project models it is desirable to choose the activities to suit the nature and purpose of the model.

A common scenario in the construction industry is as follows: a large developer with many projects treats each separate project as a basic component of its planning model; a contractor appointed on each project divides the work into activities to form the basis for construction; subcontractors produce smaller activities for their sections of work; and individual workers deal with even smaller elements in order to plan their day's work. In short, the appropriate definitions of 'activity' fundamentally depend on why we are interested in defining activities in the first place.

It follows from this that the choice of activities for planning is not an exact science. It is rather more a matter of experience. As the choice is fundamental to the success or otherwise of planning and might therefore be considered to be vital to the successful completion of a project, it is considered in some detail here.

The choice of activities depends on several factors which themselves are not independent. In particular, it is common to consider the following.

Use for which the plan is intended   A plan which is to be used for detailed control on a project would be expected to have activities which represent the project in some detail (and thus the activities would be fairly elementary). A plan which is to be used to assess the feasibility of an outline proposal of a project would probably have fewer activities than the detailed control plan but they would be compound (and thus complex) activities.

Stage of the project at which the plan is made   The further through a project that work progresses, the easier it is to see the relative importance of the individual tasks or types of work which make up the total project. The activities used should reflect this. A plan of work constructed after the completion of the project for claims purposes would therefore be expected to include not only the activities envisaged at the start of the work but also those activities which are useful to show the effects of any alterations to the originally specified work.

Amount of information available   The more detailed the information, the more realistic it becomes to plan the work in detail and hence the more realistic it is to include detailed activities.

Person or company making the plan   Several organisations produce plans of work for a project and within a single organisation, several people at different levels of the organisational hierarchy and with different responsibilities produce plans (Section 1.2). Even at a given position within an organisation, two people might well produce different activities because of their own individual background and experience.

Complexity of the project   The more complex the project, the greater the importance of producing detailed plans of work since a hold-up in one area can have wide-ranging repercussions throughout the project. Unfortunately, it is often in the most complex projects that the least information is available at the planning stage.

Planning method proposed   Not all planning methods demand the same level of detail and indeed, some planning methods are not able to make use of as much detail as others.

As mentioned earlier, not all of these considerations are independent: the use to which a plan is to be put is closely related to the person who is drawing it up; the stage of the project at which a plan is being made is related to the amount of detail available; the person making a plan is related to the planning method proposed and the complexity of the project should also be related to the planning method proposed. Despite these obvious inter-relationships, it is important to consider all the above points when selecting activities.

In addition, anyone making a plan should also consider the following points when choosing activities, depending on the level at which the plan is intended to be used.

Geographical area   It is useful to have activities which are not spread out too much in a geographical sense. The concept of 'not spread out too much' is rather difficult to define exactly and depends on the project and the use to which the plan is to be put. In the example of Tingham Bridge (Appendix A), work on the two opposite sides of the river may be considered to be in different geographical areas because the access problems may be different for the two sides. In the case of a road project it is likely that stretches of road between the various structures would be considered as different geographical areas.

Structure   Different structures should be in different activities. Thus, on a road project with many structures (for example, the Tingham Bypass (Appendix A)) it would be sensible to put structures into separate activities. This is because different structures are often done at different times and have different problems.

Structural element   Within a single structure, it is useful to split the work down by structural element. Thus, for Tingham Bridge, the foundations would be treated differently from the piers which are different from the deck.

Work type   Within a structural element, it is useful to split the work down by type of work. For Tingham Bridge, the types of work would be such things as *erect formwork*, *fix reinforcement* and *place concrete*.

Predominant resource   It is often useful to divide up the work on a project into that done by different major resources. This allows for easier planning and control of resource demand (Section 4.4). Some of the resources for Tingham Bridge might be *piling gang, formwork gang, steel fixing gang*, and *concreting gang* or, if the work is to be subcontracted, the different subcontract gangs. Clearly these may all contain some *general labourers* but the suggested resources have been chosen as the important ones. If more than one gang of any particular type is to be used on the project, it is useful to separate their work to enable proper control action to be taken.

Responsibility   It is very useful to have a plan in which the person responsible for an activity can be identified. Thus, if a particular piece of work includes piling, with a subcontractor being responsible for it, this work should be included as a separate activity (or activities) in the plan.

Contractual considerations   If there are conditions in a contract, such as information about payment in stages defined by completed elements of work, the planners should be prepared to consider these conditions by choosing suitable activities to enable the payment stages to be identified.

Considering the level of detail in a plan, the preceding discussion on geographical area assumes a level of detail in a plan which it may be wrong to assume. For example, if the bridge formed a small part of a much larger project, the overall project plan may consider the foundations of the bridge to be a single activity. Only at greater levels of detail might it be considered necessary to separate the work on the two sides of the river. By not separating the work in the outline plans, problems might be missed. In this example the access problems across the river exist, no matter what the scale of the plan, but such problems are easily missed in outline plans. For this reason, outline plans are often optimistic, which may or may not be a good thing.

Considering for whom a plan is produced, the client's representative is more likely to be interested in the structure or structural element than the contractor is; the contractor is more likely to be interested in the type of work or major resource than the client's representative is. Inconsistency in the activity requirements of the various parties can lead either to unsatisfactory plans or to more than one plan. Neither of these situations is good. Many of the problems can be overcome by use of summary activities, or hammocks, introduced at the start of Section 4.2.

4.2.2   Example of activity selection

To illustrate the choice of activities, consider the construction of Tingham Bridge on the Tingham Bypass (Appendix A).

Two scenarios are considered for the completion of this bridge:

(1)   The bridge is to be considered as a complete project in its own right.
(2)   The bridge is to be considered as part of the bypass project consisting of several kilometres of dual carriageway road and associated structures. Most of these involve a similar type of *in situ* reinforced and precast concrete construction but generally have only one span.

Following the guidelines presented in Section 4.2.1 gives activities similar to those contained in the following lists for Scenario 1 of the example:

Planning for the client or the contractor's directors   These people would probably use a Gantt chart (Section 2.2.1) for planning the project and the activities might be:

- *set up site*
- *foundations*
- *substructure*
- *superstructure*
- *finishes.*

Planning for the project manager employed by the contractor or for the engineer employed by the client to oversee construction   These people would use either a Gantt chart or a network (Section 2.5) and the activities might be:

- *set up site*
- *foundations west*
- *foundations centre*
- *foundations east*
- *substructure west*
- *substructure centre*
- *substructure east*
- *superstructure east*

- *superstructure west*
- *surface*
- *finishes.*

Planning for the manager employed by the contractor with responsibility for the daily control of the bridge This person would use both Gantt charts and networks and would produce several plans as the project progressed. At the start of the project, the activities might be:

- *set up site*
- *piling and cap for foundations west*
- *piling and cap for foundations centre*
- *piling and cap for foundations east*
- *formwork for foundations west*
- *formwork for foundations centre*
- *formwork for foundations east*
- *reinforcement for foundations west*
- *reinforcement for foundations centre*
- *reinforcement for foundations east*
- *concrete foundations west*
- *concrete foundations centre*
- *concrete foundations east*
- *reinforcement for substructure west*
- *reinforcement for substructure centre*
- *reinforcement for substructure east*
- *formwork for substructure west*
- *formwork for substructure centre*
- *formwork for substructure east*
- *concrete for substructure west*
- *concrete for substructure centre*
- *concrete for substructure east*
- *strip formwork to substructure west*
- *strip formwork to substructure centre*
- *strip formwork to substructure east*
- *reinforcement to east wing walls*
- *formwork to east wing walls*
- *concrete to east wing walls*
- *strip formwork to east wing walls*
- *reinforcement to west wing walls*
- *formwork to west wing walls*
- *concrete to west wing walls*
- *strip formwork to west wing walls*
- *drainage behind east abutment*
- *drainage behind west abutment*
- *backfill to east abutment*
- *backfill to west abutment*
- *erect beams to west deck*

- *erect formwork to west deck*
- *reinforcement to west deck*
- *concrete to west deck*
- *erect falsework to east deck*
- *erect formwork to east deck*
- *reinforcement to east deck*
- *concrete to east deck*
- *strip formwork and falsework to east deck*
- *waterproof membrane to decks*
- *surface bridge*
- *install parapets*
- *slabs under deck*
- *landscaping*
- *tidy site.*

These activities would be augmented as the project progressed and as more information became available. For example, there is no activity for the fixing of the bridge bearings so they would currently have to be assumed to be in either the substructure activities or the deck activities.

For this level of planning, it might be argued that the amount of detail is either too great or too little. It may be too great because of the large number of activities, especially towards the end of the project, whereas it may be too little because of obvious omissions (as for 'bearings' above).

If the same procedure were followed for Scenario 2, the lists of activities would be similar but there would be an extra layer in the contractor's organisation, perhaps requiring a plan for someone in charge of all the structures on the project. At this level, the detail of the bridge manager's plan would be too great, whereas the detail of the project manager's plan might be insufficient.

4.2.3 Activity descriptions

Planners have a tendency, especially when using computers or working on large projects, to abbreviate activity names or use activity numbers in their place. Since one of the main purposes of a plan of work is as an aid to communication between the parties involved in a project, excessive abbreviation detracts from the usefulness of a plan. All the people involved with a plan (at all levels in all the relevant organisations) should understand the same thing when an activity name is used. In order to make a plan clear and unambiguous it is therefore important to use full activity names, to make the name as descriptive as possible, and never to use just a number no matter how advantageous or convenient it might appear to be on small projects. Indeed, even full activity names might not be sufficient and an explanatory note might be useful for reference purposes.

As an example, taking the project manager's activities from the Tingham Bridge project (Section 4.2.2), a typical note might be:

Activity	Note
Substructure west	Activity to include all fabricating and fixing of reinforcement and formwork; concreting; curing concrete; removal of formwork; positioning of bridge bearings; backfill around the base of the structure; slabs on the earthwork under the deck; and all necessary drainage.

Anyone questioning what had happened to the falsework for the West substructure of the bridge would be able to check whether an omission had occurred by checking the other activities and their descriptions.

Throughout this book, as in real life, compromises have had to be made. Activity names such as 'Activity A' should be avoided because they have little or no meaning by themselves. However, where it would be too verbose to illustrate a simple general point using full activity names, abbreviations have occasionally been used.

4.2.4 Size of activities

Although activity size is often used as a measure of a plan, what is perhaps rather more important than the absolute size of an activity is the relative size of all the activities in a plan. In bills of quantities (see Preface) it is common for about 80% of the cost/value to be contained in 20% of the items, giving 20% of the items as large and 80% as small. If this were to be repeated in the activities, it would give problems with using the plan for control purposes since an activity is not as easily split into parts as the items in the bills of quantities are. This is because the larger the activity is, the less likely it is to be made up of homogeneous work, whereas in a bill of quantities all the work in a single item is of a single type (by definition).

Large activities are most common in broad, long-term plans used by clients for evaluating projects and by clients and contractors for monitoring overall progress. They are often expressed in terms of whole structures or major parts of structures and would usually contain many different types of work.

Small activities are most common in short-term programmes used by contractors for detailed planning and control. The activities are usually defined in terms of single work types or single resource types.

If large and small activities are mixed within a single plan, it can be expected that the plan would vary in usefulness and not be continuously useful to a single person or group of people. The only time when it is sensible to have activities of varying sizes is when using a *rolling programme* approach to planning. In this, activities are planned in more detail as the time for working on them approaches. Thus, activities a long way in the future are very large, multidisciplinary, and rather uncertain, whereas those for immediate operation are small, often single work type and considerably more certain. Despite this, planners may be forced into producing plans with varying levels of detail by the different amounts of information available for different areas of the project when the plans are being produced. In such a situation, the plan should be updated and the detail included as soon as it becomes available.

The size of an activity can be measured in many ways but, although there is a rough correlation between them, there is no strict rule that if an activity is small by one measure then it will be small by another measure.

The most common ways of measuring the size of activities are by duration and by value.

DURATION

The duration of an activity can in theory be almost any size. It is important to consider the uses of a plan in deciding the minimum and maximum desirable duration. In most forms of construction, except in exceptional circumstances, it seems unreasonable to have very small duration activities (less then half a day) because the impracticality and uncertainty of collecting and correlating information on such a time-scale would not generally allow any use to be made of it. Exceptional circumstances would be where the activity related to a key event, the importance and organisation of which might otherwise be overlooked. It also seems unreasonable in the vast majority of situations to have activities with durations which are a large proportion of the duration of the project. This would appear in most situations to limit durations to the range of 1 day (or shift) up to 30 days (or shifts).

Such general advice should not stop very detailed planning of specific areas of work. For example, it might be sensible to plan to much greater detail the operations to be carried out during a single night shift when a major road is to be closed. In this case activity durations of 30 minutes might not only be acceptable they might actually be essential.

VALUE

This can be further divided into any number of different values including, for example, profit, cost, income, material cost and labour cost. There is often a strong correlation both between these values and also between them and the activity duration. It should not be assumed, however, that there is always a correlation. Typical examples where correlations do not exist are the following:

- With expensive procurement activities where the cost may be great but the profit small.
- With activities like 'ground consolidation' where the duration may be very long but the value (in all the cases above) is small
- With small activities where work (such as *switch the traffic from one lane to another* in a road construction project or *connect house drains*

to sewer in a building project) is sometimes given very small or even zero duration. In reality this is often not a sensible tactic because the preparation for the actual switch often takes a considerable time and this should be included in the activity duration or a *preparation for switch* activity included. It is impossible to monitor the progress of a zero duration activity; it is either complete or not started.

Furthermore, it is impossible to assign a resource to a zero duration activity in any sensible manner. Resources assigned to a zero duration activity would not appear in a project resource aggregation. Equally, when considering financial information, it is impossible to assign finances to a zero duration activity in any sensible manner since the spend and earn rates become infinitely large. Nevertheless, there are occasions in practice when finances are attached to zero duration activities to represent, as *point* costs, the often quite large sums of money for insurance payments, land rental payments, council tax and so on. Bearing in mind the above points, planners should try to avoid this practice, for example, by choosing a small but finite duration for the activity.

4.3 RESOURCES AND ACTIVITIES

4.3.1 Initial considerations

Resources are all those things that are required for work to progress. The term includes everything, such as space and finance, as well as labour, plant and materials. It is important to understand the relationship between resources and activities when deciding on the activities for a project plan. This should be apparent from Section 4.2. A plan of work needs to allow for the conflicting resource demands of activities. The resources which exist in construction work need to be defined and allocated to activities which can then be scheduled to meet the objectives of a plan. These points are considered in the rest of this chapter.

A construction activity is normally a complex piece of construction work requiring several different types of resource to carry it out. A project needs a much greater number of different types of resource. The project or site resources are allocated to activities to carry out the work and the progress of an activity is defined by, or is at least related to, the quantity or number of each resource allocated to it. This process can become very complex and have considerable influence on the plan and cost of the project.

When considering the allocation of resources to activities, many points should be borne in mind. The two outlined here are of fundamental importance and are discussed in detail as the chapter progresses.

Number of resource types There are usually a large number of types of resource which could be involved on a construction project. The greater the number actually used, the more complex the problem of both allocating the resources to an activity and subsequently scheduling the activity. Furthermore, any schedule produced with a large number of types of resource runs the risk of being very poor because of the difficulty of matching the demands made by the activities on the resources. It is common for a manager to recognise a small number of key resources to work with and either group together the remainder as money or assume the remainder are unimportant and always available in the required numbers.

Activity durations   When considering the duration of an activity it is important to remember that the actual time taken to carry out the work is, to an extent, determined by the number of resources allocated to the activity. The more resources, the smaller the duration (in general), although in extreme circumstances over-crowding could occur and retard progress. If there are a number of different types of resource, the changes in duration can be unpredictable. In performing planning calculations, it is usual to assume that an activity duration is fixed at the optimum and that the resources assigned to it are also optimum. In reality, the numbers of resources on a single activity often vary from day to day and hence the rate of progress of the activity also varies.

4.3.2   Resource management

The resources required on a project range from obviously important resources such as large cranes and skilled craftsmen to rather less important and more readily available ones such as nails and hammers. Theoretically, it is possible to consider each and every resource independently and produce a plan of work which makes best use of all of them.

The choice of resources is a matter for the manager and is a matter of skill and experience. The choice is complicated further because decisions normally have to be made very early in the life of a project when not all information is available. It is important to identify the resources which are likely to cause problems through the project. In many cases the important resources change as the work proceeds from one phase to the next and this should be reflected in the choice of resources. For example, at the start of a road construction project, earthmoving equipment will generally be a key resource; in the middle of the project, the key resources will probably be those used for the structures; at the end of the project the key resources will be those for paving, signs and other finishes tasks.

The difficulty in selecting resources is most acute at a change-over between phases, such as between the earthworks and structures or between the structures and finishes (in the above example). Where phases overlap, the key resources also overlap and the number for the manager to consider is increased. Similar difficulties occur when using a plan for control purposes since more records must be kept and analysed as the type of work changes.

Although it is usually better to treat each project on its merits, it is important to realise that a project is part of a company's business and that it has to be planned and controlled as such. For this reason, it is sometimes necessary to consider using resources which would not be used if the project were planned and controlled in isolation.

Planners should always consider matters relating to resource management.

4.3.3 Resources and finances

Resources and finances can be allocated to activities in a number of ways depending on the type of resource under consideration. Indeed the manner of allocation is sometimes used as a classification system for resources (Chapter 3). Resources are usually allocated in one of two ways:

- *Rate constant* resources are those such as labour for which it is normal to define the number of resources required in each time unit of the activity's duration.
- *Total constant* resources are those such as materials for which it is normal to define the total amount of the resource which is required for completion of the activity.

It is possible to define labour requirements in total constant terms (for example, an activity requires 20 worker-days of steelfixer work) and a material in rate constant terms (for example, an activity requires fill material to be delivered at $1500\,m^3$ per day over its duration). The manner in which resources are allocated to activities is important for planners to understand.

4.3.4 Comment

The above points (Sections 4.3.1–4.3.3) raise many questions about the assignment of resources to activities and the determination of the duration of activities. Section 4.4 discusses fundamental issues concerned with modelling the resource demand of activities. Some practical considerations affecting decisions about resources and activities are then discussed in Section 4.5. These are:

- gang sizes
- project resource totals
- money
- overhead resources.

Following this, Sections 4.6 and 4.7 provide a detailed examination of the relationship between resources and activity durations.

4.4 ACTIVITY RESOURCE DEMAND

The activity resource demand is the pattern of resources (Figure 2.22) required by an activity over its duration. Throughout this discussion, the term resource could equally mean finance. The main points to consider are:

- first, whether an activity uses only a single type of resource or whether several resource types are required;
- second, whether the resources assigned to an activity are allowed to vary over the duration of the activity or whether they must remain constant.

4.4.1 Single or multiple resource activities

Single resource activities are much simpler to use in most planning techniques and indeed are the only ones which can be used in some instances (such as time/cost trade-off problems). Unfortunately, it is common in construction to have work which cannot be carried out by a single resource type. For example, whereas *erecting formwork* is referred to as a task for carpenters, they need assistance from labourers to speed up the work and sometimes from cranes to lift the formwork to make the fixing work possible. Recognising that such activities exist, there are three options as to how to model the activity resource demand:

(1) Ignore all the resources except the key one on the activity (in this case carpenters). This is a dangerous option since the key resource may well turn out to be another resource, such as cranes, and this would not be apparent from the plan.

(2) Combine all the resources into a single resource (in this case a carpentry gang). This is potentially a good idea if the resources always work together. However, it is common to have resources moving from one gang to another or at least working with several gangs throughout a day. An example of this is a crane which can assist with formwork erection, reinforcement provision and concrete placing on the same structure on the same day. The more detailed the activities, the more likely this is to occur.

(3) Accept multiple resource activities. On most projects there are large numbers of resource types, ranging from workers and large plant to small tools and nails. Consideration of all of these separately would complicate the planning process to the point of making it impractical. It is therefore imperative to be selective in the resources to be considered, even if multiple resource activities are to be allowed. It should be remembered that, when doing this, the resources which have not been considered in the planning process are assumed to be there in as large a number as required and this must be allowed for during the execution of the project.

4.4.2 Constant or variable resource demand activities

For all the resources considered in the planning process, there are two possible assumptions for the resource demand of activities. The 'resource demand' is here defined as the quantity of resource (or resource level) of a given type assigned to an activity. Figures 4.1 and 4.2 show constant and

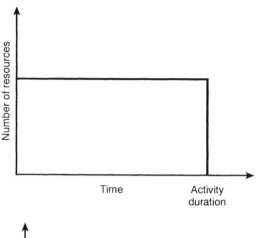

Figure 4.1 Constant resource demand activity.

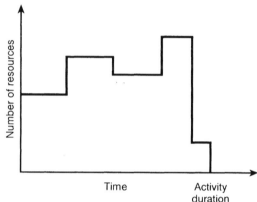

Figure 4.2 Variable resource demand activity.

variable resource demand respectively for a single resource activity. The variation in the resource level in Figure 4.2 is caused by the different quantities of the resource required for the various types of work comprising the activity. By far the simplest form of resource demand to use is the constant resource demand but, especially on large activities, the variable resource demand is by far the most realistic model.

The differences in the two models only really become apparent when the resources are aggregated for all the activities to provide the project resource demand, or when attempts are made to provide resource-limited or time-limited schedules. When either of these is attempted, the complexity of the planner's task is considerably increased with a variable resource demand activity model. Further complexity is added when considering variable duration activities along with variable resource demand (Section 4.6).

Since the variable resource demand model is more realistic, it might seem best to use this. In practice, if large numbers of activities are involved and the constant resource demand model is used (taking the resource use on an activity as the average of the levels in the variable demand model), the model gives results (after analysis) which are very similar to the results obtained by using the variable resource demand model. The decision must be made as to whether the effort and time in producing a better model (variable demand) is worth it. It can be difficult to make this decision because in certain situations, perhaps unknown when the plan is produced, extra detail may be required but is not afforded by the constant demand model. The following information should help with this decision.

Although there is no easily determined rule as to what constitutes a large number of concurrent activities, a number as small as 10 gives reasonable answers. This means that only plans with a very small number of very large activities need the variable demand model and it is questionable whether such broad plans should be used for resource planning.

It is therefore advised that the constant resource demand model should be used at all times.

If the planner suspects a problem with this model (from the answers obtained) then the activities in the problem area should be split into smaller simpler ones which are closer to the constant demand assumption. This simulates a variable demand model by using smaller activities which can have different constant resource levels. The problem of what to do when an activity has more than one resource type, each with a variable demand is now discussed.

If an activity has a long duration or is complex, it may require different quantities of different resource types at different times. This is an extension to the single resource, variable resource demand activity model introduced above, but now several key resources are considered on one activity. An example of this is shown in Figure 4.3 which represents the simplified labour resource demand for the construction of a typical reinforced concrete structural element. The resources considered are:

Labour	carpenters
	steelfixers
	labourers (assisting trades and serving as a concrete gang)

For labour, it can be seen that the resource demand changes several times throughout the duration of the activity and that the changes do not always occur at the same time for each resource.

Similar diagrams could be produced for the other resources such as:

Plant	crane
	concreting equipment
Materials	formwork
	reinforcement
	concrete

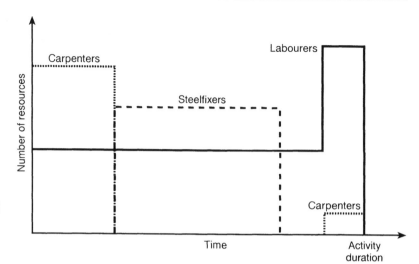

Figure 4.3 Resource demand for typical reinforced concrete activity.

The less detailed the activities in a project, the more likely it is that changes in resource demand will occur. If very detailed activities such as *supply formwork, supply reinforcement, fix formwork, fix reinforcement* and *pour concrete* are used to replace the single activity, then each of the more detailed activities could be modelled with a constant resource demand throughout its duration.

Further, the effect of changes to a project on the duration and resource demand of variable resource demand activities is very difficult if not impossible to model with any accuracy. Since changes to projects are a fact of life in construction, this also suggests that detailed plans should if possible have constant resource demand activities.

There is also a problem (similar to the one of assigning many resources to a single activity) with the allocation of a single resource to more than one activity at a time. This was alluded to in Option (2) in Section 4.4.1. Most resources can and should be assigned to individual activities since they work on one activity at a time. A few tend to work on several activities during the course of a day (for example, cranes) and these need to be treated differently. This is mentioned in Section 4.5.4 for the example of cranes.

Computer packages which offer complex activity resource demand facilities should be treated with some care because of the difficulty of defining the demands with great enough accuracy. There are also considerable problems inherent in their use when scheduling (Chapter 7) and trying to determine the resource–duration relationship for the activity (Section 4.6).

4.5 OTHER PRACTICAL CONSIDERATIONS

4.5.1 Gang sizes

Resources usually work in groups of set sizes. In labour terms these are gangs of workers under the direct supervision of a ganger. It may seem convenient in planning to consider the gang to be the smallest unit of resource but, as labour levels change, this may give very misleading results because of the changes in productivity. If changes in gang sizes are anticipated, the planner can consider using individual resources which have a constant productivity. The disadvantage of working in terms of individual resources is that the particular gang which works on site is not represented in the plan. Whichever method is used it is essential to check the realism of the plan.

4.5.2 Project resources

It must be remembered that the sum of the resources allocated to the activities and the idle resources must equal the sum of the resources on the project. There is a tendency, especially on large projects, for sections to plan independently and for this obvious check to be omitted. It is common to find some of the project resources not allocated to sections or to find some of the resources allocated twice.

One particular situation when this can occur without it being realised is when considering large items of plant on a geographically dispersed project. Some resources cannot move around a large construction project instantaneously. Typically, large cranes have to be ordered for specific dates and take several days to erect. This must be allowed for in the activities and the resource plan.

Another common example is the movement of crawler cranes between bridge sites on road projects. A project manager might correctly ensure that a crane is only used by one section at any one time but the time for travelling the crane from one position to another is sometimes neglected. Such problems can significantly affect the validity of a plan and its cost-effectiveness.

No ideal way of tackling this problem exists and it is up to the planner to ensure that resource movement is realistic. This may be one situation in which it may be advantageous to use resource constraints in a project network and include specific activities for large resource movements. Resource constraints should, however, be used with care since they greatly reduce the ability of resource schedulers to produce good schedules.

4.5.3 Money

As discussed in Chapter 3, money can be used as an aggregate resource. If this is done the problems of even spread still need to be considered. For example, the provision of material may require a large amount of money compared with the installation of that material. An approximation to the money demand for the reinforced concrete activity described in Section 4.4.2 is shown in Figure 4.4. It can be seen that it is far from even, and many of the same problems in scheduling and in the resource-duration definition (Section 4.6) still exist.

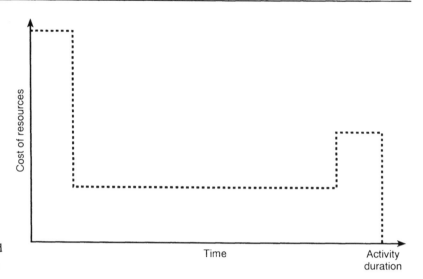

Figure 4.4 Money demand for the reinforced concrete activity.

In addition to the considerations of resource type, time lags should also be considered if the money is modelling payment into or out of a company (Sections 2.7 and 3.3).

4.5.4 Overhead resources

Overhead resources are not usually allocated to activities since by definition they are not provided uniquely for particular activities. They must not be forgotten, however, and the following points may be useful when they are considered.

Some resources such as cranes can be considered as real resources rather than overhead resources. They can work on many activities of the project in a single time period (such as a day or week). Thus if a crane can work on up to 10 activities in a day, it is useful to consider it not as a single crane but as 10 part-cranes, each of which can work on a single activity at a given time.

Summary or hammock activities (Section 4.2) can be used for the allocation and monitoring of overheads. The duration of such activities is not defined but instead is calculated as a result of compounding the timing of others. Care should therefore be taken when allocating resources to them. Total constant resources (Section 4.3.3) change their rate of use as the activity duration changes and thus may give a misleading indication of resource demand especially when aggregated. Rate constant resources (Section 4.3.3) change in total quantity used as the duration changes. This may lead to unpredicted changes in project costs, the reasons and sources of which would not be immediately apparent.

4.6   RESOURCE–DURATION RELATIONSHIP

It is important to remember that the time taken to carry out work is related to the quantity of resources allocated to an activity. The question of what to do about different rates of working (rather than variable resource demand) is now discussed.

Several computer packages allow activities to change duration and call them *stretchable*. The decision to model different rates of working becomes particularly important when using networks with several types of connections (Chapter 5) or when scheduling to give a resource-limited programme (Chapter 7). Summarising the discussions: when using networks, the assumption of stretchability affects the results from the timing calculations used to determine the critical path and the project duration; when scheduling, it might be considered advantageous to vary the number of resources and hence the activity duration to make fuller use of project resources.

If such alteration of the resource level is to be done automatically during the procedure, the way it is done must be defined. This means a resource–duration relationship for each activity is needed.

4.6.1   Variable duration activity models

For single resource activities, defining the resource–duration relationship may be possible and in fact is done in time/cost trade-off problems when the relationship can be assumed to be as shown in Figures 4.5 and 4.6. Figure 4.5 shows what is assumed to be reality.

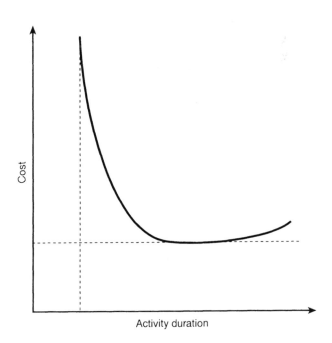

Figure 4.5 Variable duration activity model.

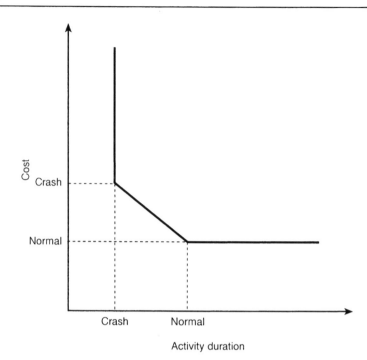

Figure 4.6 Idealised variable duration activity model.

Here the duration of an activity gets harder to reduce (or crash) despite increasing the amount of resources and money allocated to it and indeed cannot be reduced indefinitely; similarly, an optimum duration can be identified where resources are used efficiently and the cost minimised, before further reduction of resources leads to less efficient working and an increase in the cost of the activity.

Figure 4.6 shows a simple model which is used as an approximation of the reality; it can be seen that the model assumes a linear relationship between resources (here expressed in terms of cost) and activity duration over the activity durations which are of interest.

It is quite reasonable to ask a project manager or planner to define the ends of such a line and use this model for single-resource, constant resource demand activities (Section 4.4.2). For variable resource demand activities (Section 4.4.2), the model depicted in Figure 4.6 is not valid since the section of the resource demand being considered also has to be defined. In effect, the only time this model would work with variable resource demand activities would be if the segments of the resource demand profile represented smaller activities which must be carried out in series. This is rarely the case since large activities are more likely to be complex and could be viewed as mini-projects in their own right.

Considering multiple resource activities, these are exceedingly complex even if the constant demand model is assumed. In this, the resources which make up an activity are considered to be balanced when they are assigned in the optimum manner to provide the optimum activity duration. The activity would behave differently to a change in availability of each resource not only because that resource would be able to work at a different speed but also because the balance of the resources on the activity has been changed. An example would perhaps be useful at this point!

4.6.2 An example

Consider a simple, although perhaps extreme, example of two painters painting a wall and needing a ladder to reach the top of the wall. The normal activity duration and resource requirements are shown in Table 4.1.

It might be inferred from this information that there are 20 painter-hours required for the activity. For simplicity at this stage, assume that this can be achieved by 1 painter in 20 hours, 2 painters in 10 hours or 4 painters in 5 hours. This is only possible provided all tools and equipment are available as required, so the resources shown in Table 4.2 give an activity duration of 5 hours and those in Table 4.3 give one of 20 hours.

However, these same durations may be achieved in other ways. For example the resources shown in Table 4.4 give the same duration as those in Table 4.2 assuming two painters work from ladders and two do not.

A reduction of other resources would have a rather different effect. Table 4.5 shows the effect of a reduction in paint brushes. This returns

Table 4.1 'Optimum' resource requirements for small activity.

Activity: paint walls
Normal duration: 10 hours

Resource	Number
Painter	2
Paint brushes	2
Ladders	2
Pots of paint	2

Table 4.2 'Sub-optimum' resource requirements (1) for small activity.

Activity: paint walls
Normal duration: 5 hours

Resource	Number
Painter	4
Paint brushes	4
Ladders	4
Pots of paint	4

Table 4.3 'Sub-optimum' resource requirements (2) for small activity.

Activity: paint walls
Normal duration: 20 hours

Resource	Number
Painter	1
Paint brushes	1
Ladders	1
Pots of paint	1

Table 4.4 'Sub-optimum' non-balanced resource requirements (1) for small activity.

Activity: paint walls
Normal duration: 5 hours

Resource	Number
Painter	4
Paint brushes	4
Ladders	2
Pots of paint	4

Table 4.5 'Sub-optimum' non-balanced resource requirements (2) for small activity.

Activity: paint walls
Normal duration: 10 hours

Resource	Number
Painter	4
Paint brushes	2
Ladders	4
Pots of paint	4

Table 4.6 'Sub-optimum' non-balanced resource requirements (3) for small activity.

Activity: paint walls
Normal duration: 6.25 hours

Resource	Number
Painter	4
Paint brushes	4
Ladders	4
Pots of paint	2

the duration to the same as that obtained by the use of 2 painters since a paint brush is an essential piece of equipment for a painter.

Table 4.6 shows a different situation again. This is obtained by a reduction in the number of pots of paint. By doing this, the painters can all still work but because they need to share the paint their productivity is reduced. Here it is assumed to be only 80% of normal productivity and the activity duration is therefore 6.25 hours.

4.6.3 Comment

The above examples show the possible effects of increasing the resources in the multiple-resource, constant resource demand model. In general they range between having no effect at all and having a significant effect because of the balance between the resources on the activity. Every activity is different so resource reduction would have similarly complex effects.

The effects of resource alteration on the multiple-resource, variable resource demand model are even more complex. Indeed, the effects on the constant resource demand model are so difficult to take into account and so difficult to specify that it is recommended that such a model is not used. If a computer package offers these features, it is important for the planner to run small test projects to ensure that it behaves in a manner that the planner understands and is willing to accept.

When using multiple resource activities, the results of analysing a plan should be examined and if the planner or project manager suspects that benefit could be gained from altering an activity's duration or resources, then the alteration should be made by the planner and the effect on the project determined.

4.7 ACTIVITY RESOURCE DEMAND AND DURATIONS

4.7.1 Splittable and non-splittable activities

Having suggested that stretchable activities are potentially dangerous in planning because of the difficulty in defining the resource–time relationship for all but the simplest single-resource, constant resource demand activity model, there is one other type of activity model which must be considered. This is the splittable activity.

A *splittable* activity is one that is allowed to stop and start throughout its execution. A *non-splittable* activity is one that must be completed in a single piece once it has started. Most construction tasks can be stopped and started and could therefore be claimed to be splittable. Some activities in construction remain essentially non-splittable by their very nature, an example of this being the slip-forming of concrete structures.

At the planning stage it should be asked if the activity model to be adopted should allow splitting or not and, if it is permitted, several questions should consequently be answered.

It has already been stated that most construction activities are splittable by nature but being splittable and being allowed to split are two different issues. As described in Section 4.6, the normal duration and resources assigned to an activity are (or should be) optimum. If the resources have to stop work on an activity and move to another, to carry out some or all of that activity before returning to the first activity to carry out more work, it is highly unlikely that the resources will be able to achieve the productivity assumed for the initial optimum durations.

The true productivity achieved depends on several factors:

Type of work Complex activities which require considerable organisation achieve lower productivity than the simple ones.

Resources involved Multiple resource gangs and gangs brought together especially for an activity achieve lower productivity than single resources.

Amount of time that the resources are allowed to work on an activity If a resource is only allowed to work for a very short time, the productivity is low because of the organisation required at the start of work and the slow productivity which can occur towards the end of a period of work on an activity.

Amount of time between visits by the resources to an activity If a long time elapses between visits, the productivity is likely to be lower than if there is a short time between them. Additionally, if the time is very long, extra work may be required on some activities either to make them suitable for leaving or to return them to the state they were in when they were left. Both of these can be significant amounts.

Physical distance of an activity from other activities If an activity is a long way from the other activities which the resources move to, the movement time may be significant, especially for activities using large plant.

The activity split imposed by a planning technique is different from the natural splits which occur at the end of each day or at weekends. Natural splits are allowed for (planned for) both by the managers and by the workers. Imposing other planned splits could be like incorporating disruption, studies of which have indicated that it can easily lower productivity to 40% of planned values.

Consequently it is recommended that activities should always be assumed to be non-splittable at the planning stage.

If the non-splittability of an activity appears to cause problems, it should be physically split, by the planner, into more detailed but still non-splittable activities. This assumption does not necessarily produce the most efficient plan despite the individual activities being performed in their most efficient manner. This is because of the potential difficulty in utilising all the resources fully. If the plan is drawn up in this way, alterations of sequence and other variations can be significant (Chapter 14).

4.7.2 Learning effect

The more times workers repeat an operation (up to a point) the better they become at doing it. This effect is known as the learning effect, although psychologists might wish to differentiate between the learning which has taken place and the translation of the learning into behaviour which improves productivity or performance. It has been studied in depth by several researchers. Unfortunately, there is also a forgetting effect which means that if workers are taken off a particular task they will not be as efficient at it when they return.

In construction, the learning effect covers not only the psychological learning effect but also the organisational aspects of work. If a worker or gang of workers is set to carry out a task, their performance at it will increase as time goes by because they become more organised in performing the task. They know more accurately what has to be done and where the resources and tools for carrying out the task are best placed. If the task is changed or interrupted, learning becomes forgetting and productivity decreases.

As well as applying to individual tasks, learning also applies to whole projects and it is normal for productivity to increase as a project progresses even if the work changes. This is perhaps due to the organisation rather than to true learning but it has the same effect.

Learning and forgetting effects can be large and one study showed an overall improvement in time of 17% brought about by proper consideration of learning effects on a project. It is therefore important that learning effects be taken into account in assessing the duration of activities. It is suggested that an average productivity for a resource on a task be used and this can be varied as more information becomes available.

4.7.3 Calculation of activity durations

It is important wherever possible to calculate the durations of activities from all the information which is available at the time of planning. At the very least this should be from the quantity of work involved in the activity, the resources assigned to it, and the productivity for the resource, as shown in the equation.

$$D = Q/NP$$

where D is the activity duration, Q is the quantity of work to be done, N is the number of resources (of a type), and P is the productivity (quantity of work per unit time per unit resource).

It is relatively simple to understand and apply this equation. A planner would usually take a standard figure for P, check and amend it as necessary for the project, would select N for the project and, together with Q from the project documents or supplementary calculations, would then determine D. From preceding sections of this chapter it may be deduced that the main problem with the equation is what to take for the value of P and how to select N.

As the productivity of a resource varies throughout an activity, it was suggested in Section 4.7.2 that an average value for productivity is taken until more information becomes available. This average value should, if possible, come from company records (rather than other sources) of similar resources performing similar work.

All too frequently the lack of detailed information and the complexity (not always appreciated) of the activity models used in a plan gives the planner an excuse to estimate activity durations based on 'experience', with no knowledge of the productivities that such durations must imply for each resource. This estimate is likely to be significantly in error and the practice is to be discouraged at all times. It is again recommended that only simple constant resource demand activities should be included in plans whenever reasonably possible.

Even when planning the procurement activities (for example *order materials*), it is advisable to ask suppliers for information which allows monitoring of progress to be performed and control action to be taken if necessary.

The calculation of activity durations from basic data often provides a check on the planning assumptions (and the bid documentation if appropriate) and as such is valuable in evaluating the plan (and the estimate if appropriate; Chapter 13).

Planners must always bear in mind the general observation that work will take whatever time is available for its completion. It is important to set appropriate durations for activities by taking into account both experience of how long work takes and what duration for the project would be most desirable.

4.7.4 Rate of activity completion

When updating plans or looking at them from the point of view of project control, it is often necessary to assess how much of an activity has been completed. In reality this is rather more difficult than it appears. For all the reasons discussed above (learning, forgetting, multiple resource activities and variable resource demands), the rate of completion of an activity is usually not linear.

It is therefore necessary to define exactly what is meant by completion of an activity. Two measures of completion are normally considered:

- duration
- work content.

Figure 4.7 shows a simple model of the relationship between these.

There is some reason to suppose that an activity might behave like a project (which is usually taken to follow an S-curve, as shown in Figures

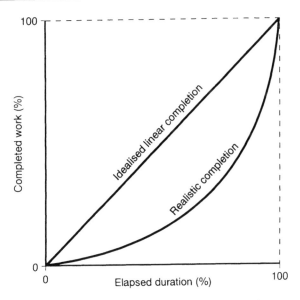

Figure 4.7 Measures of activity completion.

2.23, 2.26 and 2.27, from commencement to completion). However, for a well-defined detailed activity (ideally but not essentially a single-resource, constant resource demand activity) the shape shown in Figure 4.7 is preferred. It can be seen that there is a significant difference between the completion at a particular time for the idealised and realistic curves. This may not matter but it could have large financial implications. In some forms of construction contract, for example, the payment system is based on the completion of activities. If the measured work were used as a measure of completion of an activity, contractors might be paid less than that calculated from the amount of effort expended (because learning effects slow the rate of production but represent effort expended). Conversely, using an elapsed duration measure would over-estimate the correct payment for the contractor if the activity contained, for example, a significant element of finance for provision of materials throughout the activity duration. In addition, using the elapsed duration measure assumes that the original duration estimate was, and remains, correct and that the learning curve effects would get absorbed later when higher than average productivity levels are achieved.

Taking all these into account, the simpler the activity the more sensible it is to assess its rate of completion and the closer the elapsed duration measure and the completed work measure are to each other.

4.8   SUMMARY

- Activities form the basis of project models.

- The choice of activities is fundamental to good planning.

- Several different types of activity exist.

- Different types of activities need different treatment in the planning process.

- When choosing activities consider:
  - the use to which the plan is to be put
  - the stage of the project at which the plan is to be made
  - the amount of information available
  - the person making the plan
  - the complexity of the project
  - the planning method proposed.

- When producing a plan consider:
  - the geographical area of the project
  - the layout of the site
  - the structures involved
  - the structural elements involved
  - the work types involved
  - the predominant resources to be used
  - the responsibility for the work
  - the contractual implications.

- If there is not enough detail in the plan, the plan may be optimistic.

- Clients and contractors require different activities since they have different uses for plans.

- If different plans are produced for clients and contractors it is essential that they integrate properly.

- Always have full activity descriptions on plans.

- Try to maintain a constant size for activities throughout a plan unless the intention is to produce a rolling programme.

- Financial value is not always a good measure of activity size for a contractor.

- Resource use is sometimes a useful measure of the size of an activity from a contractor's point of view.

- When allocating resources to activities bear in mind:
  - the number of different types of resources – the larger the number the more complex the planning process
  - the size of gangs to be used
  - the spread of the resources over the duration of an activity
  - the allocation of major items of equipment to more than one activity at a time.

- Too many resources makes scheduling, monitoring and control difficult.

- Too few resources makes scheduling, monitoring and control unrealistic.

- Twenty key resources is usually a sufficient number to consider in project planning.

- The choice of resources to be considered depends on the project.

- Company standards and the integration of projects may influence the choice of resources.

- Activities should ideally be confined to single key resources.

- Constant resource demand activities should be used if possible in planning.

- Variable resource demand activities are very difficult to work with in practice.

- The simpler the activity-resource relationship the better.

- If an activity needs to split, the planner (not a computer package) should control it.

- Learning (and forgetting) effects can be significant.

- The relationship between elapsed duration and quantity of completed work for an activity is generally not linear.

5

ACTIVITY CONNECTIONS AND CALCULATIONS

' "Contrariwise," continued Tweedledee, "if it was so, it might be; and if it were so, it would be: but as it isn't, it ain't. That's logic." '

LEWIS CARROLL
(Through the Looking-Glass)

This statement seems to be designed to confuse initially, but it actually concludes quite clearly. Plans are designed to be clear but sometimes, as is shown in this chapter, they can easily confuse (both the planner and the plan user) if specific questions are asked of the logic and the calculations. Contrariwise as they say.

5.1 INTRODUCTION

Most plans are produced in the belief that they are good for their intended purpose. It is often a lack of understanding of something fundamental that makes plans less well suited than the planner had expected. The lack of understanding can be illustrated by looking in detail at the practical interpretation of the model produced by the planner and asking what it really means. A practical interpretation of a plan involves trying to understand every part of it in the context of the work represented on it. Regardless of how a plan is actually presented (Chapter 2) various things need to be understood, including:

- the meaning of the elements (activities) of a plan;
- links between the elements of a plan;
- durations assigned to the elements of a plan;

This chapter takes a detailed look at possible interpretations of the connections, or logic links, between activities and of the durations assigned to a plan. Activities as individual components of a plan are covered in Chapter 4.

Although, hopefully, the logic of a plan will have been thought through, it is not always represented on the final programme. The form of representation which sets down the logic in detail is the network (Section 2.5), and this is the type of model used in this chapter to discuss a plan's logic links. It is important to consider networks because many plans are derived using them; this is because much project planning software is network-based. As with all technical software applications it is important to understand the results that any particular computer package gives.

This chapter shows how important it is to understand both the computer output and the method used by the computer to obtain it. The chapter uses simple examples which can be worked through by hand. The different assumptions that can be made when interpreting a model are shown to give different, and at times quite surprising, results. Different software packages use different assumptions and so can give different results. In some cases the assumptions used in writing the software are not known by the user and, indeed, the in-built assumptions are not always explained adequately in the software documentation.

The chapter starts with an introduction to the analysis of simple networks by using the Tingham Bridge project (Appendix A) with its fairly straightforward logic. A different simple example is then used to raise some concerns about the use of logic links. More complex network connections are considered and further examples are used to explore the effects of different assumptions.

To understand the issues behind the various assumptions it is necessary to understand how to analyse a network. The basics are introduced in the following section. It is hoped that most readers are already familiar with the terminology and techniques of basic network analysis (in which case the following section could be omitted). Just in case there are some planners reading this who are not familiar with basic network analysis techniques (that is, they leave much or all of the analysis to the computer and are unwilling or unable to check the output), or who prefer to use arrow diagrams or activity-on-arrow networks, the following section is included.

For planners and computer packages there is a choice between using the activity-on-node (or precedence diagram) format and using the activity-on-arrow (or arrow diagram) format, both introduced in Section 2.5. This book in general uses activity-on-node networks or precedence diagrams but the same results would be achieved had the alternative system been used.

5.2 SIMPLE NETWORKS

Simple networks are of the activity-on-node type (Section 2.5.2) or the activity-on-arrow type (Section 2.5.1). If each activity is assigned a duration, a simple network can be analysed to yield timing information about the activities and the project. The information of most use comprises:

- earliest start time (EST) of the activities,
- earliest finish time (EFT) of the activities,
- latest start time (LST) of the activities,
- latest finish time (LFT) of the activities.

This information gives (1) the earliest time that any activity (and hence the project) can be started or finished; (2) the latest time that the activities (and hence the project) can be started or finished so that the final activity can be finished at its earliest finish time (or some other specified time); and (3) the float available for each activity (and hence the project). Float, or slack, is a word which can be defined in several ways to mean the time available for an activity in addition to its duration. The most commonly encountered types of float are:

Total float (TF) The time by which an activity may be delayed or extended without affecting the total project duration.

Free float The time by which an activity may be delayed or extended without affecting the start of any following activity.

Independent float The time by which an activity may be delayed or extended without affecting preceding or following activities in any way.

Float can be a sensitive word, especially in the area of claims' evaluation, a topic which is discussed further in Chapter 14. It is also important to understand the meaning of floats on different types of activity and this is discussed in Sections 5.3 and 5.8.

5.2.1 Basic network analysis

Conventionally, networks are analysed to give the EST, EFT, LFT, LST and TF values (Section 5.2 above) for every activity. The technique employed involves performing a forward pass and a backward pass through a network, a process which yields these values directly in the case of an activity-on-node network. For an activity-on-arrow network the forward and backward passes yield the earliest and latest event times (EET and LET) from which can be calculated the other values. The results of the analysis are shown in Table 5.1 for the simple activity-on-node network (Section 2.5.2) for the Tingham Bridge project (Appendix A and Figure 2.20) which uses connections only between the end of an activity and the start of another activity.

Figure 5.1 combines the network analysis information with the network, the numbers being shown against each activity according to the

Table 5.1 Results of basic network analysis.

Activity	Duration (days)	Earliest start time (end of day)	Earliest finish time (end of day)	Latest start time (end of day)	Latest finish time (end of day)	Total float (days)
1 Start	0	0	0	0	0	0
2 Pile & cap E	10	0	10	0	10	0
3 Pile & cap C	5	0	5	11	16	11
4 Pile & cap W	8	0	8	30	38	30
5 Substructure E	23	10	33	10	33	0
6 Substructure C	17	5	22	16	33	11
7 Substructure W	20	8	28	38	58	30
8 *In situ* span	30	33	63	33	63	0
9 Precast span	5	28	33	58	63	30
10 Surfacing	5	63	68	63	68	0
11 Finishings	14	68	82	68	82	0
12 End	0	82	82	82	82	0

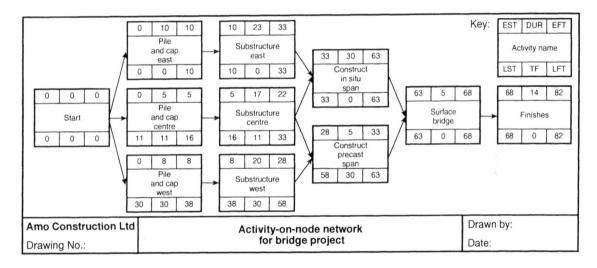

Figure 5.1 Results of basic network analysis shown on a network.

key. It is recommended to show all the calculated values in tabular form separate from the network, but standard representations do allow them to be superimposed on the network in the boxes used for the activity names. The following paragraphs may help to show where the numbers have come from.

In the forward pass, the analysis moves through the network from the start to the finish, considering activities in turn for which the EFTs of all the immediately logically preceding activities have already been determined. (The term 'immediately logically preceding' is used here to mean that there is a direct logical link in the network between the activities.) The EST of such an activity is the highest value of the EFTs of the immediately logically preceding activities. This is because the completion of the preceding activities is a precondition to the commencement of the activity in question. Its EST cannot arrive until the last precondition occurs, which is the latest EFT of the logically preceding activities. Its own EFT equals its EST plus its duration. The EST and EFT of each activity are computed in this manner. It is recommended to begin the forward pass with a single activity called *start* which has zero duration and resources and an EST value of zero (see below).

The backward pass commences at the LFT of the final activity which is normally set to be equal to its EFT, although any specified time can be used. The analysis moves through the network from the finish to the start, considering activities in turn for which the LSTs of all the immediately logically following activities have already been determined. For such an activity the LFT equals the lowest value of the LSTs of all the immediately logically following activities. Its own LST equals its LFT minus its duration. It is recommended to begin the backward pass with an activity called *end* which has zero duration and resources (see below) and an LFT value equal to either the EFT of the activity *end* or any chosen value for the project's LFT.

The TF value for the activities equals the difference between their EFT and EST, which is also the difference between their LFT and LST (but beware other types of network (Section 5.4.1) which allow splittable activities). The activities with the least total float are called critical activities and the route(s) which connect the critical activities are the critical path(s) through the network. If the LFT and EFT for the project are the same value, the critical activities in the network have zero total float. The critical path is therefore the longest path (or paths) through a network.

Networks with many start points and end points can be produced but they offer no real advantages over those with just one of each. There is a danger that an activity might appear to be a start or an end because a planner has accidentally missed out a connection. Such activities are usually called *danglers* and can obviously cause problems. Users of plans should be able to identify clearly all the activities which could commence once a project has started. This is generally easy if they are all connected to a *start* activity from which the analysis of a network follows in a straightforward way. Similarly, a network analysis is more complete if all the last activities on the network branches come together into a final activity called *end*. These actions ensure that all the activities have been tied in and any time constraints on the start or end of projects can be inserted by the planner at just one point (the start or end of the network).

5.2.2 Networks for which the programmed completion time is not the critical path duration

As mentioned above, it is possible and indeed quite common for a project to be programmed to be completed in a time other than the critical path duration (CPD). In particular, if no resource constraints are included in a network, it is usual for the CPD to be significantly less than the programmed project duration. Projects are programmed to be carried out in a particular duration for one of the following (or combination thereof) reasons:

- to satisfy the requirements of the client/owner,
- to make the most efficient use of resources,
- to make the project more beneficial financially,
- to share the risks between the parties in some manner.

When the programmed completion date is x units of time later than the the CPD, the real total floats of each activity are x units greater than those calculated when considering the LFT of the project to be the same as its EFT. The total floats of the activities on the critical path are thus x units of time. In such circumstances the critical path is no longer critical in the normal sense (that is, critical activities with zero total float) since the critical activities can now be delayed by up to x units beyond their earliest times without necessarily delaying the project completion. It can be shown, however, that such arguments contain many pitfalls and assumptions, especially when allocating liability for delay. This is discussed in Chapter 14.

5.3 ACTIVITY INTERACTIONS

Defining the building blocks of a project model (Chapter 4) is only one part of producing a plan. The interaction between the building blocks (in this case, between the activities) must also be determined. This is introduced through an example.

Consider the painting of a door, a project involving the following stages which, for ease of communication in this book, can be defined as activities A, B and C with durations shown in brackets. (See Section 4.2.3 for the dangers in calling activities by a number or letter.)

Activity	Description
A	*strip* off the existing paint (2 hours)
B	apply *paint* (1 hour)
C	allow paint to *dry* (12 hours)

5.3.1 Use of connections for different types of activities

Even the simple example of the door painting raises a fundamental issue in activity selection because there are different types of activities here: *strip* and *paint* are activities that require resources. That is, they cannot be started until the resources are available. The activity *drying* requires no resources other than the correct environment, and indeed will start regardless of what the planner wants to happen; it is therefore different to the other activities and is identified here as a chemical activity (Section 4.2). Other examples of similar activities which occur frequently when planning construction in detail are curing of concrete and hardening of glues such as epoxy resins.

Because of this fundamental difference in the behaviour of different types of activity, the planner must be careful. Particular differences are pointed out as they occur in the text (for example, Section 5.8.2) but planners ought to look out particularly for the problems in using standard network connections to incorporate chemical activities into project models.

5.3.2 Simple logic links

The relationship between the activities in the door painting example can readily be modelled as shown in Figure 5.2. Activity A must be completed before B can commence and B must be completed before C can commence. This is an activity-on-node network (Section 2.5.2) and the total duration of the project is 15 hours being the sum of the durations of the individual activities.

In this project each and every activity is interdependent. For instance Activity C cannot start before B has been completed, but neither can it commence until A has been completed. In most real projects there are some, often many, activities which are not interdependent. Consider a project involving the painting of two doors. Using the same connections the situation is represented in Figure 5.3.

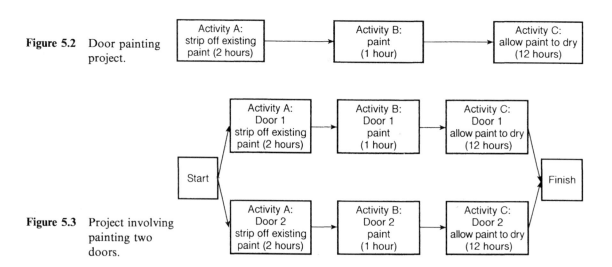

Figure 5.2 Door painting project.

Figure 5.3 Project involving painting two doors.

Note here the additional activities *start* and *finish* of zero duration in order to make the logic explicit. As described in Section 5.2.1, there is no reason why there should not be many starts and ends to a network as long as they are recognised for what they are.

5.3.3   Modelling with finish–start connections

In the two-door network (Figure 5.3), a standard finish–start (FS) connection has been used to incorporate the chemical activity (Section 5.3.1) in the network. In terms of the technology of construction, the activities relating to one door are independent of those on the other. For example the paint on door 1 can have dried even before door 2 is stripped. The planner should now ask:

Is it correct to use a FS connection between paint and dry?

In this case the answer is 'yes'. Drying cannot start until after painting, but it *must* start immediately after painting.

In many situations a standard FS connection would be inappropriate and the need for other connections arises (Sections 5.4 and 5.7). (This example is extended in Section 5.8 to introduce a situation where the answer to the above question is 'no' and a rather nonsensical situation arises in which the paint on Door 1 need not start drying until some time after the painting has finished! (Section 5.8.2))

5.4   POSSIBLE MEANINGS OF CONNECTIONS

A connection is simply a directional line between two nodes in a network. In an activity-on-node network (one with only finish-to-start connections) the links show what might be termed absolute precedence. In the example introduced in Section 5.3, the model can only be wholly correct if no element of Activity B commences until all elements of Activity A are completed. Additionally, the connections represent pure logic (they have no duration and consume no resources). On some occasions the simple FS connection is an acceptable approximation to 'wholly correct' (as in the example in Section 5.3); on other occasions this is not the case. In a precedence diagram (Section 2.5.3) different connections can be used, as discussed below.

A range of types of connection between activities can usually be implemented in computer packages. Unfortunately the connections can be interpreted in several ways. Not only do different planners sometimes assume different meanings for connections, they also frequently make different assumptions to the computer programs that they use. This can give rise to inconsistencies in plans and consequent confusion. Some of the different assumptions which can be made and the effects they can have in practice are now examined. The discussion is in terms of precedence networks but similar features of arrow diagrams have similar problems.

5.4.1   Interpreting standard connections – work or delay lags

Four common types of connection between the starts and ends of activities are defined as follows:

Start–start	(SS)
Finish–finish	(FF)
Finish–start	(FS)
Start–finish	(SF)

To represent these connections graphically it is necessary to define the start of an activity as the left side of an activity box and the end of an activity as the right side of an activity box.

An example showing all these different connections is given in Figure 5.20. When a number is assigned to a precedence connection it is hereafter called a lag, the meaning of which is discussed below.

Trying to use the above relationships to model a project needs some clarification. Consider the example of simple earthworks for a road project shown in Figure 5.4(a).

The project might be modelled as shown in Figure 5.4(b), in which two activities, *excavation* and *sub-base* are defined and two connections link them, an SS connection of duration 10 and an FF connection of duration 3. All the numbers represent durations in days. According to one interpretation, this network would be read, quite reasonably, as:

> *The sub-base cannot start until 10 days after the excavation has started and the sub-base cannot finish until 3 days after the excavation has finished.*

The earliest start and finish times for the activities are shown in Table 5.2 for the cases of non-splittable and splittable (or stretchable) activities (Section 4.7.1).

By way of explanation, the SS connection suggests that *sub-base* can start at time 10 and the FF connection suggests that *sub-base* cannot finish until time 43 (3 days after the earliest finish of *bulk excavation*). If the activities are considered non-splittable this means that *sub-base* cannot start until time 33 (i.e. $43 - 10$). If the activities are considered

Figure 5.4   Simple earthworks project: (a) small road project (longitudinal section); (b) possible network model.

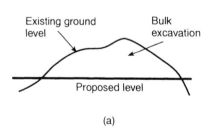

(a)

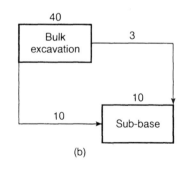

(b)

Table 5.2 Earliest start and finish times for network shown in Figure 5.4.

Activity	EST	EFT
(a) Non-splittable activities		
Bulk excavation	0	40
Sub-base	33	43
(b) Splittable or stretchable activities		
Bulk excavation	0	40
Sub-base	10	43

splittable or stretchable the earliest finish of *bulk excavation* remains at time 40 but it is now possible for *sub-base* to start at time 10, even though it cannot finish until time 43.

In reality what the planner was trying to represent was probably more like:

It will take the excavation equipment 10 days to do enough work to allow the sub-base gang to have a large enough piece of work to allow them to start. After the excavation gang has finished, the sub-base gang will have 3 days work left to do to finish the area in which the excavation gang was last working. Furthermore, the sub-base must follow the excavation at all points by some amount which will ensure that the two sets of resources do not get in each other's way.

This arguably more realistic interpretation of the network differs from the earlier interpretation in two important respects.

(1) The lag durations on the arrows relate to work, not just time. In the case of the SS connection (or lag) it is work on the preceding activity, and in the case of the FF lag it is work on the following activity.

(2) The two lags together (SS and FF) are used to denote a progress restriction imposed by the former activity on the latter. This is explained in Section 5.4.2.

5.4.2 Progress and constraint lines

One interpretation of the progress restriction implied by the use of SS and FF connections together is illustrated graphically in Figure 5.5 for the project in Figure 5.4. The solid line represents an assumed linear rate of working on the *excavation*. Assuming also a linear relationship between the start and end lags, the dotted line indicates the earliest time at which any percentage of the *sub-base* could be completed. The planner's intended interpretation (Section 5.4.1) of the interaction between the activities suggests that the latter assumption is not necessarily a reasonable one.

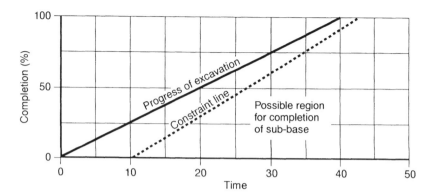

Figure 5.5 Graphical representation of SS and FF constraints.

Taking the two types of activity (non-splittable and splittable or stretchable), it is possible to show, on such a diagram of percentage completed against time, the earliest time at which the *sub-base* activity could be done.

Figure 5.6 shows the situation if the *sub-base* is considered non-splittable. This gives an earliest start time of the *sub-base* as time 33.

If, as is perhaps more realistic, the *sub-base* can be split, there is a range of possible answers (Figure 5.7) depending on the degree of splitting allowed. Figure 5.7(c), for example, shows the situation if a minimum work period of 3 days is allowed.

It can be seen from Figure 5.7 (a)–(c) that, as the sub-activity duration increases, the start time of the *sub-base* activity becomes later and vice versa. The phenomenon of being able to bring forward the start time of an activity by allowing the activity to split into smaller parts can be an advantage if the objective is to produce the shortest time for a project (provided there are subsequent activities which proceed at a slower rate than the split activity). Since the objective in planning should be to model reality, such shortening may in fact only be a sign that the plan is optimistic. Indeed, in order to achieve the earliest start time given by the assumption of pure time lags (rather than work lags (Section 5.4.1)), the *sub-base* must be considered to be splittable into infinitely small subactivities. This is usually an unrealistic assumption but one employed by some computer packages. This is the equivalent of a stretchable activity with all the implications discussed in Section 4.7.

If the network is augmented to show one other activity, as shown in Figure 5.8, this problem gets compounded. Here the duration of the third activity (*base course*) is 10 days and both the start (SS) and finish (FF) lags between it and the *sub-base* activity have a duration of 4 days. This is convenient for the explanation, as it makes this part of the network into a *balanced ladder*. If the lags were uneven or the duration different to that of the preceding activity, the calculations would be much more difficult to carry out (see below).

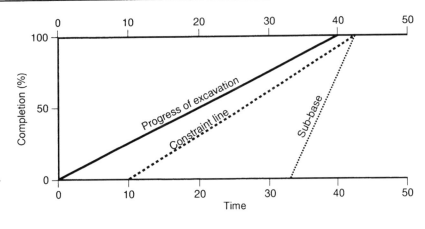

Figure 5.6 Progress of non-splittable sub-base.

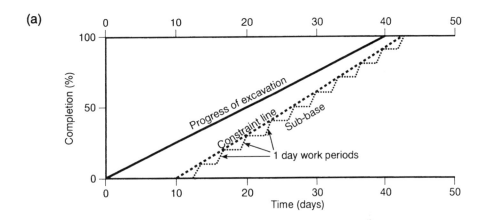

(a)

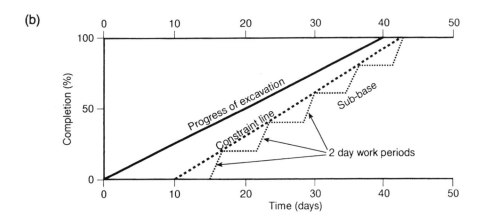

(b)

Figure 5.7 Progress of splittable sub-base: (a) Minimum split 1 day; (b) Minimum split 2 days.

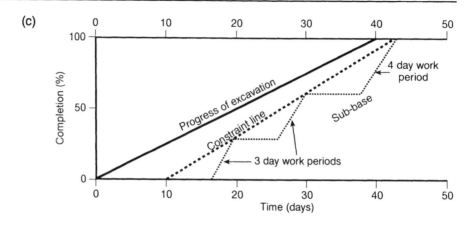

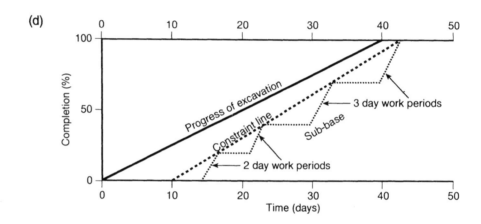

Figure 5.7 (c) Minimum split 3 days (3-3-4); (d) Minimum split 2 days (2-2-3–3).

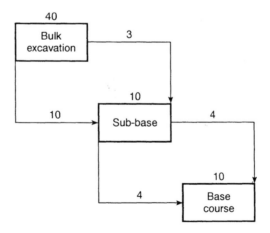

Figure 5.8 Augmented road network.

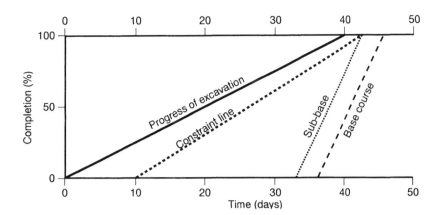

Figure 5.9 Progress with three non-splittable activities.

Table 5.3   Earliest start and finish times under different assumptions for the network in Figure 5.8.

Activity	EST	EFT
(a) Non-splittable		
Bulk excavation	0	40
Sub-base	33	43
Base course	37	47
(b) Minimum split 1 day		
Bulk excavation	0	40
Sub-base	12	43
Base course	16	47
(c) Minimum split 2 days		
Bulk excavation	0	40
Sub-base	15	43
Base course	19	47
(d) Minimum split 3 days		
Bulk excavation	0	40
Sub-base	17	43
Base course	21	47
(e) Infinitely splittable or stretchable		
Bulk excavation	0	40
Sub-base	10	43
Base course	14	47

Diagrams showing progress to completion are shown in Figure 5.9 (non-splittable) and Figure 5.10(a)-(c) (splittable) for the same assumptions used in Figure 5.6 (non-splittable) and Figure 5.7(a)–(c) (splittable). For convenience, the minimum split sizes of the *base course* activity have been taken to be the same as those for the *sub-base*.

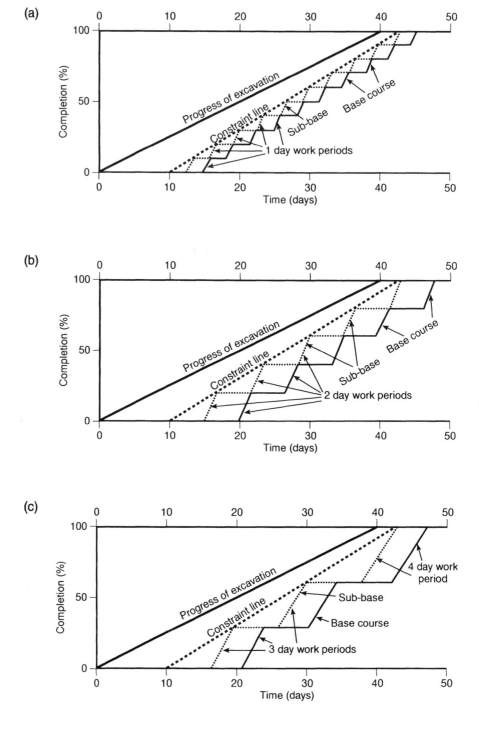

Figure 5.10 Progress with three splittable activities: (a) Minimum split 1 day; (b) Minimum split 2 days; (c) Minimum split 3 days (3-3-4).

Table 5.3 shows the earliest start and finish times of the activities under all four assumptions (non-splittable and splittable with 1, 2, and 3 days minimum split). For comparison, the results with infinitely splittable or stretchable activities are also given.

If the last activity (*base course*) were reduced in duration by 1 day to 9 days, an interesting situation arises (alluded to above) in all except the infinitely splittable case. Because the progress is faster than that of the preceding activity, the earliest start time of the following activity is in fact increased in order not to overtake the former activity. This is termed an *unbalanced ladder*, which can give rise to many problems in understanding project schedules (Section 5.6.2).

By drawing progress and constraint line diagrams, it can be seen how much work has been done on one activity before the next one commences. This can be important, as described below. It is acknowledged that the graphical approach described above can be time-consuming and difficult, but it does help to resolve problems of understanding the model that has been created and the effects of various assumptions.

It is therefore recommended that the graphical approach be used when appropriate on small parts of networks, particularly when computer analyses are not understood.

5.5   FURTHER CONSIDERATION OF LAGS

Consider a second example, the construction of a retaining wall shown in Figure 5.11, modelled by the network shown in Figure 5.12. The retaining wall is to be constructed in four sections.

In this example, the relationships are rather more complicated and might be expressed using the *construct wall* and *strip formwork* pair of

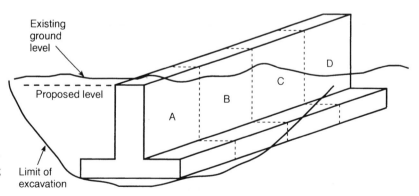

Figure 5.11 Sketch of retaining wall showing work sections.

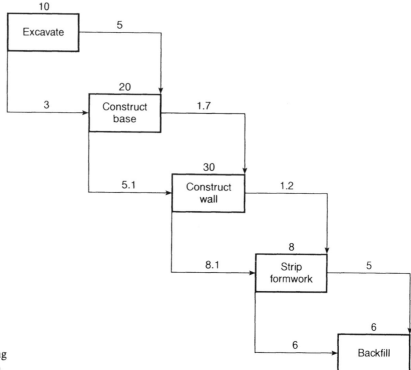

Figure 5.12 Precedence network for retaining wall construction.

activities as an example, as follows:

> *After the wall construction has started, it will take 8 days to complete the first phase and after a further day of curing the concrete, the stripping of formwork can begin. After finishing the wall construction there must be one day allowed for the wall to cure before the last phase of formwork stripping can begin, with 2 days worth of work in it.*

This combines the lags described for the simple road project with pure time lags (for the *curing*) and gives a full relationship definition shown in Figure 5.13. It is suggested that the work and delay lags always occur in the order shown. The order of work and delay on the lag becomes particularly important when considering the effect of multiple calendars, discussed in Chapter 6.

The calculations for these types of lag have the same difficulties as those described for the road project (Section 5.4) and are not repeated here. Section 5.6 shows that calculations become more complicated unless either non-splittable, infinitely splittable or stretchable activities are used. As the activity models described above make none of these assumptions the calculations are not carried out and indeed these assumptions are rarely if ever made by computer packages. This means

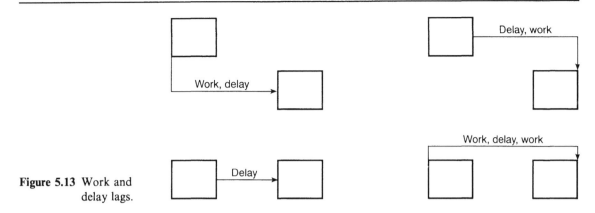

Figure 5.13 Work and delay lags.

that the most realistic assumptions which are commonly made by planners are rarely, if ever, used for the calculations.

5.5.1 Inclusive and exclusive lags

The examples described so far only have one relationship coming into or going out of the start or end of an activity. If this is not so and the lags on the arrows are considered to be of the types shown in Figure 5.13, it is important to recognise what work is occurring on the lags. Consider, for example, a road project with two bridges included as shown in Figure 5.14.

If access to the bridge sites can only be achieved by excavation along the road (from either end), the project might be represented as shown in Figure 5.15.

The question then arises as to what durations to give to the excavation activity and the two connections from it. Three possible answers are shown in Figure 5.16.

Figure 5.14 Road and bridge project (longitudinal section).

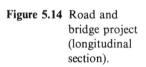

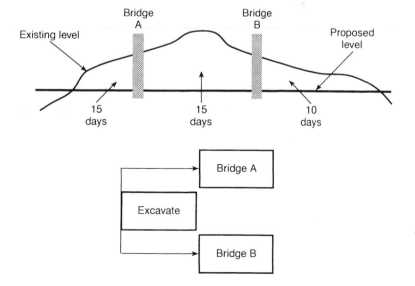

Figure 5.15 Network for road and bridge project shown in Figure 5.14.

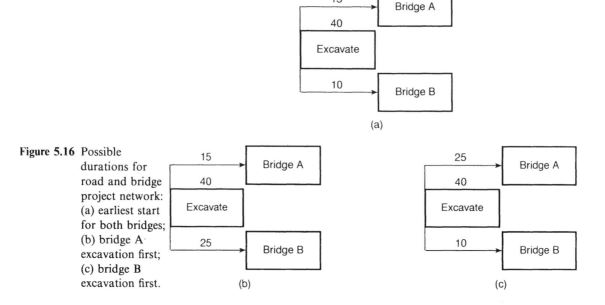

Figure 5.16 Possible durations for road and bridge project network: (a) earliest start for both bridges; (b) bridge A excavation first; (c) bridge B excavation first.

Figure 5.16(a) gives the earliest possible start time for both bridges, namely after 15 days and after 10 days of excavation at each site. With a duration for the excavation of 40, the two lags represent 15 and 10 of the 40 days and so cannot happen together under this model with these numbers.

Figure 5.16(b) models this fact with the further decision that the excavation to reach Bridge A will take place first. It gives the correct lags for this decision by making the SS lag on Bridge A equal to 15 and the SS lag on Bridge B equal to $15 + 10 = 25$.

Figure 5.16(c) models the alternative decision that the excavation to reach Bridge B will take place first. It gives the correct lags for this decision by making the SS lag on Bridge B equal to 10 and the SS lag on Bridge A equal to $10 + 15 = 25$.

In Figure 5.16(a) the work on each lag represents a totally different part of the excavation to be done, whereas in Figures 5.16(b) and (c), the work on the longer lag includes the work on the shorter. The lags in Figure 5.16(a) are called exclusive work lags and those in Figures 5.16(b) and (c) are called inclusive work lags. Exclusive lags give the earliest start time for the activities but can result in infeasible solutions. Inclusive work lags assume a method of work and give the earliest times for that method but not necessarily the shortest critical path.

The example given is perhaps extreme but the same aspects have to be considered whenever two or more connections modelling some work originate from the start of an activity.

This problem is not confined to the starts of activities, as whenever two or more work connections terminate at the end of an activity they too may be considered to be inclusive or exclusive work lags.

5.5.2   Space lags

In the above discussion it was assumed that the lags are either all work or pure delay. Four further possibilities occur, often (indeed typically) on linear projects. These are called *space lags*. They state that for two connected activities a specific part of the preceding activity must have advanced by the appropriate space lag before the corresponding part of the succeeding activity can be performed. An example of such a situation was described in Section 2.3.1 and illustrated in Figure 2.8.

5.6   NETWORK ANALYSIS WITH SPLITTABLE AND NON-SPLITTABLE ACTIVITIES

Stretchable activities (Section 4.6) and splittable and non-splittable activities (Section 4.7.1) have been introduced as ways of modelling certain aspects of a project. They have been used above (Sections 5.4 and 5.5) to illustrate the problems of modelling with networks and to demonstrate the need to make some assumptions about a model. For these types of activity to be of much use in planning it is therefore necessary to be able to analyse networks which use them. Much network analysis is performed by computer so it also necessary to know how the software handles the activities and their connections.

Some computer packages consider activities to be stretchable (able to alter in duration depending on the resources assigned to them). Since this is like most construction activities, it would initially appear to be the best assumption. On further analysis it becomes clear that most construction activities utilise many different resources and that the relationship between the number of each resource and the duration is different (and usually by no means linear). Thus, for example, if an operation such as lifting precast beams into place requires two cranes and six labourers, a reduction in the number of labourers available may increase the duration of the activity and a decrease in the number of cranes would prevent the activity being worked on at all. In general it is impossible to specify the resource–duration relationship of an activity (Section 4.7) in detail and the stretchable activity type is therefore impracticable in many situations.

In construction it might be argued that most activities are splittable with only activities like *slip forming* being non-splittable. Many computer planning packages allow splittable activities and several assume that all activities are splittable by default. If an activity is split during the calculation procedure, it is important that the minimum duration of the sub-activities and the maximum allowed gap between them are known. This information is required in order to produce an answer. Without a specific declaration of these values the analysis must make assumptions, as indeed is often the case. If the values, or splitting factors, are not declared it is recommended that planning be carried out with non-splittable activities and appropriate checks be made to assess the reasonableness of that assumption.

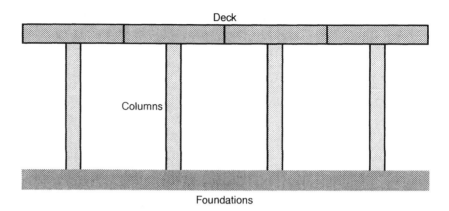

Figure 5.17 A bridge
support
and deck.

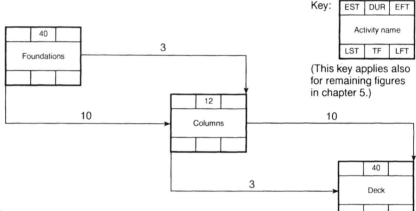

Figure 5.18 Precedence network for bridge
support and deck.

Throughout the rest of this section, illustrations of calculations are given in terms of splittable and non-splittable activities. If no other assumption is given, it is assumed that the activity is splittable into infinitely small parts and that there is no restriction on the time which can elapse between the split portions of the activities. (These are also normal assumptions for computer packages.) As mentioned above, some model of the effect of splitting an activity should be included in the calculation procedure. This might involve making an allowance for stopping and restarting by choosing sub-activity durations which give an increased total duration for an activity. No such increased duration is applied in the following examples. It is shown in the examples that the basic assumptions of non-splittability or splittability can significantly affect the outcome of normal critical path calculations.

The calculations in the following sections use the structure shown in Figure 5.17 for which the precedence network is drawn in Figure 5.18 (with durations given in days).

The calculations are relatively straightforward but it is first necessary to make the assumption that the activities are either splittable or non-splittable. The forward and backward passes in precedence networks are calculated, as shown here, in essentially the same way as for simple networks (Section 5.2). The procedure outlined here is valid if all activities are non-splittable.

In the forward pass, an activity is considered when the ESTs and EFTs of all immediately preceding activities have been determined. All connections into the activity are then considered and its EST and EFT values determined in accordance with the assumptions made about the behaviour of the activity (that is, splittable or non-splittable). In the backward pass, an activity is considered when the LSTs and LFTs of all immediately following activities have been determined. All connections out of the activity are then considered and its LFT and LST values determined also in accordance with the assumptions made about the behaviour of the activity.

There are slight differences in the above procedure if there are some splittable activities (see Section 5.6.1 for the differences) or, for example, in the complex situation when there are SF connections within a ladder. This latter point is demonstrated in the Tingham Tank Farm (Section 15.3.2).

5.6.1   Calculations with splittable activities

The calculation assuming infinitely splittable activities is readily carried out using the example in Figures 5.17 and 5.18. The procedure discussed at the end of the preceding section is re-examined and modified for splittable activities after the following example.

The activity *foundations* commences at 0 and ends at day 40. (Note that days 0 and 40 mean the ends of days 0 and 40. In fact it does not matter what they mean for the purposes of analysis as long as all references to days mean the same thing.)

Columns starts when it can, that is day 10, being the start time for *foundations* plus the SS lag of 10 days. *Columns* ends at the later of:

(1)   start of *columns* plus minimum duration of *columns*
      $= 10 + 12 = $ day 22
(2)   end of *foundations* plus the FF lag *foundations–columns*
      $= 40 + 3 = $ day 43.

Consequently *columns* ends at day 43. Because it starts at day 10 its duration of 12 days is split somehow into the 33 day period.

Deck starts when it can, i.e. the start time of *columns* (day 10) plus the SS lag (3 days), namely day 13. It ends at the later of:

(3)   start of *deck* plus minimum duration of 40 days
      $= 13 + 40 = 53$
(4)   end of *columns* plus the FF lag *columns–deck*
      $= 43 + 10 = 53$

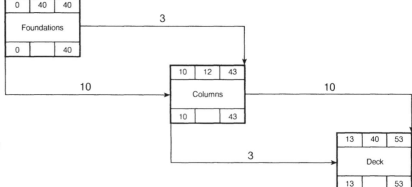

Figure 5.19 Calculation for the network shown in Figure 5.18 assuming splittable activities.

Consequently both routes (3) and (4) between *columns* and *deck* are equally critical.

The above analysis is essentially a forward pass through the network. A backward pass in this case yields the same figures for the latest start and finish times as for the earliest, giving the completed analysis shown in Figure 5.19.

With the activities being splittable (or stretchable), the calculation procedure (end of previous section) is made rather more complicated due to the splittability of activities. The calculations proceed as expected but with extra considerations which are now outlined:

In the forward pass, (a) the timing of the earliest end *of an activity can be defined either by the earliest start plus its minimum (non-split) duration or by the completion of connections coming into the end; and (b) the timing of the earliest* start *of an activity can only be calculated from the completion of the lags into it and* not *from the earliest end.*

It is sometimes possible to carry out the calculation of the forward pass or part of the forward pass by calculating all the start times and then all the finish times. In a similar manner the backward pass can sometimes be calculated in two phases. These two features allow correct networks to be drawn which appear to contain loops. An example of such a network is shown in Figure 5.20 which represents a situation in which there is a limited space available for storage of excavated material and some backfill has to be done before the excavation can be completed.

Here the calculations can be carried out only if the activities are considered splittable or stretchable but not if they are considered non-splittable. A practical illustration of this is shown in the Tingham Tank Farm project in Chapter 15.

In the calculations normally carried out on simple activity-on-node or activity-on-arrow networks, certain activities emerge as critical (Section 5.2). The concept of activity criticality is rather difficult to apply to precedence networks using SS, FF, FS and SF connections with

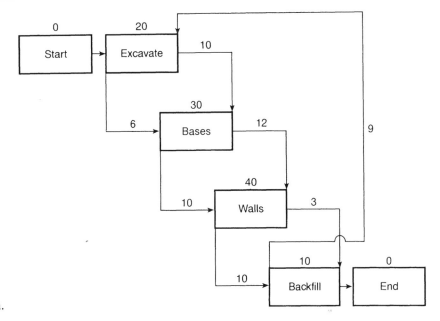

Figure 5.20 Precedence
network with
SF connection.

non-zero lags. For example, in Figure 5.19 all activity starts and finishes are constrained to be where they are if the project is to be completed in 53 days. This means, for instance, that the start and finish of *columns* are, in this sense, critical. Yet the time between its start and finish is now 33 days rather than the 12 days minimum indicated; so, in what sense is the activity critical? This matter is mentioned later (Section 5.6.3) for it is not a mere conundrum but of fundamental importance in the interpretation of precedence networks.

The assumption of infinite splittability always produces a lower bound solution for the overall project completion time, as illustrated in Figure 5.7. To visualise what splitting actually means, the graphical approach demonstrated there should be used. The lower bound results achieved with infinite splittability may be infeasible or impractical. If splittability is restricted, for example when using a non-splittable assumption, the overall completion time may increase due to differential rate effects; that is, different connected activities proceeding at different rates, also illustrated graphically in Section 5.4.2.

5.6.2 Calculations with non-splittable activities

The assumption of non-splittable activities implies that once an activity has commenced it must continue until completion. Thus, for instance in considering *columns*, once it starts it will finish 12 days later.

The result of the network analysis is shown in Figure 5.21. As was noted at the end of Section 5.6.1, the elimination of splittability may increase the project duration. In this case it is increased by 21 days to 74 days.

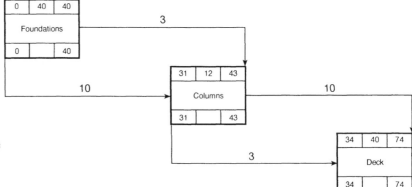

Figure 5.21 Calculation for the network shown in Figure 5.18 for non-splittable activities.

What has happened here is that *columns* cannot finish until 3 days after *foundations* is completed. Therefore its start must be moved to day 31 as shown. This has a knock-on effect for *deck* which is dragged forward to day 34 because of the SS lag of 3 days between *columns* and *deck*. This in turn pushes the finish of *deck* to $34 + 40 = 74$. The completed solution gives a project completion time of 74 days, 21 days longer than the splittable solution.

SIMPLIFIED ANALYSIS

With the non-splittable assumption, it is possible to simplify a network to assist in its analysis. The general principles are indicated in Figures 5.22 and 5.23 in which two activities X and Y with durations x and y are connected by SS and FF lags of duration s and f, respectively.

The technique requires identification of the critical route, which is the longer of $(s+y)$ or $(x+f)$. If $(s+y)$ is the longer route then the FF connection is redundant (in terms of criticality) and can be eliminated. Otherwise the SS connection is redundant and can be eliminated.

Using these ideas it is possible to redraw Figure 5.21 in the simplified form of Figure 5.24 which enables the calculations to be performed very quickly as the number of connections is reduced significantly.

UNBALANCED LADDERS

The above calculations can be extended to show something rather interesting about the non-splittable model. Using Figure 5.21 as an example and giving *columns* a longer duration, say 33 days, the network can be reanalysed as shown in Figure 5.25.

The commencement of *columns* has moved earlier, to day 10, and the commencement of *deck* has also moved earlier, to day 13, enabling it to finish at day 53. This phenomenon of decreasing the duration of the network by lengthening the activities can perhaps be understood by referring to the critical connections introduced in Figures 5.22 and 5.23. The critical connections for Figure 5.21 are shown in Figure 5.24. These

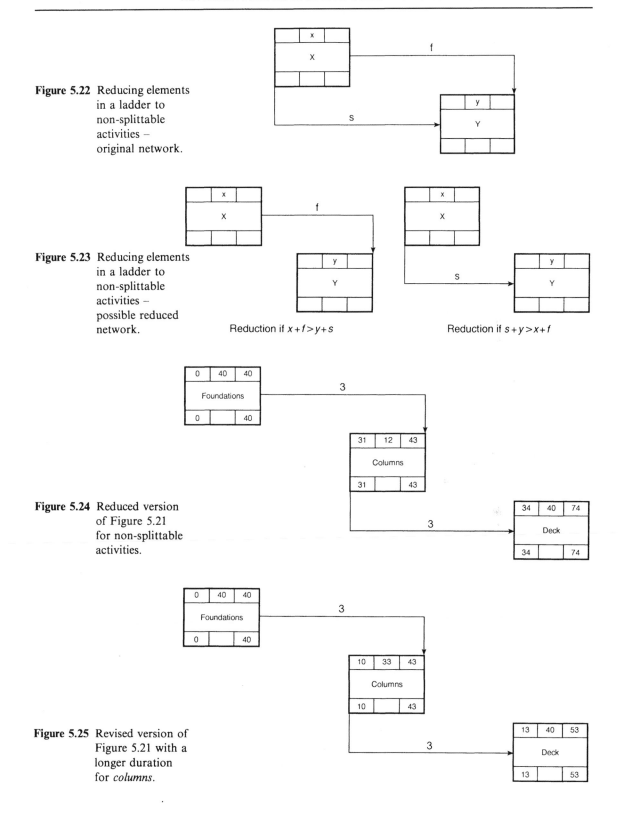

Figure 5.22 Reducing elements in a ladder to non-splittable activities – original network.

Figure 5.23 Reducing elements in a ladder to non-splittable activities – possible reduced network.

Reduction if $x+f>y+s$

Reduction if $s+y>x+f$

Figure 5.24 Reduced version of Figure 5.21 for non-splittable activities.

Figure 5.25 Revised version of Figure 5.21 with a longer duration for *columns*.

remain the critical connections until the duration of *columns* becomes 33 days at which point $(x + f) = (s + y)$ from Figure 5.23. Thus for any duration of *columns* up to 33 days, the position in time of *columns* after analysis is set by the end-time constraint on *foundations*, as shown by the critical connections in Figure 5.24. Accordingly, if the duration of *columns* were longer, its start time would be earlier (because its end time is fixed). This would enable *deck* to start earlier (owing to the critical SS connection in Figure 5.24) and hence enable the network to finish earlier. In the extreme, shown in Figure 5.25, increasing *columns* from 12 to 33 (that is, by 21 days) reduces the network duration by $74 - 53 = 21$ days. This happens because the ladder in the network in Figure 5.21 is *unbalanced*, and it does not become balanced until the duration of *columns* becomes 33 days, unless some other values (such as f or s values) change.

Although it seems paradoxical that increasing the length of any activity could reduce the project duration, this is the logical result of the assumptions used and highlights their shortcomings – unless, of course, the activity is truly non-splittable. The unbalanced ladder problem occurs when $x + f > s + y$ for the connections coming into an activity, and when $s + y > x + f$ for the connections going out of an activity. (It should be noted that, in this explanation, x always refers to the first activity duration, y the second activity duration, s the SS lag and f the FF lag for two connected activities.)

The fundamental issue with the lack of balance concerns the fact that activities which are dependent upon each other are progressing at different rates. It does not mean that the numbers for the activity durations are wrong but the analysis, in the non-splittable case, is not what the planner intended when the durations and lags were assigned. For example, considering what Figures 5.17 and 5.18 actually mean, the situation might be interpreted as follows:

(1) there are four foundations each of which takes 10 days;
(2) the column construction can commence as soon as the corresponding foundation is finished;
(3) the columns each take 3 days to complete;
(4) the deck units can be installed when the corresponding column is complete;
(5) the deck units each take 10 days to install.

However, this is not the mathematical translation. Instead of the work corresponding to (2) and (4), the SS connections on the diagram specifies that, for example, Column 1 can commence at day 10 (which is acceptable) and that the next column can begin at day 13. But it is clearly impossible for this to occur since the foundation upon which it is to be constructed will, at day 13, be only 30% complete.

This effect is caused by a subsequent activity proceeding at a faster rate than an earlier one, so that the later one's start has to be delayed to prevent it from overtaking the earlier one. The paradox outlined above arises if a fast rate activity is both preceded and followed by slow rate activities. The one preceding it delays the start time of the fast activity (because it can't start until it can all be done in one go); the one succeeding it is delayed because of the delayed start time of the fast activity.

The assumption of splittability avoids this difficulty because, although the rate of working on individual parts of a subsequent activity may be higher than that of the corresponding part of the preceding activity, the subsequent activity can then stop while the preceding activity catches up.

Whenever calculations are performed assuming that activities are non-splittable, it is important to check for the balance along ladders. Section 5.6.4 summarises the situation!

5.6.3   Comment on criticality

During the discussion of basic networks using only FS connections (Section 5.2) it was possible to identify for every basic configuration (that is, networks with no activity representing external constraints on the network except for the start and finish activities) at least one critical path through the network. Along this path lay critical activities, constrained to commence at their earliest start time and, failing which, the project would be delayed beyond its minimum duration.

In practice, however, it is likely that:

- projects are rarely scheduled to be completed in their minimum duration;
- projects frequently have external constraints such as fixed dates which affect the criticality issue (Chapter 6);
- networks often include resource constraints (Section 5.8);
- the durations of activities can be changed by altering resource levels (Section 4.7).

When dealing with precedence networks, the concept of criticality becomes even more indistinct and bound up with often unrealistic assumptions (Sections 5.6.1 and 5.6.2). This is because the connections represent the links between the starts and finishes of activities only and not the links at other points in the activities. This makes the graphical approach covered in Section 5.4 very useful for considering the implications of network connections.

5.6.4   Summary

The planner's dilemma regarding splittable and non-splittable activities is perhaps best summarised at this point:

Experience in construction suggests that most work is splittable. This suggests it would be best to use splittable activities. Network models use connections into or out of the starts and ends of activities and, for ease of calculation, assume infinite splittability. The infinite split is not particularly practical as it does not allow for any restart effects on the duration of the activity. It also gives the shortest possible project duration which may not be achievable with the activity splits the planner had in mind. The question of the meaning of critical paths passing through split activities is also raised.

Ideally, construction work should not be stopped and restarted so it would be sensible to try to avoid this happening by producing plans with non-splittable activities. In trying to make the plan as practical as possible the durations and lags used, for the best of reasons, may introduce the problem of unbalanced ladders. Non-splittable activities give a maximum critical path duration for the project.

The main options are planning with non-splittable activities or infinitely splittable activities, both of which have limitations. Both can give reasonable plans.

Through their own experiences the authors recommend that non-splittable activities be used. However, the planner should always consider each case on its merits and be prepared to split activities to avoid problems such as unbalanced ladders.

5.7 TIED ACTIVITIES

All the logic connections discussed so far are of the enabling type which allow Activity B to start or finish after the start or finish of Activity A. There is one other type of connection which is likely to occur in a deterministic project model. This is the sort of connection which exists when Activity B *must* start or finish within a fixed time of the start or finish of Activity A. A simple example of this type of situation was alluded to in Sections 4.2 and 5.3 when introducing the modelling of chemical processes such as paint drying. Activities in this sort of situation have been referred to as *tied* activities with the accompanying recommendation that they are to be avoided because of the excessive complication incurred in the network analysis calculations. This solution seems unreasonable since this type of connection might occur frequently and the results obtained without them would have to be interpreted carefully and could be meaningless.

To illustrate this, consider the situation which might exist in a motorway refurbishment contract. The traffic has to be moved off one area of the motorway before the work on that stretch can be done and the traffic moved back. There will perhaps be several such areas on the project and each one must be completed within a fixed time. Such situations arise quite frequently and non-compliance with time limitations could prove very expensive depending on the contract conditions. The network shown in Figure 5.26 could be used and analysed, and the results checked to see that no problems exist with the *must* restriction. Any problems would have to be eliminated by an iterative approach involving adjustments to the network data.

This approach is only satisfactory when the network is simple (as in Figure 5.26). If however, as might be the case, the section of network

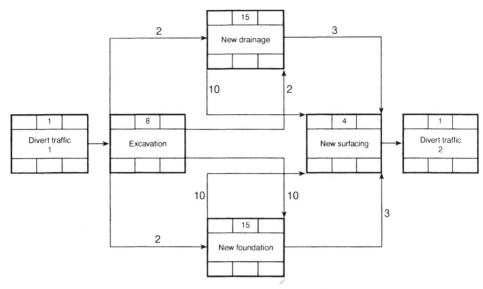

Figure 5.26 Section of motorway refurbishment network.

were repeated many times in a project with many interactions between the various parts, the iterative procedure may become very complicated to carry out manually and errors in the plan would be likely.

The proposed solution using tied activities would be to insert an extra connection as shown in Figure 5.27. This would not be a standard SF connection but a *must* connection.

The difference between these and standard connections can be illustrated by means of the following definitions (the first has been introduced in Section 5.4.1):

SS	The second activity cannot start until t time units after the start of the first activity (t is the lag duration)
Must SS	The second activity must start within t time units of the start of the first activity (t is the lag duration)

This section investigates the use of *must* lags to model tied activities.

5.7.1 Representation

Although standard documents recognise tied activities they do not in general recognise the ways of connecting them into the network, so there is no standard representation.

Here it is suggested that the *must* connections are represented like normal connections (activity-on-node format) but with the arrows drawn using dotted lines.

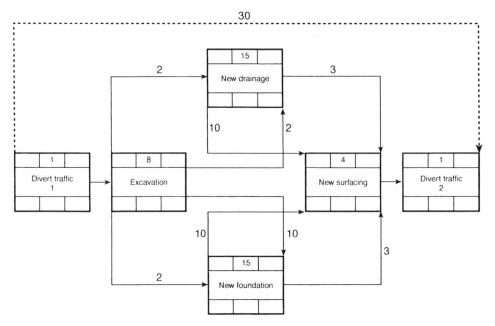

Figure 5.27 Section of motorway refurbishment network showing tied activities and *must* lag.

5.7.2 Calculation

Since these connections are not part of basic network analysis, the calculation method which has to be employed with them is covered here in some detail using non-splittable activities.

Calculation proceeds in the same manner as in normal networks but without considering the *must* connection at its origin activity. It is only considered at its destination activity. Thus in the example network, the calculation of the forward pass is as shown in Figure 5.28 until the last *divert traffic 2* activity is reached. At this stage the earliest finish time of this activity is checked to verify that the *must* connection has not been violated. In this case, the earliest finish time of the activity is 23 which is within 30 days of the start of *divert traffic 1* activity: the constraint is therefore not violated and the earliest finish time is correct.

Consider, however, the augmented network shown in Figure 5.29. The extra activity and its connections mean that the earliest finish time of the last activity is 40 which is not within the 30 limit specified by the *must* connection.

In this case, it is necessary to return to the start of the *must* connection and increase the earliest start time of the originating activity. Here it is obvious that the earliest start time of the originating activity must be increased by 10, but in general this may be rather more difficult to determine especially if there are several *must* lags in the network. Having returned to the originating activity with a new estimate of its

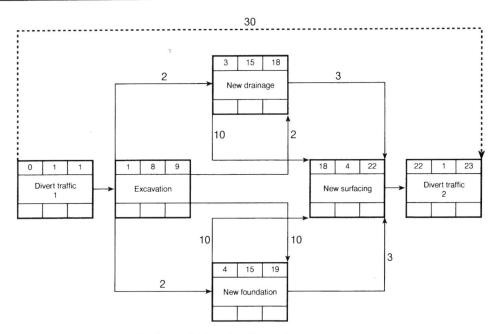

Figure 5.28 Status of network with tied activities after forward pass.

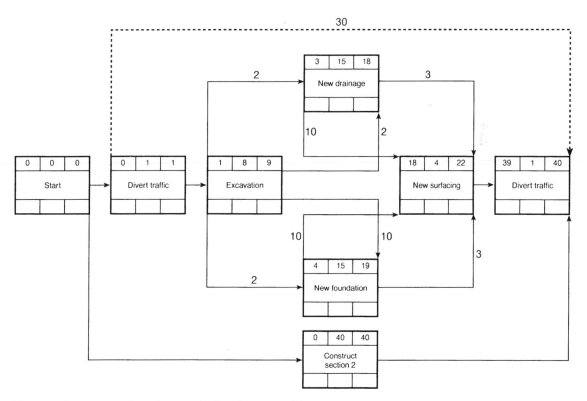

Figure 5.29 Augmented road network after first pass of forward pass.

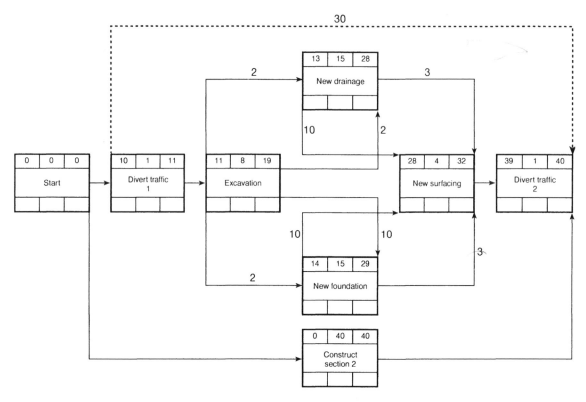

Figure 5.30 Augmented road network after second pass of forward pass.

earliest start time, the forward pass must be repeated for *all* activities following, not necessarily immediately following, that activity (in this case nearly all the activities). The analysis of this is shown in Figure 5.30.

The backward pass is performed in a similar manner but considering the *must* lags only at their origin rather than their destination. In both the situations shown, the calculations are done in a single pass but this is not the general case. The results of the calculations are shown in Figures 5.31 and 5.32.

5.7.3 Problems

The main problem with tied activities, apart from finding a computer package that handles them, is that the critical path calculations cannot in general be done in one forward pass and one backward pass. Both passes need in general to be iterative and the search for a solution may be very time-consuming even with fast computers.

Furthermore it is relatively easy to draw a network which is impossible to solve because it contains an internal contradiction. For

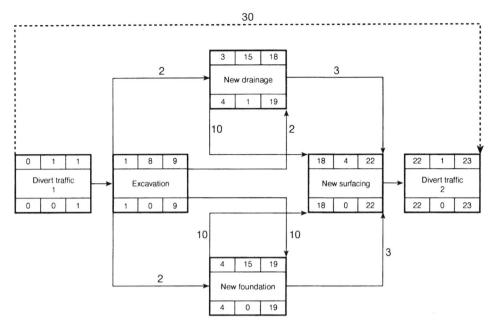

Figure 5.31 Road network after backward pass.

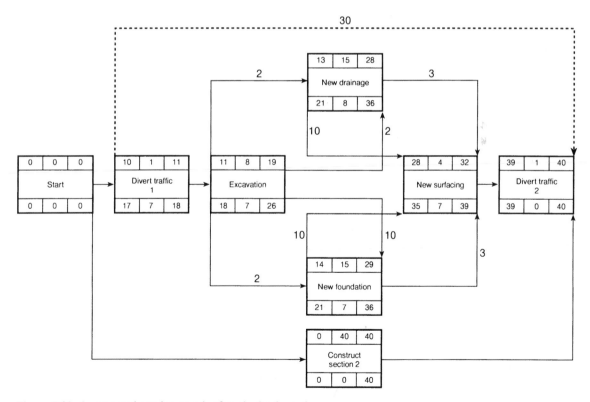

Figure 5.32 Augmented road network after the backward pass.

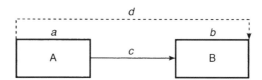

Figure 5.33 Tied connections.

instance in the network shown in Figure 5.33 the duration through the Activities A and B is $(a+b+c)$ but the tied connection requires the completion of B to be within d of the start of Activity A. Thus if $(a+b+c)>d$ there is no feasible solution.

Tied activities can sometimes be modelled using fixed dates. This is so for the basic network analysis procedure (using activity-on-node networks with FS connections without lags), although the model produced is not an exact representation of tied activities and the procedure requires the planner to operate the network analysis procedure iteratively. It is not possible, however, to use this approximation when resources and resource scheduling are applied (Chapter 7).

5.8 OTHER CONSIDERATIONS

5.8.1 Resource dependency

In terms of the resources to be used on activities there may be a dependency. Consider, for example, the project shown in Figure 5.3. The minimum duration of this project is 15 hours because both doors can be undertaken simultaneously as shown by the network. To do this it would be necessary to have two painters because the stripping and painting activities on each door would have to be programmed to occur simultaneously. If the available duration were relaxed to 18 hours, it would be possible to use just one stripper/painter who would spend three hours stripping and painting door 1 then go on to door 2. If resource-constrained in this way, the project network could be redrawn as shown in Figure 5.34.

This example serves to show that networks should be constructed using only 'technological' constraints. If this were not the case, any change in resource loading on the project or any out-of-sequence work would destroy the basis of the network. For example, on arriving on site to commence the project shown in Figure 5.34 and finding that door 1 was not available for an hour, the decision would probably be made to start by stripping door 2 and then going on to strip door 1. The logic of the resource-constrained network has been disobeyed but the original unconstrained network in Figure 5.3 remains valid as it shows only technological constraints in its logic.

Resource constraints are common in practical networks because of the apparent correctness which they give to the network as a model of a method of work. In reality, however, the problems which they generate when it is necessary to change the original method of work suggests that

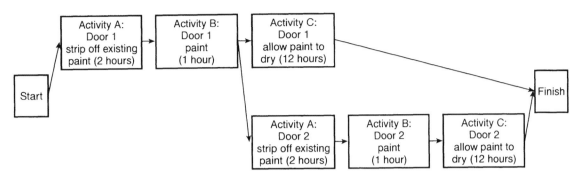

Figure 5.34 Two-door project using resource constraints.

they should be used with caution and avoided if at all possible. A discussion on the benefits and problems of resource constraints is contained in Section 7.2.1. Chapter 7 introduces resource scheduling techniques; networks should generally be left to show when work can be done irrespective of resource considerations.

5.8.2   The meaning of floats for 'chemical' activities

The example depicted in Figure 5.34 also illustrates another problem, alluded to in Section 5.3.3. The network path through door 2 is 3 hours longer than the network path through door 1 *paint drying*. This suggests that there is some float on the drying of door 1 which need not therefore start till 3 hours after door 1 has been painted. This theoretical float is meaningless in practice since the activity could not be delayed after the completion of the painting of the door.

In general, where this type of 'chemical' activity (Sections 4.2 and 5.3) is included in a network both the free and the total floats (Section 5.2) cease to have any meanings for those activities but the total float continues to be meaningful for the path through the network. This can lead to problems with interpretation of results and their use in practice.

5.8.3   Time and the critical path

A matter of some importance in planning is the determination of the permissible times at which activities can take place. The following terms are generally used and were introduced in Section 5.2.1: *critical path* and *critical activity*. From these the critical path duration can be determined as the minimum time in which a project can be undertaken assuming only technological constraints (that is, no restrictions on resources).

It is often assumed and stated that there is always at least one critical path through a network. Unfortunately, using any of the recognised definitions of criticality, this is not true in practice.

Aspects of planning which affect the critical path include:

- splittable activities (Section 5.6)
- tied activities (Section 5.7)
- multiple calendar working (Chapter 6)
- fixed dates (Chapter 6).

In searching for a critical path in a network with splittable activities it can be useful to show the total floats on both the start and the end of the activities (as these may be different). It is possible for an activity's start or end or both to be critical.

Splittable and tied activities have been discussed in this chapter. The next chapter continues the discussion in respect of multiple calendar working and fixed dates.

5.9   SUMMARY

- Choice of activities is fundamental to good planning.

- Several different types of activities exist.

- Different types of activities need different treatment in the planning process.

- Different types of activity interact differently.

- Four standard connections are defined:

  - SS   Start – Start
  - FF   Finish – Finish
  - FS   Finish – Start
  - SF   Start – Finish.

- These connections work well between normal activities but should be used with care on other types, especially chemical and process activities.

- The connections restrict only the start and ends of activities.

- SS and FF connections together between two activities are often used to mean that the whole of the progress of one activity is controlled by the whole of the progress on the other.

- Assuming activities are either splittable, non-splittable or stretchable changes the answers to network analysis calculations.

- To perform the calculations, it is usual to assume that the activities are splittable into infinitely small parts.

- Assuming infinitely splittable activities gives an infeasible lower bound to the duration of a project.

- As the size of the individual portions of a split activity increase, so the duration of the project increases.

- Assuming non-splittable activities gives an upper bound to a project's critical path duration.

- The standard assumption for a connection is that it represents an elapse of time between the starts and ends of activities.

- Planners use lags to represent work which must be done between starts and ends of activities.

- A lag can in reality be used to model both work and delay between the starts and ends of activities.

- Assuming either work or delay lags, this changes the answers to the critical path calculations when activities are assumed to be splittable or stretchable.

- When assuming work for the lags, it is important to define what work each lag actually represents.

- Calculations assuming non-splittable activities can give misleading results for unbalanced ladders – the critical path duration can decrease as the duration of the activities is increased.

- Connections are enabling devices in that they only give the allowable sequence of work.

- Tied activities help to model such situations as Activity B must start immediately after the finish of Activity A other types of connection are required.

- Network calculations for tied activities cannot usually be carried out in a single pass and sometimes there is no feasible solution.

- Floats mean different things for different activity types.

- The float on chemical activities is meaningless.

- The float on a normal process activity can be different at the start to the end if the activity is splittable.

- Process activities can have sections which are critical and other sections which are not.

- Resource constraints in networks should be avoided or considered very carefully before inclusion.

6

CALENDARS AND FIXED DATES

'But how is it
That this lives in thy mind? What seest thou else
In the dark backward and abysm of time?'

WILLIAM SHAKESPEARE
(The Tempest)

A few things to think about were introduced in Chapter 5. It is hoped they will live in the mind. A further problem in modelling arises in understanding time and the fact that work on some activities stops for the weekend or a winter season but work on others continues for 7 days a week. Multiple calendar models can help model these effects but can also give results that plunge the planner/user into a dark, backward abyss, unable to understand what is happening in the time model. It is hoped that this chapter will show how to handle these models and will also live in the mind.

6.1 INTRODUCTION

Work takes time to complete but, whereas time passes continuously, work is usually carried out intermittently. It might be stopped, for example, at:

- the end of shifts
- the end of the day
- weekends
- holiday time

- times of occasional inclement weather
- the end of a working season.

The time which can be utilised for work is defined by the calendar operating for that work. Since different types of work use different calendars, the coordination of these calendars is essential for proper planning. This is covered in Section 6.2.

Plans are also required to allow for external time constraints (termed fixed dates). This is covered in Section 6.3.

The concepts described in this chapter regarding calendars and fixed dates are very much related in the real world to the points raised in Chapter 5 for splittable and non-splittable activities and for tied activities. On some projects the topics covered in both chapters are used together and the reader is warned to be wary of the effects which each has on the other.

6.2 MULTIPLE CALENDARS

In an attempt to achieve greater accuracy in project plans, planners have introduced the concept of multiple calendars. This is done to allow some activities to work when others cannot. For example, different calendars are frequently used to allow:

- concrete curing to operate seven days a week when the site works only five;
- activities such as slip-forming concrete to operate three shifts per day whilst the rest of the site works only one;
- earthmoving to occur only within the earthmoving season;
- activities such as traffic switches and railway possessions to occur only at specific times (such as weekend nights).

In manual planning methods, the models are not sufficiently sophisticated to enable these things to be done with any certainty and multiple calendar effects are achieved by rules-of-thumb applied by planners. In computerised planning methods, the developers of the techniques have sought to provide, within their techniques, methods to enable planners to plan with many calendars. These innovations can have profound effects on the simple calculations which are normally carried out and it is therefore important to understand the possible effects when using the methods or when interpreting the answers produced by the methods.

6.2.1 Implementing multiple calendars on computers

There are two main methods of implementing multiple calendars in computer-based planning.

In the first, and most common, method it is possible to divide the time-scale into shifts and define for each shift whether an activity is 'working' or whether it is 'holiday' (not working). As many calendars as are necessary can be set up in this way (subject to any limitations in the software used), with different calendars having different combinations of working and non-working shifts. Furthermore, this can be done for either the activities or the resources. This method is ideal for modelling such things as concrete curing and traffic switches when the shifts for working are known. It is rather less satisfactory to model weather patterns several months in advance. This is the situation for which the second method was developed.

In the second method, the planner specifies working and non-working periods on one calendar and defines the percentage of normal working rate to be achieved by the different types of work. This second method is not satisfactory for the traffic switches and concrete curing.

Computer packages rarely offer both these options, the most common being the first method. The implications of multiple calendar assumptions on critical path calculations are therefore described, in the following section, in terms of this method. Similar effects are generated by the second method.

6.2.2 Critical path calculations

Consider, as an example, the small network in Figure 6.1, where the activity durations are given in days. Network analysis and the terminology associated with it have been introduced in Chapter 5, to which reference should be made if terms used here are not understood.

From inspection, it is obvious that the critical path is through all the activities and the project duration is 12 days. Now consider the site working 5 days per week. The curing activity will naturally be done on a 7 day week since, as with most chemical reactions, it does not recognise weekends.

The calculations of the earliest and latest, start and finish times of the three activities in this model are best illustrated by means of the bar charts shown in Figure 6.2.

It can be seen that starting on Thursday or Friday makes the project finish on the same day as if the project started on the following Monday. Thus, defining a critical activity in the normal way as 'one which cannot be delayed without delaying the end of the project', only the last activity *strip formwork* is critical if the project starts on Thursday or Friday. Further, since the critical path is 'the path of activities through the network from start to end containing all critical activities', there appears to be no critical path by this definition.

The total float of the first activity for a Thursday start can be given in several ways (Section 6.2.3). In terms of its own calendar it is two days;

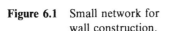

Figure 6.1 Small network for wall construction.

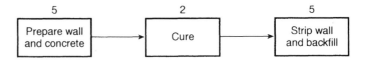

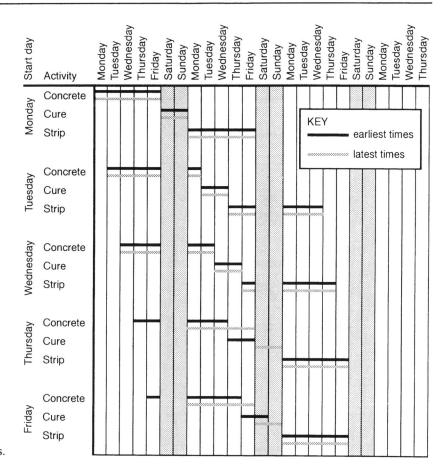

Figure 6.2 Bar charts for network in Figure 6.1 for project starting on different days.

in terms of the underlying calendar (that is, a 7 day week), it is 4 days. It is important that the planner should be aware of which figures are being quoted. In general, the float generated in a single path through a network purely by the inclusion of multiple calendars is a function of the number of changes of calendar along the path and the difference in work periods between the calendars.

Defining the critical path as 'the path with least float' and a critical activity as 'an activity on the critical path' appears to overcome the problem with the previous definition. In reality, the situation is generally compounded, giving many short critical chains scattered through the network starting and stopping when consecutive activities have different calendars. This can make a nonsense of the importance often associated with 'critical' and can cause embarrassment to planners who cannot explain the results of their analyses.

Studies of the multiple calendar/float effect have concluded that in any network using multiple calendars, a critical path from start to end is unlikely to exist. The float generated can give a maximum value equal to the sum of the changes in the activity calendars along the path. The change in the activity calendar is measured in the smallest time unit of the underlying calendar and is the number of these time units in one cycle of one of the calendars minus the number in one cycle of the other.

As well as total float, any activities on the apparent critical path can have free float and this can also increase at any calendar change by the difference in work durations of the two calendars.

It is recommended that network calculations with multiple calendars be treated with considerable caution.

6.2.3 Definition of float in multiple calendar networks

The float in a multiple calendar network can be defined in terms of any of the following:

- the number of working days in the activity's own calendar;
- the number of working days in the project main calendar;
- the number of working days in the underlying Gregorian calendar (that is, the 24 hours a day, 365 days a year – plus the leap day every four years – calendar)

None of these is ideal and all may cause problems of interpretation as follows:

Own calendar This appears to give the smallest floats of the options available but is not easily traceable through a network.

Gregorian calendar This gives the largest floats and may actually give a float to a truly critical activity.

Project main calendar Somewhere between the other two options, this has the drawbacks of both whilst having the benefit of being the normal terms of communication between project managers.

In practice it is unlikely that any of these would be sufficient by itself.

It is therefore advised that if multiple calendar techniques are used, an activity's float should be quoted in both the activity's own calendar and the Gregorian calendar.

If the computer package being used does not allow the user to define the floats, it is important to make sure that the meaning of the floats provided is understood.

6.2.4 Multiple calendars and splittability

When multiple calendars are used to give periods in which activities cannot work, it has to be decided whether to deem a break in the progress of an activity, caused by this calendar effect, a 'split' in the activity working. If it is a split, then particular attention has to be given to non-splittable activities and the minimum split considerations in splittable activities (as discussed in Chapter 5).

Consider, for example, the network and bar charts (Figure 6.1 and 6.2) used when discussing multiple calendar, critical path calculations in Section 6.2.2.

If these activities were non-splittable and a break caused by the 5 and 7 day calendars were considered to be a split, the only day the project could start (to avoid the split happening) would be a Monday. If the activities were splittable with a minimum split of 2 days, the project could start on a Monday or a Thursday.

In general, considering calendar breaks as splits makes the calculations nearly impossible and their results very difficult to understand for all but the simplest of networks.

In reality what the planner may wish to convey is that natural breaks should be allowed to occur in otherwise non-splittable activities. This might be to consider an activity as being non-splittable in a 5 day week or similar calendar. The productivity figures used in the plan would be based on average values with rises and falls in productivity around a weekend being absorbed in this figure. The resources would still be organised for working on the activity and it would be reasonable to assume the productivity would not decrease in the same way as for normal splits.

6.2.5 Uncertain calendars

The calendars used by planners are often wholly predictable. However, when working within a time frame stipulated by natural events such as temperature, river levels and the like, an additional degree of difficulty is introduced into the planning exercise. This problem is briefly discussed in Section 8.5.

6.3 FIXED DATES

Construction work almost invariably requires coordination between parties either involved in or affected by the works. Examples include:

- work at a railway station requires coordination with the railway operator
- work on the highway requires coordination with the regional and/or highway authority
- an electrical specialist subcontractor working in a restricted area in a building must coordinate the work with other trades working in the area.

In many cases, a contract may specify a procedure for coordinating the activities. In some cases, there will be a firmer constraint, known as a *fixed date* constraint. In these first two examples the railway operator and the highway authority may specify a fixed calendar within which work may proceed. Furthermore they may specify that particular items of work must be completed on or before a specified date, which forms an absolute constraint on the programme. In the case of the electrical specialist subcontractor, the coordinating contractor may stipulate that access to a particular part of the site is available within strictly fixed dates.

The greatest constraint imposed by a fixed date is one where a specified activity *must* take place on a specified date. An example of this is the closure of a motorway for 12 hours to build a new overbridge. In such cases the 12 hour possession period would be specified many months in advance so that all interested parties could be informed and plan accordingly. In such a case the activity *build the new bridge deck* is almost devoid of any latitude in planning, give or take the odd hour.

A situation sometimes occurs in which one activity must be completed within a specified time of another activity. This might be considered to be a special case of the situation described above for the bridge deck. Another more extreme example of this, in very detailed programming, might occur when concrete must be poured within a fixed time of mixing. The activities linked in this manner are said to be tied. Tied activities are covered in detail in Section 5.7.

In addition to the uses described so far, fixed dates are sometimes used to define the window in which resources are going to be available. This is an apparently practical solution to a difficult problem but can cause problems with the interpretation of the results. A discussion of these points can be found in Sections 6.3.1 to 6.3.5 with examples of calculations in which fixed dates are involved.

In summary, the types of fixed dates which may be imposed on an activity are:

- special calendars
- on-or-after fixed dates (fixed earliest start dates)
- on-or-before fixed dates (fixed latest finish dates)
- tied activities.

Additionally for splittable activities, the planner might need the earliest finish times and latest start times fixed (see Section 5.6.1 for the use of start and finish times in calculations for splittable activities).

6.3.1   Fixed earliest start date

The objective of a fixed earliest start date for an activity is to ensure that it cannot start before a predetermined date. As outlined above, this might be for such things as ensuring that access is available to a particular area of a site or that there is a minimum of interference between various contractors on a project.

For all practical purposes, it is the start of the activity which is fixed although it might be theoretically feasible to produce a model in which any element of an activity cannot happen until at least the given date. Such a model might be considered useful when splittable activities are used in a network. In a situation in which it might appear advantageous to fix the earliest time of anything *other than* the start of an activity, the activity should be split into more activities such that the start of one of them can be fixed.

The most common start date to fix is that of the start of the project. If there are several starts to a project, all of the 'start' activities should have their earliest start dates fixed to the allowed start date of the project.

6.3.2 Fixed latest finish date

This is used and handled like the fixed earliest start date, as the following words imply. The objective of a fixed latest finish date for an activity is to ensure that it must be completed by a predetermined date. As outlined above, this might be for such things as ensuring that required sectional completion dates are achieved or that there is a minimum of interference between various contractors on a project.

For all practical purposes, it is the end of the activity which is fixed, although it might be theoretically feasible to produce a model in which any element of an activity must have been completed by the given date. Such a model might be considered useful when splittable activities are used in the network. In a situation where it might appear advantageous to fix the latest time of anything other than the finish of an activity, the activity should be split into more activities such that the finish of one of them can be fixed.

The most common finish date to fix is that of the end of the project in order to ensure that the work is completed on time. If there are several ends to the network and no sectional completion is required, all the *end* activities can be fixed to the required project finish date.

6.3.3 Effect of fixed dates on timing calculations

FIXED EARLIEST START DATES

These tend to increase the floats on all the activities in the paths through the network up to the activity for which the start date is fixed. This is because they are equivalent to the addition into the network of an extra (dummy) activity which only operates during the forward pass calculations. This activity is connected with an SS connection (Section 5.4.1) between the start activity and the dummy activity and an FS connection between the dummy activity and the fixed date activity. Lags to and from the dummy have zero duration. The duration of the dummy activity is the difference between the allowed start date of the project and the fixed date. This would usually be calculated in the Gregorian calendar (Section 6.2.3).

As an example of this on a simple activity-on-node network consider the Tingham Bridge project (Appendix A and Figure 2.20). The critical path calculations for this are shown in Figure 5.1. This shows the critical path going through the *construct in situ span* activity. If a fixed date constraint is stipulated for the *construct precast span* activity, which corresponds to a commencement of the activity on day 60 of the project,

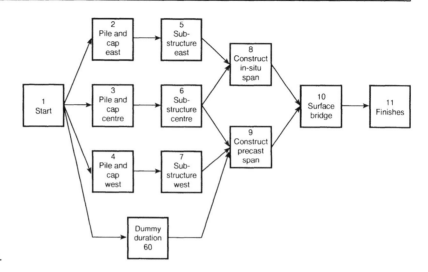

Figure 6.3   Tingham Bridge network with dummy duration.

the critical path switches to *construct precast span* and disappears for all preceding activities. The reason for this is that the inclusion of this fixed date is equivalent to introducing a *dummy* activity with duration 60 in the network, as shown in Figure 6.3.

FIXED LATEST FINISH DATES

These tend to decrease the floats on all the activities in the paths through the network from the start activity up to the activity for which the finish date is fixed. This is because they are equivalent to the addition into the network of an extra activity which only operates during the backward pass calculations. The dummy activity is connected from the fixed date activity with an FS connection and to the end of the project with an FF connection. The lags to and from the dummy activity have zero duration. The duration of the activity is the difference between the fixed finish date of the activity and the finish of the project. This would usually be calculated in the Gregorian calendar (Section 6.2.3).

Fixed latest finish dates can give rise to negative float which would indicate that an activity is already late if the fixed finish date is to be met. If this is the case, the planner must reduce the length of the path between the start of the project and the activity with the fixed finish date either by decreasing the durations of some or all of the activities along the path or by changing the method of work implied by the activities and connections.

Fixed dates obviously interact with one another and it can often be a complex matter to determine which fixed dates are having the most detrimental effect on the critical path calculations. Because the fixed dates alter the float of activities, they can cause the critical path to fragment and, in effect, disappear. Even defining the critical path as the path with least float does not always clarify the situation since in general this is rarely a continuous path through the network from start to finish.

6.3.4 Fixed dates and multiple calendars

Fixing dates when using multiple calendar techniques is similar to fixing them in a single calendar technique, except that it is essential to ensure that the fixed date is a working date in the calendar of the activity on which it is operating.

Because both fixed dates and multiple calendar techniques independently act to remove the critical path, it is most unusual in a multiple calendar network with fixed dates to have a critical path. Any critical path that is produced should be examined very carefully before it is relied upon.

6.3.5 Fixed dates in practice

The application of many fixed dates to a network can make it difficult to understand the effects of the individual restrictions. In order to minimise this it is suggested that the following steps might be applied.

- observe how the network behaves with no fixed dates applied initially;
- impose fixed dates either individually or in selected batches;
- observe how the network is affected and which fixed dates have the most effect.

Fixed latest (finish) dates should be imposed last (after checking to see if the network fits the contract period) due to their effect on critical paths.

Fixing dates for activities restricts the time window in which they can be carried out. The consequence of this is that there are fewer options for a resource scheduler to level out the use of resources.

6.4 SUMMARY

- A considerable amount of construction work is affected by calendars.

- There are two types of calendar effects – predictable and non-predictable.

- The apparent increase in accuracy obtained by using multiple calendars may be bought at the expense of clarity and usefulness.

- Computers typically implement multiple calendars in one of two ways.

- In general the introduction of multiple calendars changes the floats of activities.

- Every time an activity working on one calendar is followed by an activity working on another calendar, float can be generated.

- The float of an activity can be defined in terms of its own calendar, the underlying calendar or any other convenient calendar.

- It is unusual to have a critical path in a network with multiple calendar activities.

- Activities taking longer than 5 days and working only 5 days per week might, by definition, be splittable. However, the effects of weekend splitting are not as severe as the effects of splitting discussed in Chapter 5.

- Network calculations with multiple calendars should be treated carefully.

- Fixed dates are useful for modelling project constraints.

- The effect of a fixed earliest start date on an activity can be modelled in activity-on-node networks by using dummy activities.

- Fixed dates alter the floats of activities and the critical path.

- The influence of fixed dates which interact can be difficult to understand.

7

PLANNING RESOURCES AND FINANCES

'Many hands make light work'

ANON.

'Too many cooks spoil the broth'

ANON.

These two proverbs offer words of wisdom to the planner who is interested not just in planning activities but also in scheduling the resources to carry out the work. It can be a fine line between optimising the resource usage and adversely affecting productivity – but the attempt at scheduling should be made. Planning the resources can save time later because, after all, a stitch in time saves nine.

7.1 INTRODUCTION

Resources and finances were introduced in Chapter 3. Subsequent chapters have concentrated on understanding what the activities of a project plan are and how to handle them, without considering in great detail the resources that would do the planned work. Certainly, resources have been considered in so far as activities need to model work as realistically as possible, and the resource implications of network analyses need to be understood and accepted as reasonable. It is, however, worth giving more attention to planning the resources and finances as they are so important to the successful outcome of a project. When to do this is debatable; that it should be done is beyond question. The management of planning is discussed further in Chapter 11. The simplified diagram of the resource planning process (Figure 11.1) shows that the whole of this process is somewhat circular and the order in which the stages are tackled is therefore somewhat arbitrary. It is felt that now is the right place in the book to discuss techniques for planning the resources and finances.

In this chapter the application of techniques for the planning of resources and finances is outlined. Although many of the techniques could be applied manually using Gantt charts or linked bar charts, the amount of time and effort required means that this is rarely done, except in retrospect for claims' analysis (Chapter 14) or on very simple projects. The nature of the techniques, being generally iterative and repetitive, lend themselves to computerisation. As networks generally form the basis of computer planning packages the discussion in this chapter is based on networks (Section 2.5).

7.2 RESOURCE SCHEDULES FROM NETWORKS

7.2.1 Earliest start and latest start schedules

In project planning, a network naturally produces, when analysed, an overall project duration. This is the critical path duration and is the shortest time in which a project can be completed using the method assumed in the construction of the network, given that the individual activity durations are correct. Simple network analysis and the appropriate terminology are introduced in Chapter 5. Two possible *schedules* (lists of dates showing the start and end times of activities) can also be produced from the network analysis. The schedules show:

- all the activities at their earliest start time – an EST schedule
- all the activities at their latest start time – an LST schedule.

If a network (or other project model) contains only technological constraints, the general effect (in project resource demand terms) of using the schedules produced from the critical path analysis is shown in Figures 7.1 and 7.2. Figure 7.1 shows the number of resource required in each time unit or period; Figure 7.2 shows the cumulative number of resources required since the start of the project.

The project resource budget shown in Figure 7.2 is not a true budget unless the resource is one of the financial resources, earning or liability (Section 3.3). The similar curves for other financial resources are shown in Figure 7.11 and the effects of the delays between earning and income and between liability and expenditure can be clearly seen. This is discussed in detail in Sections 7.8 and 7.9.

One common use of such charts is in fact with an abstract resource of *activity days* (Section 3.2.8) rather than a financial resource. If all

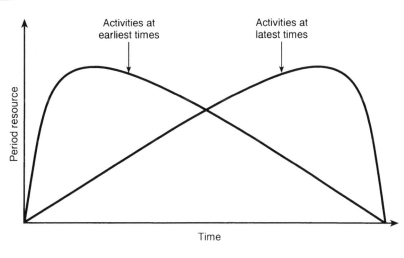

Figure 7.1 Idealised resource profiles for EST and LST schedules.

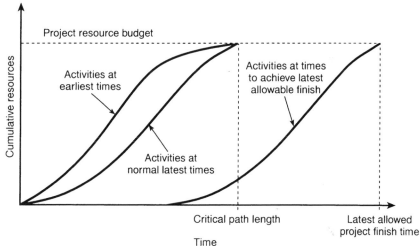

Figure 7.2 Idealised cumulative resource use for EST and LST schedules.

activities are similar in size (Section 4.2.4) the chart can be used to monitor progress (Section 12.3.5).

If a network has been produced using only technological constraints a series of graphs like those shown in Figures 7.1 and 7.2 would be obtained, one for each resource. They would not have any particular resource considerations for the network taken into account except considerations appropriate to each activity in isolation (Chapter 4). Since it is usually not desirable to carry out the work with all the activities at either their EST or their LST, planners often attempt to include a large number of *resource constraints* in the original network. These are constraints which look like, and have the same effect as, technological constraints but which are there solely to ensure that the resource use in the EST schedule is reasonable and practical. For example, where there are two independent activities which need the same resource (such as a crane), the planner may use a link between them to ensure that they may not be constructed simultaneously. Such a link is not necessary on technological grounds; it simply ensures that the resource demand is sensible. Such a link is known as a resource constraint. Section 5.8.1 illustrates and discusses this point. As another example, it is common in a major road project to see constraints ensuring that the bridges are constructed in a specific order because that is the order in which the planner envisages the cranes moving. In reality, if the need arose, it would be possible to construct the bridges in almost any order, and indeed it may even be possible to construct them all at the same time.

If resource constraints are used, the network itself may be close to being a plan of work and, as such, conceptually useful. It is, however, very difficult to update or alter it when changes occur in the work during the execution of the project.

Since changes caused by such varied things as client decisions, contractor preference, weather and unforeseen conditions occur very frequently in most construction projects, it is recommended that resource constraints are not included in networks.

If technological constraints only are included in the network (as theory suggests), it is often not necessary nor even preferable to programme to finish the project in the critical path or shortest duration and the network is certainly not a plan, programme or schedule of work. There is however a need to schedule the resources in order to produce a reasonable plan.

The two schedules discussed so far (the EST and LST schedules represented in Figures 7.1 and 7.2), are the most commonly used, but are rarely if ever the best schedules to work with. The EST schedule usually gives a high early peak resource demand which causes inefficient resource usage. The LST schedule causes inefficient resource use later in the project and increases the risk of late completion because of variations to the work affecting activities which are being carried out as late as possible (see the example in Section 7.3).

The schedule employed should be somewhere between the two extremes. It cannot have activities starting before their earliest start times since this would imply that the technology of the construction method as modelled by the network is wrong. Similarly, it should not have activities starting after their latest start time as calculated from the desired project completion date since this would mean that the project would finish late. It is thus necessary to choose the start and finish dates for all the activities to produce the desired schedule. They may be chosen for many reasons, for example to:

- earn money as early as possible,
- smooth the use of resources
- provide continuity of employment for subcontractors.

7.2.2 Allocating and scheduling resources

A schedule is simply a list of activity start and finish times. The techniques designed to produce a schedule by considering resources are collectively known as resource scheduling.

Creating a plan of work involves determining resource levels, and allocating and scheduling resources with due regard to smoothing or levelling. These terms can be defined as follows:

Scheduling   This is the process of determining activity start and finish times or dates subject to constraints of logic, time and resources.

Resource allocation   This is the scheduling of activities and the resources required by them within predetermined constraints of availability and/or project time.

Resource aggregation   This is the summation of requirements for each resource, for each time period.

Resource smoothing   This is the scheduling of activities (within their limits of float) so that fluctuations in individual resource requirements are minimised.

Resource levelling   This is the process of producing a schedule that reduces the variation between maximum and minimum values of resource requirements.

The main ways of displaying information are the resource histogram (or profile) showing the quantity of resource required in any time period (Section 2.6) and the bar chart showing when each activity will be operating (Section 2.2). These are both used frequently in the rest of this chapter.

Construction is seldom simple and diagrams such as Figures 7.1 and 7.2 are usually not particularly easy to construct unless the advice of Chapter 4 has been rigidly followed. Most activities require more than one type of resource and also probably require resources different to those of other activities. There may well be limits on resource levels and the project duration may be specified. These points complicate the resource scheduling problem, but some attempt at scheduling should be made as it should help with:

- smoothing resources,
- ensuring maximum resource limits are not exceeded,
- setting resource levels to meet contract completion dates.

7.3 RESOURCE AGGREGATION AND SMOOTHING

7.3.1 Resource aggregation

Resource aggregation is the simplest and most useful resource planning technique. It provides the number of each type of resource required during each time period in order to achieve any given schedule. The results of this procedure are useful in determining the resources required:

- on a project
- in resource smoothing
- in resource scheduling.

If the resources under consideration are financial resources, the result is a budget (Sections 7.8 and 7.9).

There are several ways of performing resource aggregation. The most straightforward for manual application is based on the assumption that the resource demand over the activity duration is constant, although it could be altered to work with any or all of the activity models discussed earlier (Chapter 4). If the resource demand is in terms of total resources, it should be converted to a rate demand (Section 4.3.3) by dividing the total activity resource requirement by the activity duration for each activity. The method, which is independent of the planning method used initially to provide the activity start and finish times, is outlined below:

> *From the schedule of activities (the activities with their start and end times), take each activity in turn and add the number of resources required per time unit into the aggregation.*

Each resource should be treated separately and the result is a resource histogram or resource profile for each main resource on the project. Examples of these are shown in Figures 7.3 and 7.4 for a single resource in the Tingham Bridge project. Using the EST values from the

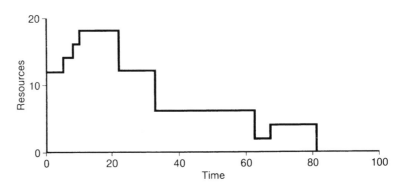

Figure 7.3 Earliest start time resource profile for Tingham Bridge project.

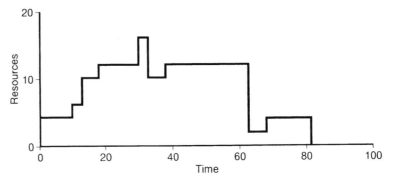

Figure 7.4 Latest start time resource profile for Tingham Bridge project.

critical path analysis (Figure 5.1) to define the schedule of work, the resource requirements of the project can be found by using *resource aggregation*. The EST resource profile is shown in Figure 7.3. The resource profile for a latest start schedule is shown in Figure 7.4.

The smoothness of the curves depends on the number of activities in the plan. Usually, the greater the number of activities in the plan, the greater the number which can theoretically be executed concurrently and the greater the number of time periods in which the resource demand changes. This leads to a smoother resource demand profile, though not necessarily a more level one. Such smoothness is particularly evident when the resources under consideration are financial, since individual activities can have a much greater range of values than if the resources were labour and plant (when the numbers used on individual activities tend to be small integer values).

It can be seen that in general the histograms give high peak resource demands and uneven resource usage. The EST schedule peaks early and tails off slowly, whereas the LST schedule builds up slowly to a late peak and drops off rapidly. This is to be expected because in general there are a lot of activities that can be started early in a project and a lot that can be finished later if so desired.

The resource profiles would be similar in shape if the critical path analysis were performed with the project's latest finish date fixed to be the latest allowable finish date rather than the earliest possible finish date. This is illustrated in Figure 7.2.

It can be seen that the two profiles for the Tingham Bridge project (Figures 7.3 and 7.4) show the trends that are typical of earliest and latest schedules (Figure 7.1) but that there are obvious changes (steps) in the resource level because of the small number of activities.

Particular attention should be paid to the aggregation of resources such as materials, for which the constant resource demand model of an activity (Section 4.4) may not be realistic. For example, it is common to have construction activities with the demand for concrete concentrated at the end and the need for formwork at the start. If there are not enough activities which are able to be carried out approximately concurrently in the schedule, the assumption of constant resource demand may give a misleading profile. In such a case, it is necessary either to model the varying resource demand on an activity or, more usually, to split the activities down until the constant resource demand assumption becomes acceptable. This is a matter of judgement and experience.

7.3.2 Resource smoothing

Where activities have float, their timing can be scheduled within these floats to help smooth resource demands, thereby improving the efficiency of the operation without increasing its duration. Resource histograms can be drawn and problems of peak or intermittent demand identified. In the example of a network for the Tingham Bridge project (Appendix A and

Table 7.1 Activity durations and resources for Tingham Bridge project.

Activity	Duration	Resources
Pile and cap east	10	4
Pile and cap west	8	4
Pile and cap centre	5	4
Substructure east	23	6
Substructure west	20	6
Substructure centre	17	6
Construct *in situ* span	30	6
Construct precast span	5	6
Surface bridge	5	2
Finishes	14	4

Figure 2.20), the resource requirements of the activities are as shown in Table 7.1.

It has been noted that the shapes of the resource profiles for both the earliest and latest schedules are rather pointed and therefore not ideal to work to. The resource demand for the project can be smoothed by moving the activities within their float time and the results of this process are shown in Figures 7.5 and 7.6. Figure 7.5 is a bar chart showing when the activities will be carried out within the allowed period. Figure 7.6 shows the consequent resource profile obtained by aggregation.

There is no set way to achieve a smoothed resource profile manually; it is a matter of trial and error. To obtain the result shown (Figure 7.6) the first step was to produce a network for the project followed by a Gantt chart showing the activities at their earliest start times and indicating the available total float for each. Resource aggregation was then used to provide a resource profile. These are the basic tools used in the smoothing process which was carried out graphically. The sequence for smoothing which was used in this instance to generate the schedule in Figures 7.5 and 7.6 is as follows:

> *A peak demand was identified from the resource profile and from the Gantt chart. The activities in progress at that time which contributed to that peak were identified. One of those activities,* substructure west, *was selected and moved within its float time in order to reduce the peak. The new position (in time) was chosen by identifying a dip in the resource demand and relocating* substructure west *there. In the example it was decided to move it to finish as late as possible at time 58.*
>
> *The next step was to check on consequential changes, that is, what other activities had to be moved because of the relocation of* substructure west. *The network diagram is used to ensure that the connections in the network are maintained. In this example, the activity* construct precast span *could not be started until after the completion of* substructure west, *and so was moved. The best place to move it to was*

determined by consideration of the resource profile and the allowable timing, identified from the Gantt chart. It was noted that the activity could not start until its latest time so it was moved to there.

It is obvious that it is quite possible for the consequential effects of moving an activity to make the resource profile worse than it was originally. In this case the planner would have to decide whether or not to return to the original state and choose another initial move.

On small projects with few resources, it is reasonable to attempt manual smoothing. On normal construction projects the smoothing process is quite difficult to do since there are usually a large number of activities and many different types of resource which tend to interact with one another. It is not unusual to cause peaks and troughs in the resource profile for one resource by smoothing another. For these reasons managers, if working manually, usually select key resources to work with and smooth only these; the non-key resources are simply aggregated and the peak demand checked for acceptability.

If a computer is available, the process is more manageable on a real construction project but the techniques available still do not easily handle

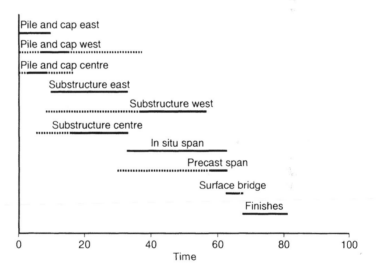

Figure 7.5 Bar chart showing smoothed schedule for Tingham Bridge project.

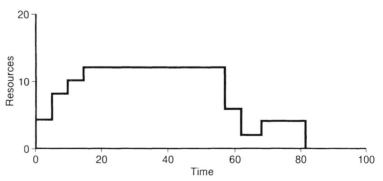

Figure 7.6 Smoothed resource profile for Tingham Bridge project.

a large number of activities and large numbers of resources and do not claim to give the best answer for any project. The techniques employed are heuristic (based on rule-of-thumb) algorithms and work by following a series of rules which have been found to provide a better than average answer in most situations. A similar result is obtained by heuristic resource scheduling procedures (Section 7.4.2), but since different resource scheduling procedures have different objectives, a typical computer-based heuristic resource smoothing process is described (Figure 7.7).

In this, the computer program first calculates the resource profile for a given schedule (usually the earliest start schedule). It then selects a resource peak and tries to reduce it by moving an activity from that time period to another. The activity chosen and the amount of movement attempted help to define what is a good scheduling method or a poor one for any given project. The way the activities are selected is by using heuristics. For example, the technique might choose the activity which has the greatest total float because that activity is the one that can be moved the most; alternatively, it might choose the one with the greatest free float (Section 5.2) because that is the one that can be moved the greatest distance without having to move any other activities and thus affect the resource profile so much.

Recognition of where in time to move the activity to is relatively simple if the activity model being used has a constant resource demand and a fixed duration. If this is allowed to change to a variable resource demand and/or variable duration model (Sections 4.4 and 4.6), the difficulty in positioning the activity is greatly increased. If there are resource constraints in the network as well as technological constraints (Section 7.2.1), the performance of the process is usually adversely affected because of its inability to recognise that it can over-ride some of the constraints if it has to. For these reasons it is advisable to use simple activity models and only technological logic if resource smoothing is to be employed on the project.

No single heuristic will always provide the best answer and it is sensible to try several in the computer package which is being used (if possible).

If a critical path analysis fixes the finish date of the last activity at the latest allowable project finish date, there are usually no activities with zero total float and all activities can thus be moved in the smoothing procedure outline above. In this case the result is very similar to that obtained by the resource scheduling process described below.

7.4 RESOURCE SCHEDULING

The results of the critical path analysis give the ESTs and LSTs of the activities. If the resources available to undertake the project are unlimited the EST or LST schedule may be used. If, as is usual, resources are limited, it may not be possible to run these schedules owing to insufficient resources. The planner must decide which activities are to run and when.

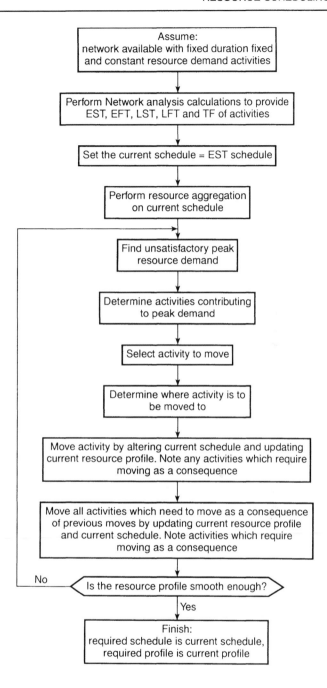

Figure 7.7 A typical heuristic resource smoothing procedure.

This is *resource scheduling*. Different authors have classified resource scheduling techniques in many ways. For most purposes, it is sufficient to classify them into the two types:

- optimal
- heuristic.

7.4.1 Optimal resource scheduling

Optimal resource scheduling techniques seek to define the problem as a mathematical programming problem and solve it to produce the best answer. There is some debate as to what the best answer is. For example, some techniques say that the best solution is the one that gives the shortest project duration; others assume that the best solution is the one which provides the smoothest resource profile; yet others combine these two objectives trading off an increase in project duration with a decrease in resource level variation. It would appear that no answer will ever gain universal acceptance, each project having its own peculiarities and hence its own definition of best. This is especially true when activity models other than the simple fixed duration and constant resource demand type are considered.

The optimal solution may be defined as the solution which optimises the *objective function* (as it is called in mathematical programming). Thus if the objective function is defined as a function of time only, the optimum solution is, by definition, that which minimises the project duration irrespective of all other considerations. Since the balance of objectives is different on different projects, the objective function varies from project to project. In some cases, minimum time will be of greatest importance; in other cases, minimum cost (which will often result from the smoother resource profile) will be most important; in other cases, factors other than overall time or cost may be considered or a combination of simple objectives may be required.

To some extent, the debate is unimportant at present since none of the proposed techniques can be used on real projects because of the numbers of variables and constraints. As computer power becomes more cheaply available, and resource scheduling becomes a more accepted part of the planning process, this situation may change and planners should be aware of the techniques available.

7.4.2 Heuristic resource scheduling

Heuristic resource scheduling techniques aim to produce better-than-average resource schedules by simulating a project manager and using rules-of-thumb to choose between activities when they are competing for the use of a scarce resource. The techniques are all based on network models of a project. The word 'heuristic' means aiding discovery or proceeding by trial and error, and the rules-of-thumb are referred to as 'heuristics'. They are also called priority numbers since they define the importance of the activities which are competing for scarce resources. Many heuristic techniques have been developed and are available in computer packages. Each technique, just like each project manager, works slightly differently and is consequently likely to be better in some situations than in others.

If the priorities of the activities are defined before all the allocation of resources to the activities is carried out, the technique is called a *serial*

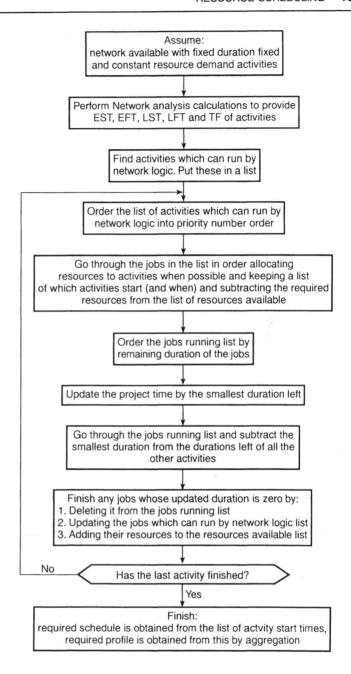

Figure 7.8 Outline flowchart for an heuristic resource scheduling procedure.

The flowchart boxes read:

Assume:
network available with fixed duration fixed and constant resource demand activities

Perform Network analysis calculations to provide EST, EFT, LST, LFT and TF of activities

Find activities which can run by network logic. Put these in a list

Order the list of activities which can run by network logic into priority number order

Go through the jobs in the list in order allocating resources to activities when possible and keeping a list of which activities start (and when) and subtracting the required resources from the list of resources available

Order the jobs running list by remaining duration of the jobs

Update the project time by the smallest duration left

Go through the jobs running list and subtract the smallest duration from the durations left of all the other activities

Finish any jobs whose updated duration is zero by:
1. Deleting it from the jobs running list
2. Updating the jobs which can run by network logic list
3. Adding their resources to the resources available list

Has the last activity finished? No / Yes

Finish:
required schedule is obtained from the list of actvity start times, required profile is obtained from this by aggregation

technique. If the priorities of the activities are calculated during the scheduling process (at the same time as the allocation of resources) the technique is called a *parallel* technique.

Figure 7.8 is a representation of a simple heuristic resource scheduling procedure using parameters from network analysis as heuristics. This is a *serial sort* method in which the network calculations are performed once only, at the beginning, and not repeated even though many of the activities may be delayed, causing the earliest and latest activity timing information to alter.

The flow chart is based on the idea of a certain predetermined number of each type of resource being available for use on the project. Since each activity requires a fixed number of resources for a fixed duration, several activities may be competing for the resources at any one time. If there are not enough resources to operate all activities concurrently, a decision rule is required to decide which activities get preference.

The decision rules are based on priority numbers and are frequently one of the parameters from the critical path analysis (Section 5.2):

Priority number	Meaning
EST	start the jobs which can start first
EFT	start the jobs which can finish first
LST	start the jobs which must start first
LFT	start the jobs which must finish first
TF	start the most critical jobs
Duration	start the jobs which will finish first (those with the smallest duration)

Some of these have obvious properties:

EST and EFT   These tend to produce a bunch of activities at the start of the contract, giving a large probability of choosing a poor order.

LST and LFT   These produce reasonable schedules (see below).

TF   With serial sort methods this has little meaning after taking a few steps into the scheduling procedure, since the critical path may easily change. It would however seem to be a reasonable choice.

Duration   Choosing the smallest duration gives the earliest possible chance of changing the situation and hence correcting errors.

Other parameters have been suggested including a wide variety of activity properties other than just those based on timing. Examples are:

Priority number	Meaning
Manager's choice	start the jobs which the manager chooses
Income	start the job with the greatest income first
Expenditure	start the job with the lowest cost first
Profit	start the job with the highest profit first
Resource demand (1)	start the job with the greatest resource demand first
Resource demand (2)	start the job with the smallest resource demand first
Network position	start the job which controls most activities in the network first.

The meanings for these can also be explained in physical terms:

Manager's choice   The idea of allowing the manager to intervene in the scheduling process and hence affect it is reasonable, although with large networks it is difficult to assign priorities sensibly to all activities. It is better to use this to over-ride one of the calculated priority numbers.

Income   This is intended to improve the cash flow by providing income as early as possible.

Expenditure   This attempts to maximise income by spending money as late as possible.

Profit   This attempts to improve cash flow by providing profit as early as possible.

Resource demand (1)   This attempts to remove bottlenecks in resource use by allocating to those activities with high resource demand as early as possible. This tends to give fewest activities working at the start of the project.

Resource demand (2)   This takes the opposite approach and attempts to remove bottlenecks by allocating to activities which require fewest resources. This tends to give most activities working at any one time.

Network position   This attempts to allocate resources based on the importance of the activities in the network. Those activities which control most activities have resources allocated first. This has the effect of making a maximum number of activities able to be worked on early in the project.

Compound priorities, which combine several of the simple measures, could also be considered. For example, it might be thought important to combine timing information and profit earning potential; this could be achieved by forming a priority formula, such as

$$K_1(\text{LFT}) + K_2 \times (\text{Profit})$$

where K_1 and K_2 are constants to be determined by experience or experiment on similar projects.

No matter which type of priority number is used, there will usually be more than one activity with the same number. In this case another sort is required. It is usual to use another one of the standard parameters for this. The first priority number is called the *major* sort and the second the *minor* sort.

The activities are assumed to be non-splittable in this model but the process could easily be adapted to cater for splittable ones. This could be done by altering the procedure which is carried out when an activity is finished. In the non-splittable model, only the resources from the finished activities are returned to the resource pool for re-allocation; if all resources are returned to the resource pool at this stage and the unfinished activities returned to the list of those that can run, the result would be a splittable activity scheduler. This could be further adapted to cater for reduced productivity of resources and several of the other problems inherent in the use of splittable activities if a general rule for them could be found. Unfortunately such a rule is usually not available.

A more refined heuristic scheduler can be produced using parallel sort methods. These use the same basic process as that described above but involve recalculating the activities' timing parameters (EST, LST, EFT, LFT and TF) every time an allocation is made, thereby taking account of any delays caused by the schedules. These parallel sort methods are really only applicable when the timing parameters are being used as priority numbers. Other types of parameter, such as profit, income and network position, do not vary when activities are delayed and the serial and parallel sort methods are identical in these cases.

A lot of work has been done investigating the performance of heuristic techniques and is summarised below.

The scheduling experiments to compare the performance of the different parameters in the techniques were carried out on many different networks. This was essential since in real life different projects give rise to very different networks. Two examples of where differences occur are concerned with network shape and activity timing. Network shape differences can be illustrated by considering different types of work. For example, roadworks tend to have very long thin networks in which there are few parallel paths, whereas buildings, especially at the finishings stage have short fat networks with many parallel paths. Timing differences occur because of the relative lengths of the paths through the network. A network in which most activities have a considerable amount of float would be expected to respond differently in a scheduling process to one in which there was very little float. In the network with much float, the scheduler would have much more choice of the order of activities.

By generating the networks automatically with a range of predetermined shape and timing properties, it was possible to investigate the performance of the various heuristics across the range of networks which were likely to occur in practice. The results showed that there was little difference between the heuristics and even a random rule performed almost as well as the chosen priorities. However, the small amount of evidence available seemed to indicate that the LFT sort was the best major sort and the minor sort was unimportant.

The original work was concerned almost totally with small networks of 20 activities and, as such, the results might appear to be inapplicable to projects where there are several hundred activities. However, further investigation using a small practical network of 90 activities, reinforced these results (although the latter work could not in fact be classed as generally applicable because of the small sample size). Experience of using heuristic scheduling on networks of several hundred activities appears to confirm the choice of either LST or LFT to be good and preferable to using EST and EFT.

7.5   VARIABLE DURATION ACTIVITIES

This section is included to illustrate the fact that any network is only a model of a project and the assumptions made affect the answers produced by network analysis. Some assumptions are made consciously, while others are made automatically and without thought.

The duration and the resource demands of an activity are very closely linked (Section 4.6). The duration assigned to any activity assumes a certain level of resources, which implies a cost for an activity. As the duration alters, so does the resource requirement and hence the cost. The relationship is not linear but is as shown in Figure 4.5. It is more useful when linearised into the form shown in Figure 4.6. The selection of resources thus becomes a trade-off between time and number of resources or cost.

It is assumed that the duration and resources assigned initially to activities are the optimum from cost considerations. This is the point defined by the normal duration and normal cost on the graph (Figure 4.6) and in practical terms means that the resources allocated to the activity are the smallest number required to complete the activity in the normal duration.

By increasing the number of resources, the duration is decreased but only until what is called the *crash* duration is achieved. It is not possible to decrease the duration any further but the cost can increase above the crash cost; this is equivalent to assigning resources to the activity and the productivity of the work force decreasing at the same rate as the numbers increase. Similarly it is not possible to decrease the cost below the normal cost although the duration can be increased; this recognises that there is an absolute minimum number of resources which can be assigned to an activity without which progress cannot be made.

Several important assumptions are inherent in this type of activity model. First, it must be possible to reduce a complex multiple resource activity to a single resource one where that resource is cost. Multiple resource activity models are discussed at length (Section 4.4.1) and it can be seen that it may be unrealistic to reduce all resources to cost. If this proves to be a problem, it is possible to overcome it by splitting the activities into smaller more detailed ones for which the model is more realistic. Second, there is an inherent assumption that the resources are constant throughout the duration of the activity (Section 4.4.2). If this is a problem, reducing the size of the activities should overcome it. Finally, and perhaps most importantly, it is assumed that it is possible to obtain the information for each activity. This can be a considerable problem with no easy solution.

Using the ideas of variable duration activities and treating cost as a resource, it is possible to formulate a manual technique for shortening a project's duration at minimum cost. Although the method is explained here in terms of the minimum increase in cost to accelerate a project, it could equally be used to determine the greatest increase in profit or

Table 7.2 Normal and crash durations and costs for Tingham Bridge project.

Activity	Original criticality (denoted C)	Normal duration (days)	Crash duration (days)	Normal cost (£'000s)	Crash cost (£'000s)	Extra per day (£'000s)
1	C	0	0	0	0	0
2	C	10	7	150	195	15
3		5	4	100	110	10
4		8	6	120	160	20
5	C	23	19	250	274	6
6		17	13	150	170	5
7		20	16	160	188	7
8	C	30	25	400	450	10
9		5	5	60	60	0
10	C	5	5	90	90	0
11	C	14	10	50	74	6
12	C	0	0	0	0	0

indeed the change in any variable which could be defined for an activity in the form indicated in Figure 7.10 (this figure is used in Section 7.6 for mathematical programming).

Table 7.2 shows the normal and crash durations and costs for all the activities in the network for the Tingham Bridge project (Section 7.3.2). Also shown are the extra costs per day of reducing the duration of each activity.

By performing the network timing calculations using the crash durations, the project duration can be reduced to a minimum of 66 days (easily checked in this simple network by inspecting the paths through the network and finding the longest). No further reduction is possible using this construction method since it is impossible to reduce the individual activity durations any further. If further reduction were required, the construction method would need to be changed to remove some of the technological constraints. This is a difficult proposition in this particular project but often feasible in real projects.

The cost of the project at normal rates is £1 530 000 whereas operating everything at crash rates costs an extra £241 000 namely £1 771 000. Crashing all possible activities is not usually the best or least cost method of speeding up a project. For example, in this network there is no need to crash the activities *pile and cap west* and *substructure west* since their float is so large.

The procedure for reducing the project duration at minimum cost can be summarised in the following steps:

(1) Perform the timing calculations for the network using the normal durations to identify the critical path.
(2) Examine activities on the critical path to find the one which can be shortened for least cost.
(3) Shorten this as much as possible. This will be defined by one of:

— the total amount it is possible to shorten the activity before its crash duration is reached;

— the total amount it is possible to shorten the activity before the required project length is achieved;

— the amount it is possible to shorten the activity before another path becomes critical. When this happens, several activities need to be shortened simultaneously in order to reduce the length of all critical paths.

(4) If it is necessary to shorten the project more, and all jobs on the critical paths have not been crashed, return to step 2.

As with most network techniques, this is easily computerised to great advantage with large projects.

Following these steps through, a project duration of 66 days can be obtained at a cost of £147 000 over the normal cost of £1 530 000. This is a saving of £94 000 over crashing all the activities. The time/cost curve for this reduction is line A on Figure 7.9.

If time-related overhead costs or other costs dependent on project duration are considered, the optimum duration could be changed. Line B on Figure 7.9 shows the sum of the activity dependent cost (line A) and a cost of £10 000 per day (independent of project duration). The minimum can be seen to be between 69 and 74 days.

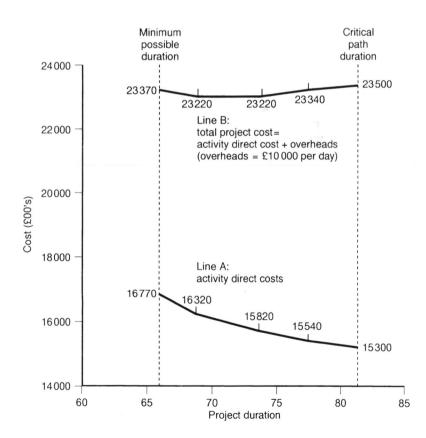

Figure 7.9 Project duration cost curve for Tingham Bridge project.

The shape of the lines on the graph in Figure 7.9 are typical of such lines for projects. The activity direct cost always increases more rapidly as the project duration gets shorter (in mathematical terms it is concave upwards); the time-related overheads (not shown explicitly in Figure 7.9) are always linear; and the total cost (line B) has a single minimum and is concave upwards. These simple rules can be used as a check and also as a mechanism to help calculation.

7.6 MATHEMATICAL PROGRAMMING FOR TIME–COST TRADE-OFF

An extension of the use of the variable duration model is to consider the problem of minimising the activity direct cost as a mathematical programming problem.

Consider the time–cost relationship shown in Figure 7.10. This is the portion of the linearised curve of Figure 4.6 between the crash and normal costs. Let the gradient of this portion of the line be G_i. Activity i must have a duration between d_i and D_i. Let this be y_i. From Figure 7.10, the cost of performing job i in time y_i is:

$$K_i - G_i y_i$$

where K_i and G_i are constants.

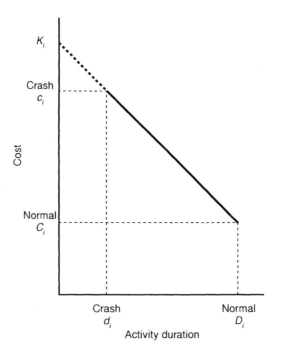

Figure 7.10 Idealised activity duration cost relationship for activity *i*.

The total direct activity cost of the whole project is:

$$\sum_{i=1}^{i=N}(K_i - G_iy_i)$$

for N activities.

Since the K_i are constants, let

$$\sum_{i=1}^{i=N} K_i = k$$

where k is another constant.

The total cost is

$$k - \sum_{i=1}^{i=N} G_iy_i$$

This is a cost function which must be minimised and the problem is equivalent to maximising

$$\sum_{i=1}^{i=N} G_iy_i$$

where the durations y_i are the variables.

There are necessarily constraints on the values which these variables can take. The value of y_i must lie within the range d_i to D_i; additionally, all logic in the network must be complied with; and finally a long-stop completion date may be set to restrict the overall project duration. These constraints may be described mathematically, whereupon the problem may be solved using the technique known as linear programming. This technique, which is described in books on mathematics and operational research, involves finding the optimum solution which complies with all the constraints on y_i.

Briefly, the constraints can be written as:

(1) Network logic: if S_i is the start time of activity i then, for all connections ij that exist,

$$S_j - S_i - y_i < 0$$

(2) Preventing the project going slower than normal:

$$y_i < D_i$$

(3) Preventing the project going faster than 'crash':

$$-y_i < -d_i$$

(4) Ensuring that the project finishes within m days

$$T_N < m$$

As in the manual calculation above, it is possible to include the overall project duration in the cost function, which then becomes

$$K_1(T_N + y_n) + K_3\left(K_2 - \sum_{i=1}^{i=N} G_i y_i\right)$$

where K_1, K_2, K_3 are constants, and can be minimised to produce a minimum cost schedule.

7.7 APPLICATION OF SCHEDULING

Resource scheduling is only one of many tools for planning and control. It is based on network analysis but contains many stages. It is essential that each stage be done correctly and in the correct order, both to produce a situation from which the next stage can proceed and to ensure the final answer is satisfactory.

For resource scheduling to produce useful answers, it is essential that the network can be drawn correctly. The main points to be observed are:

(1) The choice of activities must be detailed enough to model the planning stage for which the scheduling is to be used.
(2) No resource constraints should be drawn into the network. If they are included, the critical path may be the length of the project, but the resource schedule is unlikely to be as efficient as is possible otherwise.

It is also important to choose the number of types of resource carefully. If too many different resources are used, the schedule would probably be inefficient because of the interaction of the various constraints. Equally, if too few resources are considered the schedule may be overoptimistic. In practice, if more than 15 or 20 resources are considered, it is recommended that only the 15–20 key resources should be levelled and the rest should only be aggregated for monitoring purposes.

Having drawn the network, it is usually impractical to perform the scheduling manually. The choice of computer program should be made on similar criteria to those for basic network analysis. Whichever is chosen it is essential that the exact method for scheduling is known and understood. Even slight differences between methods can cause difficulties in operating the system for site work.

The schedule produced, even if by the most sophisticated scheduler from the most detailed data, is unlikely to be of direct use for short-term planning and control. The shortest period over which it is usually practicable to operate such a system is between 2 and 4 weeks with the latter being the most popular period.

With monthly updates and reschedules, it is possible to operate a rolling programme. Here, the network is updated and made more detailed for the first 3 months, with work further in the future being considered to be required in less detail. At the end of the month, the detail is rolled forward by 1 month to maintain the detailed programme 3 months ahead. The inaccuracies induced by doing this are usually small compared with the knowledge of events many months into the future. There will, however, always be errors or inaccuracies in an automatically produced schedule, and it should be treated like any other forecast.

The schedule is only the computer's suggestion for the plan of work. It has been produced using the rules presented but there may be improvements that can be made. For this reason the schedule should be examined critically and amended as necessary.

The resource profiles for the EST and LST schedules, in addition to the levelled profile for the schedule may be of use in this examination and, as in many design processes, it may be necessary to re-run the computer program. Without care being taken at this stage, acceptance of the planning and control system by the rest of the project organisation may be adversely affected.

Regular updating provides a total history of a project which can be of use to the site in disputes and to the company for future estimating policy. With some computer systems, it is possible to produce budgets directly from the schedule, thus forming the basis for the financial work described later (Sections 7.8 and 7.9).

7.7.1 Resource scheduling and complex networks

As a footnote to the resource scheduling discussion, it is important to warn that the more complex the project model, the more difficult the resource scheduling technique required and the more difficult it is to understand the answers.

In particular, care must be taken when using:

- multiple calendar working
- splittable and non-splittable activities
- more than one type of connection.

Explanations of some of the problems which can arise are contained in Chapters 5 and 6.

7.7.2 Measures of performance of scheduling procedure

The goodness of a schedule obtained from a resource scheduling procedure or produced manually is very difficult to define. A large number of measures are possible, many of them subjective and many applicable only in particular cases. Objective criteria are required and have been developed not only to assess schedules but also to compare the scheduling methods.

Objective criteria must be measurable, and usually should have only one possible value for any set of variables. They are called objective functions. A special set of these which are formulated in terms of cost are called cost functions. The term cost function, because it is so widely used in this work, is often applied to any objective function. Obviously the lower the value of the cost function the better. The cost functions can be determined by answering the question 'what are we trying to achieve?'.

With this in mind, several functions have been suggested:

- utilisation factor (equal to the total resources provided);
- the absolute changes in the resource levels from one period to the next over the duration of the project;
- the sum of the increases in resource level;
- the peak resource level required during the project;
- the maximum change in manpower required.

Other more complex functions have been suggested for mathematical programming solutions to the scheduling problem. These have attempted to model the costs of construction more widely, but in doing so have created many complex problems.

7.8 PLANNING THE FINANCES

7.8.1 Introduction

Despite the assertion made earlier that finance is a resource, financial planning is slightly different to resource planning. The techniques used are very similar but financial planning tends to be carried out by different people with different objectives. This section looks at the specific requirements of financial planning and relates them to resource planning. It also provides the basic information for financial control (Chapter 12). The importance of financial planning and control cannot be emphasised too much since errors can have a very serious impact on the profitability of a project and the viability of the construction company.

All organisations involved in a project are interested in finances in some way. The client, the designer, the main contractor, the subcontractors and suppliers all stand to benefit from the finances of a project, and everyone should take care that their financial involvement is correctly planned and controlled. Failure of one company on a project can have very significant effects on all the others and it is in the interests of the whole team to ensure that all the appropriate individuals are in control of the appropriate financial aspects of the project.

Money is usually a scarce resource so due consideration of a project's demand for money is essential. A company's borrowing capacity is limited either by market forces or by political pressures and the amount of work that a company is able to carry out within those constraints is controlled by its effectiveness in using money as a resource.

The borrowing requirement of a project at any time is the difference between the company's expenditure and income, up to that time, on the work. To provide the money a lender must forego the benefits and pleasures of its alternative use – and some recompense is usually required. There is therefore a duty on a project manager (or the equivalent financial manager of a project) to reduce by all available means the maximum borrowing requirement and to phase the financial demands of the project into the overall financial planning of the company. Where circumstances permit the project to run in surplus, the surplus should be maintained at the highest possible level to permit productive reinvestment elsewhere.

To determine a project's demand for money as a resource, it is necessary to plan the income and expenditure of the site, that is to form a financial plan for the work. This is obviously inseparable from the physical plan of work.

The objectives of financial planning are to:

- provide information for financial control of the project;
- set targets for financial achievement of the project;
- decide on the viability of the project;
- determine the borrowing requirements of the project in order for provision to be made;
- evaluate the risk of the project;
- ensure that the project remains within financial constraints which are either set within the organisation or passed down through the project hierarchy.

Each organisation involved gives different priorities to these objectives but all organisations should have all of the objectives in mind when planning.

The amount of detail in a financial plan is as important as the amount of detail in a physical plan. It is determined by the information available and the principal purpose of the plan. For example, in order to carry out rational control action, project managers need to know from the financial plan where they *should be* and, from information collected from the project, where they *are*. For a comparison to be made, the project information needs to be in the same form as the financial plan and the diagnostic power of the comparison is increased by more detailed planning and current status monitoring. An appreciation of the costs of planning and information collection reduce, in most cases, the level of detail to reasonable proportions.

It is sometimes the case, especially in client organisations, that the only formal planning and control systems in use on projects are in overall financial terms. This is attractive because the different quantities and types of construction can usually all be reduced to common financial terms. However, such a neglect of physical planning and control can lead to problems because improvident phasing of the work may bring agreement with financial targets at the expense of future physical delays. Overall measurement may enable large sources of loss to be hidden by the products of more profitable areas of the project and the chance of productive control action is consequently lost. Financial planning and control are necessary because of the importance of financial factors in ensuring company viability, but alone are not sufficient to run a successful project.

7.8.2 Steps in financial planning

Financial planning requires:

- a project to be split into segments or elements of work;
- knowledge of when the segments will be carried out;
- knowledge of the financial implications of the segments of work.

The way in which a project is separated and the detail and accuracy of both the phasing in time and the financial information depend on the purpose of the plan.

At tender stage a plan needs perhaps to show only the major financial demands on a company's resources and position them approximately in time. After a contract has been awarded a project would perhaps be split into segments appropriate to the company's standard cost scheme and a detailed plan of work would be available.

7.8.3 Elements of financial plans

A project may be subdivided in any convenient manner to determine a financial plan. Two common sub-divisions are:

Type of work For example, concreting, excavation, brickwork (as might be found in a bar chart model of a project).

Network activity For example *construct north pier of viaduct* which might contain many different types of work.

For convenience, the subdivision should correspond with the physical plan of the works and the method of collecting site cost information.

Other considerations and explanations of possible elements of financial plans are:

Elementary financial model quantities The total cost or revenue of an element of work is that planned for its constituents. An approximate measurement from the project drawings enables the manager to estimate the quantity of work (such as number of cubic metres of mass excavation) in an element. If a bill of quantities (see Preface) is available for the work this provides an overall check on the approximate measurements.

Revenue Revenue is used here to denote the actual or potential flow of money; it is used as a general term to encompass both earning and income (Section 3.3). The earning is the amount of money due at any time whereas the income is the amount received. In general an earning becomes an income after a passage of time since payment is usually received at regular time intervals for work carried out. For a contract arranged on a bill of quantities basis, a knowledge of the quantities involved in any element of work enables the revenue resulting from that element to be calculated directly. For other contracts which involve a relationship between work done and revenue, a knowledge of the relationship enables the revenue to be calculated.

Costs   In this context costs are used to mean both liability and expenditure (Section 3.3). The liability is the amount committed by a company whereas the expenditure is the amount paid out. A liability usually becomes an expenditure after a passage of time. They are generally calculated in at least two ways by a main contractor as described below. Other parties will work with one of these, although the level of detail and accuracy of information is greater the lower they are in the project hierarchy. The client has least detail and most uncertainty and perhaps therefore most risk; a subcontractor has greatest detail and least uncertainty and hence least risk. However, risk can be shared in an infinite number of ways depending on the contract and one of the uses of planning is to enable the risks to be quantified.

Tender estimate   If an estimate has been made against a bill of quantities it is possible that an estimate for the cost of each bill item unit will have been made. Having an approximate measure of the number of units in the element of work gives an estimate for the total cost of the element. This may be an aggregate over several bill items for both the *type of work* and *network activity* project subdivisions. An estimate for a bill item is normally subdivided into *cost heads* such as labour, plant, materials or subcontract work, and a similar subdivision may be prepared for each element of work. Where there is no bill of quantities for a project an estimate for the costs of each section of the work may be prepared using internal company standard costs. If done in full by a main contractor, this involves preparing a bill of quantities for the job and producing a full estimate. The work involved and the degree of uncertainty often means, especially at client level, that this is abbreviated to considering the standard costs of the major construction operations and allowing an additional percentage for those not formally estimated. The added percentage is set by experience.

Site estimate of costs   In making a physical plan, project management is responsible for estimating the duration of each type of work or activity and specifying the labour and plant allocated to it. The unit cost of labour and plant should be known and thus the labour and plant costs for the element of work can be calculated directly. A comparison can then be made between the site cost and the estimate cost for the element, to determine whether:

- an error has been made in either calculation;
- the proposed construction method is financially satisfactory.

Time spread of money over an element of work   In practice the cost and revenue on an activity are usually non-uniform. For example, one could envisage an activity on which most plant usage may be concentrated into the initial stages (to prepare the work) while most of the labour may be involved in finishings. Without detailed information to the contrary, it is reasonable to assume that the rate of expenditure over an activity is uniform. Over a normal monthly accounting period the errors are self-compensating for the site as a whole. The arguments for and against an even spread of finances are the same as those for resources (Section 7.4).

7.8.4   Financial model for a complete work element

The components of a model for a work element consist of some or all of the following:

- *revenue*, the total estimated revenue from the element;
- *tender cost*, the estimate of costs made at tender stage, classified into labour, plant, materials and so on;
- *site cost*, the estimate of costs made by the responsible site management before construction and agreed with top management.

Each classification of cost must include a measure (by the site) of the average time taken in transferring a sum from being a liability to being an expenditure. This should reflect the normal commercial practice for comparable work if no more practical information is available.

The time lag between submitting an account and being paid is set by the conditions of contract and should be known for the project as a whole; it may be termed the income lag.

The planned start and finish date of each activity should be known from the physical plan and additional information such as the earliest and latest start date of the activity would be of value.

7.9   ASSEMBLY OF FINANCIAL PLANS

A financial plan for a sample project can be summarised in terms of cumulative financial curves (Table 7.3 and Figure 7.11). The individual lines on the curves are discussed below. Sections 3.3 and 7.8 introduce much of the terminology appropriate to finances.

It should be noted that changes to the physical plan of work, changes in prices and time delays may all give rise to important changes in the form of the revenue pattern and the curves described below must be recalculated. This process is eased if a computer system is available but manual calculation can be minimised if the initial calculations are laid out with modification in mind.

Table 7.3   Financial figures for a sample project (£000).

Month	Period liab.	Period expend.	Period earn.	Period income	Period surplus	Cumul liab.	Cumul expend.	Cumul earn.	Cumul income	Cumul surplus
0	65	30	0	0	−30	65	30	0	0	−30
1	40	25	20	18	−7	105	55	20	18	−37
2	65	55	25	23	−32	170	110	45	41	−69
3	80	70	50	44	−26	250	180	95	85	−95
4	100	70	70	64	−6	350	250	165	149	−101
5	110	95	130	116	21	460	345	295	265	−80
6	120	105	155	140	35	580	450	450	405	−45
7	105	100	175	158	58	685	550	625	563	13
8	80	80	170	152	72	765	630	795	715	85
9	60	75	85	77	2	825	705	880	792	87
10	45	65	60	54	−11	870	770	940	846	76
11	20	40	40	36	−4	890	810	980	882	72
12	10	40	20	18	−22	900	850	1000	900	50
13		30		0	−30	900	880	1000	900	20
14		20		0	−20	900	900	1000	900	0
15				50	50	900	900	1000	950	50
16				0	0	900	900	1000	950	50
17				0	0	900	900	1000	950	50
18				50	50	900	900	1000	1000	100

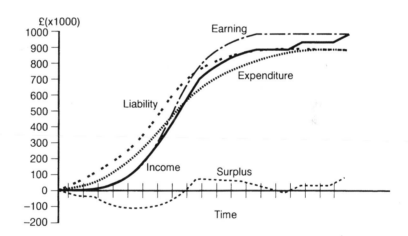

Figure 7.11 Financial plan curves for a sample project.

7.9.1   Revenue curves

If a physical plan shows that an activity is planned to be totally complete within a single interim accounting period, the income from that work arises at a date which is the end of the invoicing period plus the income lag (Section 7.8.4). For an activity covering a number of invoicing periods the proportionate incomes arise in a similar manner. The totals, over all activities, may be determined for each date and the income curve for the project may be determined. It is more usual to form a cumulative sum of income over the project time and display this as a function for time. This was described for resources and illustrated in Figure 7.2.

For comparison, or when no planned date for an activity exists, it is useful to form an income curve for both an EST and an LST schedule. These are also shown in Figure 7.2. The earliest start schedule obviously produces one outer bound amount of income but it is not uncommon to find that the completion time given by network analysis calculations is considerably earlier than the contracted or desired completion date. The difference in time is the margin which enables contractors to use their resources more effectively than would be the case in following either the earliest or latest schedule dates. The latest possible schedule income curve therefore results from delaying the latest start schedule by an amount equal to the margin for completion as shown in Figure 7.2.

7.9.2 Earning curves

The earning curves for a project can be developed in a similar manner, making no allowance for the invoicing period and income delay. In practical terms a reasonable picture can be determined by summing the weekly earning rates for activities in the weeks that the work is planned to take place. This leads to a list of weekly earnings over a project which can be produced cumulatively if required.

7.9.3 Cost curves

The treatment of costs (Section 7.8.3) derived from tender and site estimates is similar to that for earnings. In practice the tender estimate has withstood the criticism of company top management and forms a standard against which the site proposals can be judged. Once the site proposals for work method have been accepted together with their financial implications, these become the standard against which further proposals and actual performance are judged. At this stage, cost plans based on tender estimates become largely of historical interest.

7.9.4 Expenditure curves

Building up expenditure curves can be somewhat more complex than the construction of income curves because of the differences between the expenditure leads and lags for different categories of expenditure, for example 3 days for labour and 6 weeks for plant.

Complexity can be avoided by taking an acceptable average lag figure for all costs without regard to their source. Alternatively, an appropriate lag can be taken for each category of cost and the expenditure curve for each category determined. The curve for total expenditure can then be established by addition. This has the advantage that actual site expenditures are commonly collected under these categories and a detailed comparison can be made.

7.9.5 Liability curves

The complexities concerned with lags are not necessary in compiling the liability curve. These are compiled in the same manner as the earnings curves along with the appropriate cost, duration and scheduling information. As with expenditure, it is helpful to construct a liability curve for each category of cost.

7.10 SUMMARY

- Resources must be considered to provide an effective programme of work.

- Resources can be:
  - aggregated to provide the resource demands of a particular schedule;
  - smoothed to provide an efficient use of resources within the floats calculated as part of the network analysis;
  - levelled to ensure a minimum of variation of resource demand from one time period to the next;
  - allocated to produce a schedule with constraints of resource availability and time.

- Networks with resource constraints as well as technology constraints remove the freedom of choice from the techniques of resource planning and tend to provide worse plans if the project undergoes any changes.

- Redoing a resource schedule can significantly change what activities the resources carry out, but leave the plan essentially unchanged.

- Most resource scheduling procedures in use today do not claim to produce a best schedule but rather a better-than-average one.

- The greater the number of different resources considered, the more difficult the process of scheduling.

- In financial planning:
  (1) Use the system provided for physical planning to break the project into elements of work.
  (2) Determine the cost and revenue application of each element of work.
  (3) Determine the leads and lags which control the occurrence of expenditure and income for the project.

(4) Construct the curves showing the estimates of cost and revenue for the projects together with similar curves for the categories of liability and expenditure.

(5) If the financial implications of the physical plan prove to be unsatisfactory, changes to the physical plan may be attempted to reduce costs and it may be possible to renegotiate rates to improve revenue.

8

PLANNING AND UNCERTAINTY

'I am too much of a sceptic to deny the possibility
of anything.'

T. H. HUXLEY
(Letter)

A few years in the construction industry should be enough to turn
anybody into a sceptic. Construction is a particularly uncertain industry
where almost anything is possible. Planning should do as much as it
can to help the situation – such as trying to quantify the uncertainty and
measure its possible effects. Techniques are available for this. Properly
applied they should help managers allow for the possibility of some
things (if not anything) happening and might make them less sceptical.
On the other hand . . .

8.1 UNCERTAINTY IN CONSTRUCTION

Planning for uncertainty may sound paradoxical but there is much sense
in taking uncertainty into account if possible. In this book planning has
thus far been studied within a deterministic framework. Consider for
example the simple activity-on-node network in Figure 8.1 which
represents a design-and-build project, Tingham Offices (Appendix A and
Section 2.4) with a superstructure of standard design and an individually
designed raft foundation.

The activity dependencies and durations are shown as certain values.
In practice, however, neither the dependencies nor durations are known
with certainty. Take, for example, the second and third activities of the
network, relating to raft foundations which are given a total duration of

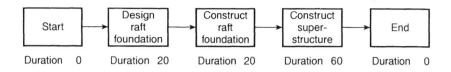

Figure 8.1 Design-and-build project network.

20 weeks each. How confident is the planner, either that a raft foundation would require exactly 20 weeks to design and construct or that, indeed, a raft foundation would be suitable? It is well known that 'in-the-ground' works are notoriously uncertain. Perhaps it would be a more realistic representation of the problem to indicate it as shown in Figure 8.2.

This is not a network in standard format, but it can be seen that both the logic of the network and the duration of activities are expressed in probabilistic terms, rather than the deterministic view adopted hitherto. The durations are given as three values defining a range of possibilities and the network contains a decision after which one of two paths will be taken but not both. With techniques available for analysing such probabilistic networks the uncertainty can be taken into account. Of course the answers would be in probabilistic terms.

Planning uncertainty can thus be categorised into distinct types:

- uncertainty of activity duration
- uncertainty of activity sequence
- uncertainty of activity resource requirement.

These are dealt with separately (Sections 8.2, 8.3 and 8.4 respectively), despite the fact that they are obviously inter-related. In addition, the question of uncertain calendars is discussed in Section 8.5.

An understanding of basic statistical techniques is important if the problem of uncertainty is to be addressed in a proper way. Readers needing more help than is included here are directed to mathematical

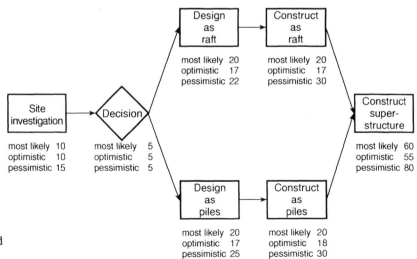

Figure 8.2 Probabilistic network of a design-and-build project.

texts for help with statistics theory. This book tries to keep the mathematics to a level which gives a good understanding of the topic without losing sight of the main engineering issues.

8.2   UNCERTAINTY OF ACTIVITY DURATION

Many studies of uncertainty in activity duration have been carried out. Two techniques are considered here. The first, programme evaluation and review technique (PERT), is one of the older scientific planning techniques which relies on many simplifying assumptions so that it can produce and analyse a project model with relatively little computational power. The second, Monte Carlo simulation, although quite old in concept, has only become a viable proposition for most normal construction projects with the advent of powerful, economically viable computers.

8.2.1   PERT

PERT is a technique which attempts to account for the inherent uncertainty of activity durations. It is based on a network model of a project and was developed shortly after the initial development of arrow diagrams. Nowadays the term PERT is often used (incorrectly) to refer to the use of arrow diagrams for project planning.

PERT was initially developed in order to assist with the planning of projects, such as research and development projects, which have uncertain outcomes with respect to time and money for each of the component activities. However, the principles are equally applicable to projects such as the construction phase of a building project. The assumptions it makes and the method of analysis it uses mean that the method has always had its detractors. It is, however, a reasonable attempt to consider uncertainty at the very early stages of a project when there is usually great uncertainty in the underlying information and when action can be taken to improve things. It should therefore be considered seriously.

The assumptions which need to be made are quite wide-ranging. They are introduced in the body of the text but their effects are not fully covered until the end of this section when the method is discussed.

PERT can best be appreciated through an example. Consider a straight-line network (Figure 8.3) with five activities A to E with finish-to-start connections and durations of 10 days each.

In this case, the project duration is 50 days. If the duration is considered to be a distributed variable rather than a certain value, approximate information about the likelihood of finishing on any particular date can be determined. There are a variety of ways in which this uncertainty information can be expressed. PERT deals with the matter by considering the distributed variable to be defined in terms of

Figure 8.3 Five activity network.

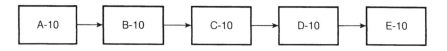

the optimistic, most likely and pessimistic duration estimates. It makes this sort of calculation easy by assuming that the activity distribution is a *beta* distribution. This is a continuous distribution which allows the distribution to be skewed so that there is a possibility of obtaining values (here for the activity duration) a relatively long way above the mean value but little possibility of getting values a long way below the mean. This is a reasonable model for a construction process which could take a long time if everything went wrong but which would rarely take much less time than the mean and could never go negative which would be possible with some other statistical distributions. Given three duration estimates, optimistic (o), pessimistic (p) and most likely (m), PERT assumes that:

(1)   the mean activity duration is

$$\frac{p + 4m + o}{6}$$

(2)   the activity standard deviation is

$$\frac{p - o}{6}$$

(3)   the activity variance is

$$\left(\frac{p - o}{6}\right)^2$$

By assuming values for the optimistic, most likely and pessimistic values, Table 8.1 shows that the best possible completion time (all optimistic) is 40 days, the worst (all pessimistic) is 63 days and the expected overall duration (all most likely) is 50 days. By no means is the answer expected to be exactly 50. There is some chance of finishing on

Table 8.1 PERT evaluation of activity durations.

Parameter	A	B	C	D	E	A + B + C + D + E
o	7	7	9	9	8	40
m	10	10	10	10	10	50
p	12	11	13	15	12	63
Mean	59/6	58/6	62/6	64/6	60/6	
Standard deviation	5/6	4/6	4/6	6/6	4/6	
Variance	25/36	16/36	16/36	36/36	16/36	

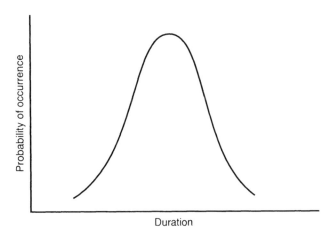

Figure 8.4 Project duration expressed as a distributed variable.

any day between 40 and 63 and it is in order to determine the likely distribution of the finishing time that PERT is used.

The mean duration of the overall project is assumed to be the sum of the mean times of the activities on the critical path. The overall variance of the project durations is given by the sum of the variances of the activities on the critical path. Thus:

- project duration is 50.5 (the sum of the mean values on the critical path);
- variance of project duration is 109/36 (the sum of the variances on the critical path);
- standard deviation of project duration is

$$\sqrt{\frac{109}{36}} = 1.74$$

(the square root of project duration variance).

It is assumed that the uncertainty associated with the overall duration is, approximately, a *normal distribution*. This derives from the central limit theorem which states that the addition of an infinite series of distributions yields a normal distribution. The assumption is reasonable in practical networks since there are usually a large enough number of activities on the critical path. We can therefore represent the project duration as a function of probability (Figure 8.4).

The network (Figure 8.3) illustrates a particularly simple situation in which there is obviously only one critical path. PERT assumes that the above analysis is carried out on the critical path alone and this in itself can be problematic. In more complex networks there are a number of paths through the network, many of which may have the possibility of becoming critical depending on the duration of the individual activities on the paths. In order to investigate this problem, all paths which have a prospect of becoming critical (because they have a low total float

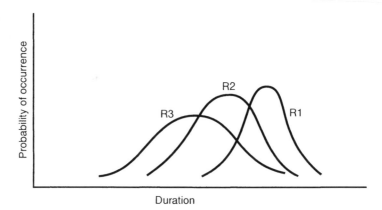

Figure 8.5 Project duration assuming a variety of critical routes through the network.

compared with their duration uncertainty) must be investigated. Figure 8.5 shows the results from a hypothetical network in which three routes R_1, R_2 and R_3 are treated as candidates for criticality. Each of the three routes gives, when analysed using PERT, its own normal distribution for the project duration (as shown in Figure 8.4). When the three are combined, the resulting distribution for the project duration is as shown in Figure 8.5. While R_1 is generally the most critical route (that is, will have the longest duration) it can be seen that there is some likelihood that either R_2 or R_3 will be.

8.2.2 Problems with PERT

There are a number of problems associated with PERT; among the most significant are:

- The form of the distribution for the activity duration has very little basis in fact but is a convenient fiction.
- On small projects there may not be enough jobs defining the project duration to make the assumption of a normal distribution for the project duration valid.
- The statistical analysis assumes that the activities are statistically independent. This is likely to be untrue. Activities following one another can be dependent, as can activities carried on at the same time, by the same resources or containing the same type of work. For example, activities performed consecutively by the same gang or subcontractor are likely to be related. Equally, similar outdoor activities going on at the same time on a project are likely to be affected by weather in a similar manner.
- The assumption that PERT is to be applied only to a single critical path means that the answers obtained always underestimate the duration of a project. The amount by which the method actually underestimates is a matter for some investigation.

- The model requires a lot more information to be provided by the manager (in terms of the three estimates of duration for each activity). Although some managers view this as a drawback, others see it as a benefit because they are able to admit their inability to predict the duration of an activity accurately.
- The method does not provide a single set of dates for control purposes and the use of the distributed results for control has problems. However, as its name suggests, PERT can be used to produce updated estimates of project durations at various stages of the project. These are useful for control and decision making purposes.
- The distributed nature of the timing of the activities means that the resource demands for the project cannot be predicted and hence controlled accurately.

The technique considers the uncertainty of the activity durations in a network-based model of a project and, despite the problems listed above, gives a reasonable indication of their likely overall effects. It is quite easy to implement even in a large network, given the critical path, and the calculations could be carried out manually if no computer program were available.

The availability of cheaper, more highly powered computers has meant that it is possible to use Monte Carlo simulation techniques instead of PERT, thereby overcoming some of the problems. It is debatable whether or not this is worth doing since the initial assumptions are probably the controlling factors in the accuracy of the forecasts.

An example of the use of PERT for a complex project is given in Section 9.4.3.

8.2.3   Monte Carlo simulation

Monte Carlo simulation is a powerful technique employing random numbers in order to combine distributed variables (such as the uncertain durations of activities in a network). The technique simulates a project by choosing at random the values for each of the variables and using these to calculate the outcome of the project. This is repeated many times, often hundreds or even thousands, to produce a distribution of the possible outcomes of the project.

The number of calculations required to perform a Monte Carlo simulation means that computers need to be employed for realistic simulations but, by way of explanation, elements of a simulation are outlined below.

Imagine that the network shown in Figure 8.3 contains activities which have durations that can be described in terms of triangular distributions (Figure 8.6). A random number X, between 0 and 99, is generated (in computer simulation a real number between 0 and 0.99 is selected). This is used to determine the value of the variable (here the activity duration) which is defined to be the value that exceeds $X\%$ of the possible values. This is demonstrated in Figure 8.6.

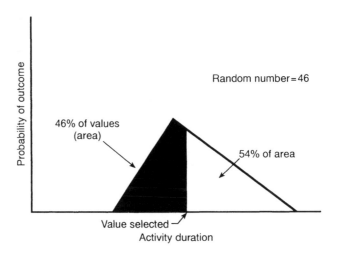

Figure 8.6 Selection of a random value from a distribution.

The first value for the duration of the first activity in the network has now been selected. In order to simulate the entire five activity network, five durations are needed. A string of five random numbers is produced using a computer-based generator, say 29, 18, 83, 34 in addition to the 46 already generated.

These five pairs of digits yield five durations. From these the overall project duration for this simulation can be computed. This is, of course, just the first simulation and does not give a true picture of the overall uncertainty. In order to do this a large number of simulations (using modern computing power) are carried out, after which the overall project duration can be represented on a probability density graph, as illustrated in Figure 8.7.

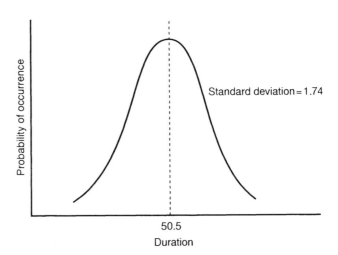

Figure 8.7 Probability distribution for the overall duration of the five activity project.

This simple principle may readily be extended to more complex networks such as the one illustrated in Figure 8.2. Here the analyst may decide that the first item for decision in each simulation is whether the foundation is to be a raft or piled. If, for instance, the planner decided that it was a 50/50 chance (a probability of 0.5) that either would be adopted, then the convention may be adopted that pairs of numbers from 00 to 49 would represent a raft option and 50 to 99 would represent a piled option. Each activity on the simulated network could then be given as an uncertainty and would be simulated using the procedure described above. This introduces the possibility of using the Monte Carlo technique for simulating an uncertain logic in the network (as well as an uncertain activity duration).

An example of the use of Monte Carlo simulation is given in Sections 9.3.3 and 9.4.3.

8.3   UNCERTAINTY OF ACTIVITY SEQUENCE

Analytical planning and programming techniques in current use are designed to work for a single, well defined method of carrying out work. They operate by modelling this pre-selected method. Their aim is to ensure that all things get done in the correct order, at the correct time, and as economically as possible. The logic of a conventional network (Section 2.5) sometimes cannot be exactly defined and this is when the generalised network technique becomes useful.

Take, for example, the problem of planning a simple linear project such as the design and construction of a long cross-country pipeline. After surveying a route, any of several things might happen. In extreme cases, it might be decided that the route is impossible and that another survey of a different alignment is required; in other circumstances, perhaps less drastic situations, it may be necessary to blast away rock. The most likely situation is that normal construction methods will be employed. Activity-on-node, precedence or activity-on-arrow networks require the actual outcome of the survey to be predicted and the network drawn for the selected one. If all three were to be considered, three networks would be required. It can be seen that, for complex projects where many decisions may be required to be made, the process of planning this type of work would be very involved (if it were possible at all).

The technique known as generalised networks is outlined in Appendix B as an analytical, rather than a simulation-based, technique for determining the characteristics of a project from a network that represents the uncertain processes in a project.

8.4   UNCERTAINTY OF ACTIVITY RESOURCE REQUIREMENT

The fact that activities and the way in which they are to be completed cannot be known in advance with certainty means that the resources required cannot be known in advance with certainty. This uncertainty may be represented using the PERT method and is most conveniently achieved by converting all resources into money. The calculations are identical to those used for uncertainty of activity duration except, of course, the project mean cost is given by the sum of all activity mean costs (not just those on the critical path); and the project variance is the sum of all the variances of all the activity variances.

By considering all resources in terms of the single resource, money, many of the difficulties of resource–duration relationships for activities are removed. However, the technique does not really produce resource plans.

The Monte Carlo technique may also be considered to model uncertain resource requirements. The use of simulation appears to allow greater flexibility than the PERT method and in particular it might appear that a variety of resource types may be dealt with independently. It is important to ensure that interdependencies, for example between time and resources, are properly modelled. The general difficulty of doing this for activities was discussed in Chapter 4 and this is greatly compounded when uncertainty is considered.

8.5   UNCERTAIN CALENDARS

Activities exist in time which is defined by a calendar. For instance an activity may have a nominal duration of 5 days, but the time it actually takes depends on the nature of the work and often also on when it is commenced. This is discussed in Section 6.2 and may be summarised as follows.

If an activity involves human labour, then it can only proceed during the working time of the labour. If, to take an extreme example, the 5 day activity commences 1 day before a 2-day industrial shut-down, the overall duration of the work becomes 2 weeks and 5 days (and maybe 3 weeks if an additional weekend intervenes). Other activities such as *curing of concrete* proceed regardless of the day of the week or the fact that everyone is on holiday.

Most work can be classified into that which may be:

- affected by an artificial calendar
- unaffected by an artificial calendar.

In the case of those which are affected, the calendar in question often relates, as in the example above, to weekends and holidays and hence are readily determined in advance so that their effect can be taken into account. A description of planning with these types of calendars is contained in Chapter 6.

Some activities are affected by calendars which are uncertain. Examples may include:

- specification clauses requiring that certain operations may not be conducted during certain types of weather;

- work that by its very nature is difficult to carry out in particular weather conditions or the productivity of which is affected by the weather;
- work, such as tidal work, which can only proceed during periods of low tide.

In both of these situations techniques are available for estimating the weather and the tidal data, but this is always (particularly in the case of the weather) subject to a great deal of uncertainty.

8.6 SUMMARY

- Uncertainty is inherent in all planning.

- Construction projects contain more uncertainty than many other types of project.

- Uncertainty can conveniently be split into:
  - uncertainty of activity sequence
  - uncertainty of activity duration
  - uncertainty of resource requirements
  - uncertain calendars.

- PERT is a simple technique to model the effects of uncertainty of activity durations (and perhaps costs).

- PERT always gives optimistic answers; however, the errors due to the method are usually much smaller than those due to the uncertainty about the underlying information.

- The Monte Carlo technique is useful and feasible if computing power is adequate.

- Generalised networks provide a way of incorporating uncertainty of method into network techniques but provide information which is only of use at a strategic level.

- The inclusion of decisions in networks means that much of the planning data cannot be calculated.

- One of the most common environmental uncertainties is the weather.

- Uncertain calendar techniques can be used to model the effects of weather.

9

FORMAL RISK MANAGEMENT

'προαιρεισθαι τε δει αδυνατα εικοτα μαλλον η δυνατα
απιθανα.'

ARISTOTLE
(*Poetics*)

A reasonable translation of this is: 'Probable impossibilities are to be preferred to improbable possibilities.' This perhaps suggests that over 2000 years ago attempts were made firstly to classify uncertain events and secondly to make decisions about them. This is risk management. Such broad classifications as 'probably impossible' or 'improbably possible' can now be extended with modern techniques which present managers with real aids to decision making. Today anything is possible.

9.1 INTRODUCTION

The fact that uncertainty exists is often of little consequence in project planning. Uncertainty becomes important only when it may affect the project objectives. In wishing to complete a project at the earliest possible time, matters which affect the completion date, and their associated uncertainties, are matters of concern. When an uncertainty threatens to affect an objective adversely there exists a risk. When an uncertainty may produce a benefit we term it an opportunity. Often we find that we have to take risks to capitalise on opportunities – nothing ventured, nothing gained.

A consideration of risks and opportunities has always formed part of the planning process in construction as elsewhere. Traditionally it has been achieved by intuition. Now the understanding of the philosophical basis of risk has improved and, with better computers being widely used, there are some sophisticated techniques available for performing risk studies. Risk management has become an invaluable management tool.

The basic thesis of risk management is that in a world presenting interdependent risks and opportunities, the task of managers is *to plan their work so that the ratio of opportunity to risk is maximised* (with the proviso that no risk is accepted which they are not prepared to run).

In this chapter, risk management is introduced and the way in which the above stated task can be achieved is discussed. The principles of risk management are still being formulated. It does, however, appear to be such a powerful tool, particularly as an aid to planning, that its inclusion as a separate chapter is justified. Examples of the use of various techniques in project planning are included.

9.1.1 Uncertainty, risk and opportunity

Uncertainty is an important part of life and business. In order to plan for the future or make decisions, the uncertainties should be assessed and used to help determine the consequences of possible outcomes and the chance of their occurrence.

This leads to a definition of uncertainty as

the quality associated with an event which results in an inability to predict its outcome accurately.

Chapter 8 shows that, although the eventual outcome cannot accurately be predicted, it is possible to assess the range of possible outcomes. The types of outcome form two distinct categories based on the nature of their consequences.

Risk The potential of an uncertain outcome of an event to *threaten* a held objective.

Opportunity The potential of an uncertain outcome of an event to *promote* a held objective.

9.1.2 Balancing risks and opportunities

If it were possible, managers would adopt project strategies which were replete with opportunity and avoided risk altogether. However, the two are usually linked in some way, and so it becomes necessary to balance the risks with the opportunities that they bring. This balancing forms the basis of decision making which is used to plan for the future. Uncertainties are acknowledged as an integral part of life; some of the possible futures will be detrimental, some beneficial. The accepted balance between the risks and the opportunities is based on the perceived risks, the opportunities sought and the acceptable level of risk. Having balanced the risks and opportunities that a range of possible decisions offers, the course of action which will produce the future with the highest ratio of opportunity to risk is chosen.

For example, the owner of a fleet of construction plant might wish to decide whether or not to insure it. If uninsured there is a risk of theft, damage to the machines or third party claims. But there is a saving on insurance premiums which represents an opportunity. This decision is influenced by the balance between the risk and the opportunity. The approach adopted by virtually all plant owners is to insure; this may be more expensive in the long run but the additional security brought with the insurance premium is considered by most to be worth more than the premium.

9.2 RISK IDENTIFICATION, ANALYSIS AND RESPONSE

9.2.1 Introduction to risk management

Risk management is the name given to the formalised process of balancing the risks and opportunities that a decision may lead to; this involves taking action to produce an acceptable balance between the two. The way in which this process is usually performed, by using experience and intuition, is largely unknown or is specific to the individual. Certain procedures though seem to be an integral part of the process:

(1) The objectives of the project or venture are identified.
(2) The uncertainties in the project or venture which threaten or promote these objectives are identified.
(3) The significance of each uncertainty is assessed in terms of the consequence and the chance of occurrence.
(4) The balance of risk and opportunity is computed for the range of decisions available.
(5) The balance produced is judged against acceptability criteria to determine the need for actions to produce an acceptable balance.
(6) The actions needed are determined with reference to the significance of each uncertainty.

Such a process is illustrated in the flowchart of Figure 9.1, which shows that the process is iterative, being repeated after the responses have been formulated to determine the effect, and hence suitability, of the responses. The flowchart shows the feedback to the start of the process if the balance proves unacceptable. It may be thought that the identification stage need not be repeated but new uncertainties may be recognised in dealing with others.

Although there is confusion as to the meaning of some of the terms used and the objectives of the process, there is consensus on there being three distinct stages to risk management. They relate to:

- risk (and opportunity) identification
- risk (and opportunity) analysis
- risk (and opportunity) response.

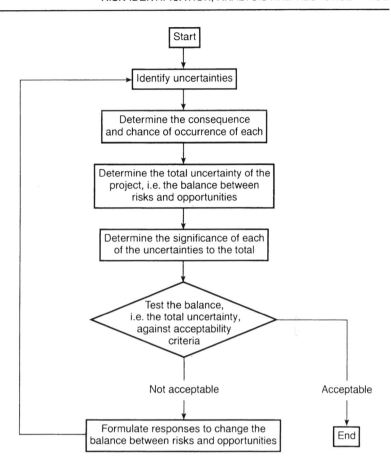

Figure 9.1 Flow chart of risk management system.

The titles of these stages are often referred to as simply risk identification, analysis and response. This should be because the process is known generally as risk management; it should not be because of an intention that only the risks, as defined earlier (Section 9.1), should be managed. It seems logical that the process of risk management should be dealing with all factors and uncertainties as the purpose is to balance the two sides, risk and opportunity, of those factors and uncertainties which may affect the objectives. The purpose of these three stages is now described briefly.

9.2.2 Risk identification

The risks and opportunities associated with a project must be identified. This is by no means straightforward as the type of construction projects to which risk management is commonly applied have many unique features. Many researchers suggest a categorised system of classification to aid in the identification process. This categorisation is based on the nature of those factors and uncertainties which may affect the objectives and can be shown as a hierarchy (Figure 9.2).

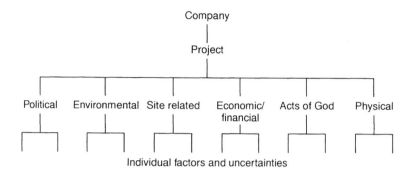

Figure 9.2 Hierarchy of uncertainty.

Standard tables and check lists may be used to assist in identifying risks and opportunities. These focus the mind and are intended to identify those factors and uncertainties which would otherwise be ignored in an informal system. The classification shown in Table 9.1 is based on the hierarchy of uncertainties in Figure 9.2.

Using a formalised system involves a rigorous analysis process rather than intuition and so all factors and uncertainties which could affect the objectives of a project should be identified and documented. The analysis stage of risk management requires time and money, the cost increasing as the number of factors and uncertainties to analyse increases. The number of factors and uncertainties identified should be limited to reduce the cost of analysis. This is a problem only acknowledged by a small number of researchers, but their solutions seem unsatisfactory. Ranking factors have been suggested, given in terms of their significance, from which are selected a number of factors stopping when an additional one would make a negligible contribution to the total. (Determining the significance of each factor is actually the role of the second stage of risk management, but this serves to highlight the feedback nature of the risk management process.)

At the present time, it is suggested that common sense is the best guide to limit the number of factors to be used. Thus if an extremely accurate detailed model of the project is to be used then many factors would have to be included. In construction a detailed model is seldom used, the managers needing only an idea of the expected outcomes. This helps to limit the number of factors and uncertainties to be identified.

Table 9.1 Uncertainty classifications.

Classification	Examples
Physical	ground conditions, plant productivity
Political	legislative changes in employment, safety
Environmental	local pressure groups, site accidents
Site related	industrial relations, plant availability, acceptance of detailed design
Economic/financial	inflation, interest rates
Acts of God	earthquake, flood

This first stage of risk management is obviously of great importance as the subsequent stages are of little use if the factors are not correctly identified. Unfortunately the importance is linked to the difficulty of the process. Despite the importance of this stage, little research has been targeted at it, the majority of research being directed to the second stage of risk management, the analysis. The reason for this can only be that the second stage is mathematically based, whereas the first stage uses judgement.

9.2.3   Risk analysis

Once a list of factors likely to affect the objectives has been compiled it must be analysed. The objectives of this second stage are:

- ascertain/assume/estimate the effects and chance of occurrence of each factor;
- determine the overall relationship between those factors which threaten and promote the project objectives (that is, the balance between risks and opportunities);
- determine the significance of the individual factors to the total outcome.

Analysis can be undertaken using a variety of techniques, Monte Carlo simulation (Section 8.2.3) being the preferred technique of the authors. In addition, techniques such as *sensitivity analysis* (Section 9.3.3) and *tolerance analysis* (Section 9.4.3) provide the manager with useful information about the potential effects of risks and opportunities.

9.2.4   Risk response

Risk management is an attempt to change the possible futures to maximise the ratio of opportunity to risk and to bring them in line with what is considered acceptable. It is the response to the computed risk regime that allows the possible future to be affected. There are many response mechanisms available, but they may be classified as follows:

- accept
- reject/avoid
- transfer
- insure against
- reduce
- share
- investigate.

The connection between risks and opportunities should be remembered when decisions are made as to which mechanism to employ. If a risk is transferred, the corresponding opportunity is often transferred with it. Insurance is a mechanism of transferring only the consequence of the risk and maintaining the opportunity; this will usually result in a greater cost. The purpose of risk management is to produce an acceptable balance between risks and opportunities, and not to minimise the former while maximising the latter.

Sections 9.3 and 9.4 show the application of the techniques to two examples relevant, respectively, to preliminary planning (a feasibility study of a privately financed toll road) and construction project planning (a study of the construction phase of a project).

9.3 RISK MANAGEMENT IN PRELIMINARY PLANNING

9.3.1 Privately funded road scheme

A feasibility study on a privately financed road scheme involves an appraisal based on the net present value (NPV) of the scheme. (Net present value is covered in many standard texts on both construction and project management and in accounting texts. Other aspects of financial planning are covered in Sections 2.7, 3.3, 7.8 and 7.9.) The scenario for the road scheme is as follows:

> *The Government is considering granting a 20 year concession period to a construction company to design, finance, build, operate and maintain a road. The company can charge a toll for vehicles wishing to use the road. The Government has imposed a maximum construction duration of 5.5 years, beyond which damages will be charged at £250 000 per week over-run.*

The purpose of this example is to determine, approximately, the profitability of the project and a measure of the uncertainty associated with it, suggesting ways in which this can be reduced, if necessary.

9.3.2 Risk identification

The first stage of the risk management process is the identification of the uncertainties which will be analysed. As the measure of the project outcome in this example is the NPV, the identification consists of determining the variables to be included in the calculation.

The money involved in the project includes the construction cost, maintenance cost and toll revenue. Variables which affect the amount of these payments or their timing include:

- time to construct,
- number of vehicles using the road,
- toll price charged.

The annual discount rate is also required in order to convert payments in the future to their equivalent amount today. These seem to be the components needed to estimate the NPV of the scheme. The principal uncertainties are thus identified as:

- construction cost
- time to construct

- annual maintenance cost
- toll price
- number of vehicles per year
- annual discount rate.

9.3.3   Risk analysis

The second part of the risk analysis is to determine the expected values for these uncertainties. This is a task for estimators and economists, whose training and experience enable them to estimate the expected values of the various identified variables and their likely ranges. After calculations, the following estimations of expected values for the variables are produced:

Variable	Expected value
Construction cost	£10m
Time to construct	4.5 years
Toll price	£1.50 per vehicle
Number of vehicles	4 million per year
Annual maintenance cost	£1.5m
Annual discount rate	10%

These values can be represented on a cash flow diagram (Figure 9.3) in which it is assumed that all incomes and expenditures are concentrated at year-ends. It is assumed that the cost of construction is distributed equally over the construction period and the number of vehicles, the toll price, the maintenance cost and the discount rate remain constant throughout.

For the first 4 years £2.2m is spent with no income. In year 5, the expenditure drops to £1.1m, and the net income equals the difference between the revenue from toll collection and the maintenance costs for half the year. In the remaining 15.5 years the net income is £4.5m.

The NPV of the scheme can be evaluated by discounting the sums back to their equivalent value today. This shows a positive NPV of £15.5m, which suggests the scheme is a worthwhile venture. However, the values used in the calculations were *expected* outcomes of uncertainties. If the managers who provided these values were questioned further, the

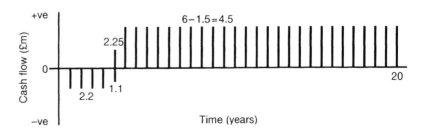

Figure 9.3   Cash flow diagram.

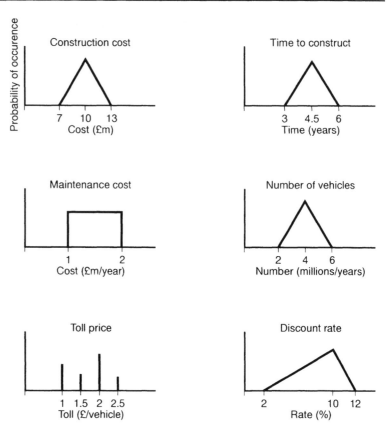

Figure 9.4 Distributions of possible values.

expected limits of variation could be ascertained. This information can be illustrated graphically using distributions of possible values for the uncertainties as shown in Figure 9.4.

The distributions shown here are not intended to portray the possible variations in such variables accurately, they are merely approximations. The shapes chosen, triangular and rectangular for the continuous distributions, are preferred for their ease of use in calculations. Many researchers suggest a normal distribution be used to represent a variable of similar nature to that depicted by the triangular; but the triangular remains a far simpler approximation to use in calculations. In the model of the project no attempt has been made to acknowledge dependencies between variables, so at this stage it should be noted that they do exist. The variables might not be independent but modelling the dependence is difficult. Sometimes, however, it might be essential.

These distributed variables provide the data required to complete the risk analysis. The first technique applied is Monte Carlo simulation, which enables the total uncertainty of the project to be determined, showing the balance between the risks and the opportunities.

MONTE CARLO SIMULATION

Monte Carlo simulation is described briefly in Section 8.2.3. The technique simulates a project by choosing at random the values for each of the variables and then uses them to calculate the outcome of the project. This is repeated many times to produce a distribution of the possible outcomes of the project.

Given the distributions in Figure 9.4, the simulation proceeds by taking a string of random number pairs, for example:

29, 18, 83, 34, 45, 64, 59, 42, 66, 79, 83, 87, ...

From these numbers the simulated values of the variables are computed in the way described in Section 8.2.3:

Variable	Simulated
Construction duration (29%)	50 months
Toll price (18%)	£1.00
Maintenance cost (83%)	£1.83m
Number of vehicles (34%)	3.64 million
Construction cost (45%)	£9.83m
Discount rate (64%)	9.1%

Using these values, the NPV of the project becomes £2.515m. This is lower than the mode but higher than both the mean and the median values (early part of Section 9.3.3) because:

- the maintenance cost is higher than expected;
- the toll priced charged is lower (perhaps due to market influences);
- fewer vehicles used the road.

However, these factors were counteracted by:

- construction cost is lower than expected
- construction duration is shorter.

Using the next six random numbers the variables take the following values:

Variable	Simulated value
Construction duration (59%)	56 months
Toll price (42%)	£1.50
Maintenance cost (66%)	£1.66m
Number of vehicles (79%)	4.71 million
Construction cost (83%)	£11.24m
Discount rate (87%)	10.4%

These values give a NPV of £17.822m.

This process is now repeated many times, a task which can sensibly be done only by computer. This gives a list of possible NPVs generated by

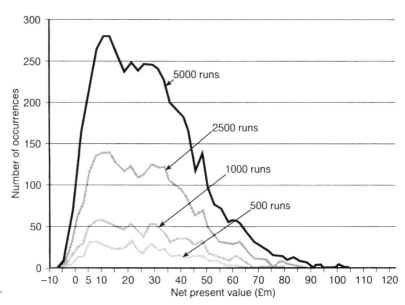

Figure 9.5 Frequency distribution of net present value.

Monte Carlo simulation. The simplest method of presenting these results is in the form of a frequency distribution. An interval of £2.5m is used to generate the frequency of occurrence in each interval. Figure 9.5 shows the distributions generated from 500, 1000, 2500 and 5000 runs of the simulation.

The expected NPV of the scheme was calculated to be £15m. The distribution of possible outcomes after 5000 runs of the simulation shows the most likely NPV to be in the region £10–15m. This trend is also shown after only 1000 runs. After the initial 500 runs, the distribution shows a highest peak to be at £20m, thus the need for a substantial number of runs of the simulation. After 5000 runs, the peaks on the graph tend to be smoothed, but they can still be detected at £10–15m, £30m and £50m. These relate to the toll price as this is represented as a discrete distribution able to take only specific values.

The distribution after 5000 runs shows a 3.6% probability of the NPV of the project being less than zero. The distribution also shows a maximum NPV in excess of £100m, and a probability of 71.2% that the NPV would exceed that expected by the original calculation.

Monte Carlo simulation has shown the balance that exists between the risks and the opportunities, the extent of each and the likelihood of occurrence. Sensitivity analysis is now performed in order to determine the uncertainties that contribute to the total uncertainty.

SENSITIVITY ANALYSIS

Sensitivity analysis is a powerful technique commonly used in industry to show the effect each variable has on the outcome of a project. As the name suggests, the technique determines the sensitivity of the project outcome to changes in a variable. The results of this type of analysis are usually presented graphically, one axis representing the percentage change in the variable with the other representing the resulting percentage change in the outcome of the project.

The first stage of sensitivity analysis is to calculate the project outcome, in this case the NPV of the project, using the expected values of the uncertainties:

Variable	Expected value
Construction cost	£10m
Time to construct	4.5 years
Toll price	£1.50 per vehicle
Number of vehicles	4 million per year
Annual maintenance cost	£1.5m
Annual discount rate	10%

Sensitivity analysis is firstly applied to the construction cost to give a tabular output (Table 9.2) which can be represented graphically (Figure 9.6).

The results show an approximately linear relationship between the construction cost and the NPV of the scheme. Similar calculations are performed on the time to construct (Table 9.3) and the toll price (Table 9.4).

All these results can now be represented on a sensitivity diagram (Figure 9.7), one axis representing the percentage change in the variable with the other representing the resulting percentage change in the outcome of the project. Sensitivity analysis has shown the prominent uncertainties in this example to be the toll price, the number of vehicles, the discount rate, and to a lesser extent the time to construct.

Table 9.2 Numeric results of sensitivity analysis on construction cost.

Change in cost (%)	Cost (£m)	NPV	Change in NPV (%)
−30	7.0	17.965	+15.6
−25	7.5	17.560	+13.0
−20	8.0	17.155	+10.4
−15	8.5	16.750	+7.8
−10	9.0	16.345	+5.2
−5	9.5	15.940	+2.6
0	10.0	15.535	0.0
+5	10.5	15.130	−2.6
+10	11.0	14.725	−5.2
+15	11.5	14.320	−7.8
+20	12.0	13.915	−10.4
+25	12.5	13.510	−13.0
+30	13.0	13.105	−15.6

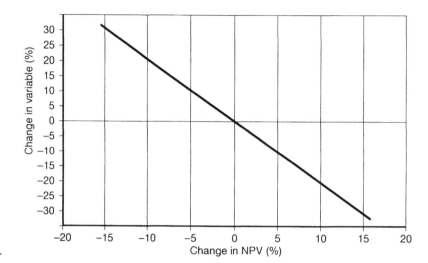

Figure 9.6 Graphical results of sensitivity analysis on construction cost.

Table 9.3 Graphical results of sensitivity analysis on the time to construct.

Change in time (%)	Time (months)	NPV	Change in NPV (%)
− 33.3	36	19.679	+ 26.7
− 22.2	42	18.228	+ 17.3
− 11.1	48	16.848	+ 8.5
0	54	15.535	0.0
+ 11.1	60	14.287	− 8.0
+ 22.2	66	13.099	− 15.7
+ 33.3	72	8.212†	− 47.1

†liquidated damages of £3.758 m (discounted) were charged due to late completion

Table 9.4 Numerical results of sensitivity analysis on toll price.

Toll price (£)	NPV (£m)	Change in toll price (%)	Change in NPV (%)
1.00	5.031	− 33.3	− 67.6
1.50	15.535	0.0	0.0
2.00	26.040	+ 33.3	+ 67.6
2.50	36.544	+ 66.6	+ 135.2

On the sensitivity diagram, limits of expectation could be added to show the likelihood of the variations shown occurring. These lines are termed contours, such as the 70% contour, and relate to probabilities. The 70% contour is constructed by marking on each line the points between which 70% of outcomes of the variable occur. These are joined to produce a continuous polygon, the 70% contour. An example of such a diagram is shown in Figure 9.8.

The addition of probability contours can confuse the meaning of the sensitivity diagram. The contour shows the region within which 70% of

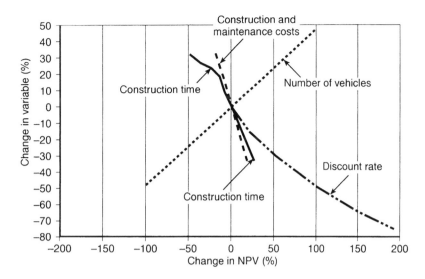

Figure 9.7 Sensitivity diagram for toll road project.

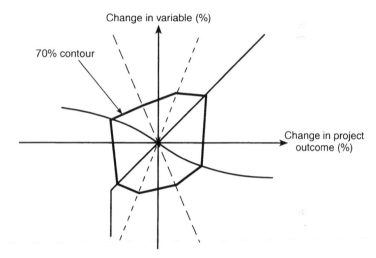

Figure 9.8 Example of a 70% contour on a sensitivity diagram.

all the possible outcomes of the variables occur. It does not show where 70% of the project outcomes, here NPV, will occur. The major deficiency of a sensitivity diagram is that it does not allow variables to vary simultaneously, which the addition of contours suggests it has.

TOLERANCE ANALYSIS

Tolerance analysis operates on a network for which the optimistic, nominal and pessimistic activity durations have been allocated, as in PERT. The tolerance referred to in the title of the technique is the difference between the pessimistic and optimistic activity durations. To overcome the problem of multiple critical paths, tolerance analysis applies Monte Carlo simulation to produce the distribution of project durations. The distribution is analysed statistically to produce the following data:

- expected project duration,
- variance of project duration,
- probability of completion within a specified time,
- most probable critical path,
- probability of occurrence of the most probable critical path,
- other critical paths.

The statistical data is then assessed against criteria and, if unacceptable, the tolerance in some of the activity durations is narrowed. Using the revised activity durations the analysis is repeated, the resulting data being checked against the same criteria.

The statistical information is the same as that generated by the use of Monte Carlo simulation in its basic form. However, tolerance analysis has the ability to change the distributions within the technique, but is restricted to using the beta distribution and can at present only be applied to the timing calculations of networks.

9.3.4 Risk response

This stage of the risk management process reflects the views and policies of the managers involved. Inspecting the balance of risks and opportunities of the toll road project gives the following information:

Parameter	Value
Minimum NPV (£m)	−7.5
Maximum NPV (£m)	116.5
Expected NPV (£m)	15.5
Probability of negative NPV (%)	3.6

A policy decision may be that this project must be profitable, and so the minimum NPV cannot be less than zero. This policy makes the project at present unacceptable. However, the results of sensitivity analysis show which uncertainties contribute significantly to the total. The decision as to which uncertainties to modify is based on the effects they have on the NPV, on the opportunities they provide and on the availability of a mechanism to modify them with. As stated earlier, risk management is a feedback process (Figure 9.1), and so it is suggested the possible modifications are made in stages.

The first stage could be to reduce the risk of the uncertainties rather than transfer them. The uncertainty associated with the time to construct could be easily reduced, but this may add to the construction costs. The uncertainty of the time to construct has been identified because it carries more risk than opportunity and, since the two are linked, reducing the risk means the opportunity is also expected to reduce. The uncertainty associated with the time to construct currently results in a possible

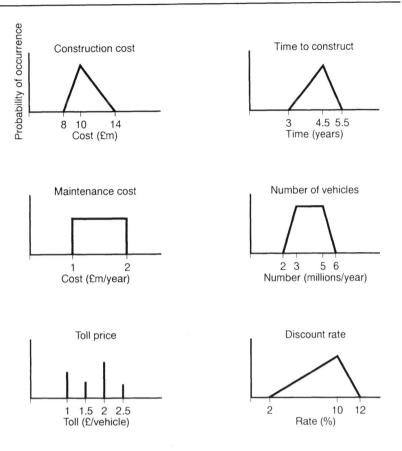

Figure 9.9 Possible modified distributions.

reduction of 47% in the NPV. This is because liquidated damages must be paid if completion is after 5.5 years. The planners could be instructed to plan the construction phase in such a way as to make completion before this time extremely likely (at present this has a 94% chance of occurrence). This may result in an increase in the minimum duration and so the resulting distribution may be as in Figure 9.9 which can be compared with Figure 9.4.

In the model of the project, dependencies between variables were not acknowledged, although they do exist. Reducing the uncertainty associated with the duration of the construction phase would probably increase the cost of the work, so this can be included in the model by creating a modified distribution of possible costs of the work (Figure 9.9) rather than by performing a different type of analysis.

Another uncertainty identified as significant is the number of vehicles that would use the road. The uncertainty can be reduced by investigating it further (as we now know this to be important). The results of such an investigation may also give modified distributions as shown in Figure 9.9.

Repeating the first two stages (identification and analysis) of the risk management process with this modified data may show the project to be still unacceptable. The second stage of formulating the response could be to investigate possible transfers of risk. If the duration of construction is still a problem this could be eliminated from the calculations as an uncertainty by transferring the problem to the contractor. This would have the effect of eliminating the opportunity of early completion as extra payments may have to be made. The extra payments would act as the opportunity to the contractor, whereas the damages for late completion would be the risk.

Repeating the identification and analysis stages of risk management may now show the project to be acceptable; this would complete the risk management process. Risk management has shown the areas of greatest uncertainty and the effect they may have on the success of the project. This information must now be used to ensure that the responses formulated in the final stage are implemented. Risk management has not ensured the project will be successful. It has merely suggested that, given the information provided, the project is extremely likely to be a success. Other uncertainties, not identified as such, could have an effect on the project, for example a political change away from privately financed infrastructure.

9.4 RISK MANAGEMENT IN THE CONSTRUCTION PHASE

In Section 9.3.4 it was suggested that the uncertainty associated with the duration of construction be transferred to the contractor. In many construction projects the contractor is engaged only for the construction phase and so may perform risk management only during that phase.

9.4.1 Householder's driveway

This example shows a network representation of the construction phase of a simple project. The project is to construct alterations to a householder's driveway. The identified activities, their planned durations and following activities are presented in Table 9.5 for the network shown in Figure 9.10. (Note how difficult it is to read the network with no names written on it (Section 4.2.3).)

As in the previous example, no attempt has been made to model dependencies between variables. These might be based on similar type of work, the productivity of a particular resource, the time at which operations are being done or something else. As an example, the activities of ordering the limestone and gravel are expected to each take 5 hours $\pm$ 30%. If these materials were ordered from the same supplier, which is likely, a delay in ordering one could result in a delay in ordering the other. If the excavation of the north foundation was easier, and hence quicker than expected, it may be that the excavation of the south foundation could take a similar time.

Table 9.5 Activity list for driveway improvement example.

Activity number	Activity name	Duration (hours)	Subsequent activities
1	Start	0	2,3,4,5,6,7
2	Excavate wall N	20	8
3	Order drains	5	14,18
4	Order walling	5	11,15
5	Remove tress	10	9,10
6	Order limestone	5	13
7	Order gravel	5	16
8	Wall foundation N	40	11
9	Excavate wall S	20	12
10	Excavate drive	50	13
11	Construct wall N	100	14
12	Wall foundation S	40	15
13	Drive foundation	20	16
14	Drainage N	10	17
15	Construct wall S	80	18
16	Surface drive	10	20
17	Landscape N	20	20
18	Drainage S	10	19
19	Landscape S	10	20
20	End	0	–

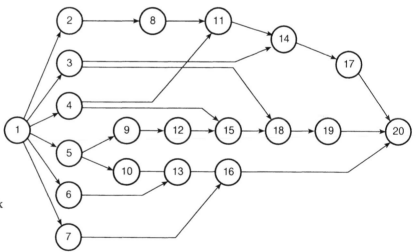

Figure 9.10 Activity network for driveway improvement example.

In standard network analysis, there is no need to identify the dependencies between activity durations. However, they must be identified when uncertainties are included and, even if not modelled, the effects should be borne in mind as they may be significant. Including dependencies in a network is difficult and also requires a modification to the Monte Carlo simulation technique, as the value of one variable dictates the distributions of others. This problem, and solutions to it, have been investigated to a limited degree only, but their importance for producing an accurate model of a project which includes uncertainties should be recognised.

If a network analysis (Section 5.2) is performed on this network, the project duration is 190 hours (23.75 days) and the critical path is along activities 1, 2, 8, 11, 14, 17 and 20. This information is of limited use as the duration used in planning the project is longer than 190 hours. The project may be planned to finish within 26 days, as the driveway is needed after that time. The purpose of the risk management exercise is to determine the risk of not finishing within the specified time limit and the activities that contribute to that risk.

9.4.2 Risk identification

The durations given in Table 9.5 are not fixed values, they are uncertainties. A risk is the possibility that the duration could be longer, an opportunity is the possibility that the duration could be shorter. The uncertainties are therefore the durations of the individual activities.

9.4.3 Risk analysis

The first stage of the risk analysis is to apply measures of uncertainty to each duration. In the toll road example (Section 9.3.3) continuous distributions are used to represent everything except the toll price. The possible durations of an activity are also represented as continuous distributions. It should be noted that this is not a particularly accurate representation as there is a limit to the increment of time used. The unit used in this example is the hour, and the smallest increment could be a quarter of an hour. It would be foolish to time an activity to the second. Despite this, continuous distributions are used to illustrate the way in which the durations can vary.

It is assumed the durations for ordering materials and excavation work have a possible variation of 30%. The other activities are taken to be less uncertain, and so subject to a possible variation of only 20% of the duration given in Table 9.5. The shape of the distribution needs to be defined. Most of the distributions used in the toll road example (Figure 9.4) were triangular. For reasons of simplicity the same is assumed for these durations.

The second stage of the risk analysis is to determine the total uncertainty in the project. Here both the PERT technique and Monte Carlo simulation are used to perform the analysis.

PERT

PERT is a continuation of critical path methods that acknowledges the uncertainties in activity durations (Section 8.2.1). For each activity three durations are provided: the nominal, pessimistic and optimistic. The nominal durations relate to those quoted in Table 9.5, that is those used

Table 9.6 PERT durations and variances for driveway improvement example.

Activity number	Pessimistic duration	Nominal duration	Optimistic duration	Expected duration	Variance
1	0	0	0	0	0
2	14	20	26	20	4
3	3.5	5	6.5	5	0.25
4	3.5	5	6.5	5	0.25
5	8	10	12	10	0.444
6	3.5	5	6.5	5	0.25
7	3.5	5	6.5	5	0.25
8	32	40	48	40	7.111
9	16	20	24	20	1.778
10	35	50	65	50	25
11	80	100	120	100	44.444
12	32	40	48	40	7.111
13	16	20	24	20	1.778
14	8	10	12	10	0.444
15	64	80	96	80	28.444
16	8	10	12	10	0.444
17	16	20	24	20	1.778
18	8	10	12	10	0.444
19	8	10	12	10	0.444
20	0	0	0	0	0

ordinarily in critical path analysis. The optimistic and pessimistic durations relate to the envisaged limits of the variations, in this example $\pm 20\%$ or $\pm 30\%$ of the nominal.

Table 9.6 shows the results of applying the PERT equations to the durations for this example. Critical path analysis is then performed using the expected durations rather than the nominal durations. In this example the distributions are symmetrical and so the two values are the same, but there can be a difference.

The critical path is along activities 1, 2, 8, 11, 14, 17 and 20, and the expected project duration is the sum of the durations of these activities; 190 hours. The variance of the project duration is the sum of the variances of the durations on the critical path; $57.778\,h^2$ or a standard deviation of 7.6 hours. This is a measure of the uncertainty associated with the project, in terms of the expected finish time.

The advantage of PERT is its simplicity and short time needed to generate a solution (which was important before computers became less expensive and easy to use). The simplifying assumptions of PERT are discussed fully in Section 8.2.2.

MONTE CARLO SIMULATION

Monte Carlo simulation was shown in the toll road example to be an effective method of determining the total uncertainty associated with a project. The technique was first applied to risk management to solve one

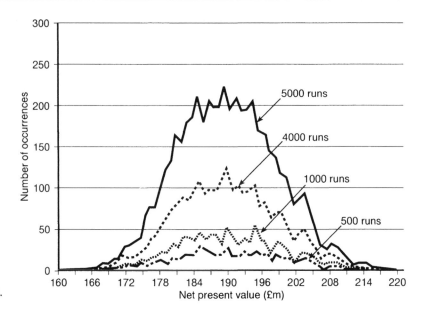

Table 9.7 Critical paths produced by the Monte Carlo simulation.

Activities in path(s)	Occurrence (%)
1, 2, 8, 11, 14, 17, 20	95.16
1, 5, 9, 12, 15, 18, 19, 20	4.68
1, 2, 5, 8, 9, 11, 12, 14, 15, 17, 18, 19, 20	0.16

of the problems identified with PERT. The methodology of applying Monte Carlo simulation to network-based timing calculations is the same as that shown in the toll road example. Values of duration are chosen for each activity from the distributions, which in this example are triangular. The project duration is calculated using these durations by applying basic network analysis. The process is repeated many times until a frequency distribution of project durations shows useful answers. Figure 9.11 shows the results of the Monte Carlo simulation after 250, 500, 1000, 2500 and 5000 runs of the project.

The graph shows that the project durations are not normally distributed as assumed by PERT. Peaks occur around 190 hours and 195 hours. This shows the interaction between the activities and the existence of multiple critical paths. Inspecting the results of the network analysis for a number of runs shows there to be two distinct critical paths, these sometimes occurring together. The activities which form these paths, and the percentage occurrence (after 2500 runs), are shown in Table 9.7.

An objective of the project is to finish within 26 days (205 hours). The distributions in Figure 9.11 show there to be a possibility (4%) that this objective will not be achieved. However, it also shows there is a probability (2%) that the project will be completed within 21.5 days. The mean duration (after 5000 runs of the simulation) is 189.66 hours with a standard deviation of 8.964 hours. This shows the balance that exists between the risks and opportunities.

9.4.4   Risk response

Risk analysis has shown the probabilities associated with completing a project within certain time limits and identified the activities which make up the critical path. If the 4% probability of not completing before day 26 is unacceptable, measures can be taken to change the balance between the risks and opportunities which, it is hoped, would also reduce this probability. The excavation of the foundations was identified as being of significant uncertainty (30% of the expected duration). This could be reduced by investigating the ground conditions or by hiring plant which would be able to deal with all envisaged conditions. Both these responses would cost money but would reduce the uncertainty. The cost of buying this security is a lost opportunity because the cost of the project would increase. This assumes the cost of the improvement is also subject to objectives, namely an expected and a maximum sustainable cost.

9.5   BENEFITS AND LIMITATIONS

9.5.1   Benefits

This chapter has demonstrated the application of risk management to two examples, a toll road appraisal and the construction phase of a small project. Both examples showed how risk management techniques determined the total uncertainty in the project and identified the uncertainties that contributed significantly to that total. Decisions regarding those uncertainties were then formulated and validated using analysis techniques.

Traditional risk management, when performed in an unformalised manner, attempts to achieve the same objectives. A formalised risk management process includes values as a measure of uncertainty rather than intuition. It is therefore expected that a formalised process of risk management is superior to the traditional method.

9.5.2   Limitations

The examples, although simple, demonstrate the limitations of risk management. Dependencies between uncertainties cannot be accurately represented at present. The objective of performing risk management is to determine the balance which exists between the risks and the opportunities, and yet some of the techniques deal only with the risks. Another problem which seems to exist in risk management is the reinvention of techniques. Tolerance analysis seems to be little more than Monte Carlo simulation as it was first applied to risk management 25 years previously. Risk mapping is similar to utility functions, an established management technique. This renaming of techniques only confuses those wishing to study or apply risk management.

These limitations can be overcome by further research, but one fundamental issue of risk management remains largely unaddressed – the purpose and target of risk management. The purpose of risk management is to determine the balance between the risks and opportunities which a decision produces in order that actions can be taken to alter that balance. A risk is the possibility that an uncertain event, the consequence of which is detrimental, may occur. Risk management assumes the target of the process is a project, with the possible consequences to that project being used as the basis for the process. In fact risk management in the construction industry can be applied to anything, not just a single project.

When risk management is applied to a project the project is treated as an independent entity. The risks and opportunities are calculated with only that project in mind, when the project itself cannot sustain losses or benefit from gains. The company must bear the losses and take the gains, the project being only a transient entity in the hopefully long life of the company. Risk management should be applied to the company, to its short- and long-term strategies, to help guide it into the future.

9.5.3 Conclusions

Risk management is a subject which has grown in popularity during the 1980s and 1990s. However, risk management, in an unformalised manner, has always been performed. The techniques used by the present process have their roots in the 1950s and 1960s (sometimes under a different name). The research which has been undertaken since the mid-1980s has refined the techniques used but there are still a number of problems. Risk management is generally applied to projects with the survival of the project in mind, rather than applying it to projects with the survival of the company as the basis of the decisions.

The two examples in this chapter were shown to have an acceptable balance of uncertainties when considered independently, but if the two projects were to be undertaken at the same time this may not still be true. To the projects, the uncertainty would remain acceptable but to the company the risks could be too great. Maximising profits on projects is important but decisions made at project level should not be against the good of the company. Obviously, before the uncertainties of projects can be combined a good understanding of the uncertainties in each project is needed. The techniques used in managing the risks to a project, whether misguided, still need enhancing to produce more accurate models of the interactions between uncertainties.

The traditional risk management process is based on experience and intuition, while many of the techniques are objective in nature. The experience and knowledge of those involved in the decision making process must not be ignored in a formalised process. A combination must exist between the objective nature of the mathematical techniques and the subjective methods of the experienced.

9.6 SUMMARY

- Uncertainty is inherent in all construction.

- Uncertainty can have both a detrimental and a beneficial effect.

- The potentially detrimental effects of uncertainty are called risk.

- The potentially beneficial effects of uncertainty are called opportunities.

- Risk management is about balancing risks and opportunities.

- Risk management should be carried on throughout a project but can have most effect at the project's inception.

- There are three main stages in risk management: identification, analysis and response.

- Identification is the most important and potentially the most difficult.

- It is often the unforeseen risks which cause the greatest problems.

- Systems have been developed to ensure that identification is carried out in a formalised manner and that omissions are therefore minimised.

- Techniques have been developed for the analysis of the identified risks. They concentrate on determining the effects on the overall project of either the individual risks or all the risks collectively.

- One of the most useful techniques is Monte Carlo simulation.

- It is important to know the sensitivity of the project outcome to the individual risks and for a combination of them.

10

SHORT-TERM PLANNING

'If you can keep your head when all about you
Are losing theirs and blaming it on you, . . .'

RUDYARD KIPLING
(If)

In construction work, people sometimes seem to lose their heads and
blame others. The problem could be to do with planning (or lack of it).
Having considered uncertainty in planning it would be surprising if
things did not change over time. Short-term planning offers the
opportunity to accommodate change and organise the work. Some
useful techniques, not used often enough, are available to help
planners and managers. Things will still go wrong but hopefully less
often, so there should be more cause for praise than blame.

10.1 INTRODUCTION

This chapter is, in its entirety, about short-term planning, an area of
planning requiring different techniques to those covered in earlier
chapters. It concentrates on short-term programming techniques and
suggests a method for incorporating financial information at site level
and the subsequent use of the information produced. A worked example
of this is included. Specific organisational requirements for the proper
implementation of short-term planning are also discussed in this
chapter, the broader organisational context of planning being covered
in Chapter 11.

Throughout this chapter, labour-only subcontractors and hired plant are treated as own labour and plant. It is very important to programme the work of all subcontractors in order to ensure the smooth operation of the project. Although the managers of a project may not have as much direct control over subcontractors, it is important to realise that control can be exercised in both a pro-active and re-active manner. The mechanisms by which this can be done are described and the need for such an approach is discussed.

10.1.1   Short-term planning

Short-term planning is needed to help overcome problems and uncertainties that arise and affect construction work. Such influences are:

- the weather
- the supply of information
- variations
- mistakes
- ground conditions
- defaults and liquidations
- machine breakdown
- defective materials
- poor coordination.

The period covered by short-term planning is arbitrary and debatable but should enable a particular type of planning to be carried out. Short-term planning should enable work over a period of time to be planned for resource utilisation, allowing for flexibility and appropriate future short-term planning of resources. From experience it is difficult to do this for a period of more than 1 or 2 weeks. The period of 4 to 8 weeks ahead falls into a middle ground, requiring a mixture of techniques identified for short- and long-term planning. It could reasonably be called medium-term planning.

Short-term planning offers many benefits and has many uses. For example, a thoughtful approach to short-term planning:

- enables early corrective action to be taken,
- allows a forward view to be taken,
- enables progress to be monitored,
- enables progress to reflect detailed planning objectives,
- gives an opportunity to check resource requirements,
- enables close coordination of related operations,
- provides a basis for operational instructions,
- provides a means of communication,
- helps to coordinate different sites,
- enables better planning for safety.

In this book, short-term planning is taken to mean the detailed programming of work to be done on a week-to-week and a day-to-day basis. It is different to all other types of programming undertaken in construction in three important respects:

- the work to be carried out and the resources available are known with much greater certainty and hopefully greater accuracy than when longer term programmes are developed.
- the objectives are closely defined by the longer term programmes.
- short-term programmes are often produced by people lower in the hierarchy of an organisation than those who produce the long-term strategic plans.

10.1.2 Finances and short-term programmes

For the reasons mentioned above, short-term programmes usually take a different form to other programmes; they are often so concerned with programming the work that their financial implications are often ignored. This is a mistake. The financial implications should be considered whenever possible, since they provide information for pro-active control (Chapter 12) and for ensuring that the company is paid its full entitlement for the work carried out.

In addition, and perhaps even more importantly from the point of view of the people actually carrying out the short-term planning, the inclusion of finances enables the accuracy of the plans to be monitored and better plans to be produced. Many more short-term plans are produced than any other type of plan during the course of a project. The information gained from them can be used not only for the improvement of future short-term plans but also for improving longer term plans and even feeding back information to estimators and tender planners. This use is described and discussed.

10.2 OBJECTIVES, TECHNIQUES AND PROCEDURE

10.2.1 Objectives

The differences described above between short-term and longer term planning mean that the principal objective of short-term planning is different to the objectives of longer term planning. The objectives of the management team, as set out in the longer term programmes, are complex and cannot be assumed just to be the attainment of the work defined by the programme activities. The objectives are influenced by resource use, productivity, quality and the personal objectives of the various personnel involved. The objectives also depend on what variations, errors and omissions might have arisen since the last update of the longer term plans.

It may further be necessary to interpret the longer term programmes in order to determine the real objectives. This is for three main reasons:

- the programmes will not have the amount of detail that is required
- the programmes may not include the reasons for doing things in specific orders
- the programmes may not be in a form suitable for the shorter term work.

Regarding the short-term period under consideration, the resources available for use on a project are likely to be known with comparative certainty at the start of the period. In the short term, for example, the number of labourers on the project is likely to deviate only by illness, absenteeism or other unpredictable events. Items of plant and equipment deviate from planned numbers only by breakdowns or emergency orders.

The objective of short-term planning can thus be stated as being:

To ensure that all the available resources are used in an efficient manner to achieve the objectives as set out in the medium- and long-term plans and as modified by changes, errors and omissions to the project.

In this definition it is important to realise that *resources* includes:

- labour, plant and materials
- subcontractors
- finances.

Regarding labour, plant, materials and subcontractors it is desirable to:

- identify what resources are available and allocate them to available work elements
- aim for continuity of use
- identify future requirements.

Regarding finances it is desirable to:

- aim for satisfactory profit and cash flows
- analyse planned and actual expenditure and income (or liability and earning).

Regarding time (which may be considered by some to be a resource) it is desirable to:

- identify and sequence the work to do in a particular time
- use time effectively and efficiently
- coordinate the work in time.

10.2.2 Information required

In order to produce a programme it is desirable to have certain information available and to have appropriate ways of representing a programme for communication to interested parties. This section considers the information required for short-term programming.

A considerable amount of information and knowledge is required since it is essential that the short-term programme is realistic. A clear, workable programme that can be understood by those who need to understand it contains considerable detail. Nevertheless it does not show everything that went into its preparation.

In devising a short-term plan consideration should be given to:

- tying in with the master programme
- important dates
- unfinished work
- sequencing and method
- continuity of work
- continuity of resource utilisation
- planning realistic productivity of resources.

It is suggested (Section 10.1.1) that a short-term programme should plan in detail for 1 week only and in outline for a second week so that resource levels can be adjusted if necessary.

Fundamental information required for the programming task for this period includes:

- the medium- and long-term objectives for the period in question
- project details
- the on-going work
- the resources available
- potential effects on future progress
- future work envisaged.

10.2.3 Sources of information

The sources of the required information are generally obvious. Clearly the project documents (such as drawings, specifications, bills of quantities, schedules of rates), subsequent variation orders and the master programme provide much of the information. Other sources include:

- past records (for example, performance on previous projects, manufacturers' data, company productivity figures);
- current records (for example, last week's or last month's performances);
- work study records (perhaps specially commissioned where repetitive elements are being carried out);
- weather forecasts (when weather critical activities are to be carried out);
- weather records (to help in the interpretation of past records and work study information).

Past records need to be properly documented so that applicable figures can be accessed and used. Current records need to be kept (frequently they are not kept in a conveniently usable form and sometimes no formal records are kept at all).

Some specific information, such as that listed below, is also desirable for good planning and should be established or calculated:

- the length of the working day and week,
- plant available and working capacities,
- resource productivity figures or output rates,
- resource constraints,
- delivery times,
- financial information,
- bonus scheme details.

Most of this may be imagined to form part of a manager's background knowledge of the company and the project in particular and is collectively sometimes called 'experience'. However, it is essential to record and catalogue such information to ensure that it can be verified and communicated to all concerned.

10.2.4   Planning techniques

Particular techniques for preparing and representing short-term programmes are available. The most well known techniques are the activity bar chart (or Gantt chart) (Section 2.2.1), the simple list (Section 2.9) and the sketch diagram.

One other technique, the resource bar chart (Section 2.2.4), is strongly recommended for short-term programming as it goes a long way to ensuring that the planner has adequately considered how the work is to be done. Work is traditionally identified by activities and represented on Gantt charts, yet the work is actually done by the contractor's (or subcontractors') resources. It would seem sensible to plan those resources to improve the chances of meeting the principal objective of short-term planning (Section 10.2.1) and the best way to represent the scheduling of resources is on a resource bar chart.

Site planners frequently claim not to have sufficient time to produce a resource bar chart. If considerable time is required to produce a resource bar chart this, itself, suggests that the resource utilisation may not properly have been considered at the short-term programming stage. This is because short-term programming exists principally to allocate the available resources to activities and the production of a resource bar chart should simply be a summary of what has been decided. Identifying the activities the resources could work on is only the first part of the procedure outlined below and can in no way be used as the best basis for controlling the work and the money (Section 10.3). Resource bar charts are discussed again in Section 10.5.4 after an example has been used in Section 10.4.

10.2.5   Procedure

Procedures to be followed for short-term planning can be considered in two parts. First, the basic elements of work to be undertaken must be identified from the overall project plan. Second, a detailed schedule must be determined in order to produce the programme itself.

The overall process of short-term planning involves:

- communication between the team (site manager, project manager, senior engineer, foreman, ganger, subcontractor, etc.),
- identifying the work to be done (from master/medium-term programmes and project details),
- determining resource output levels (from records, taking into account previous bonuses, quantities, etc.),
- deciding resource levels and plant requirements for identified work elements,
- scheduling the resources to the work elements,
- producing a programme.

A detailed procedure for producing the short-term programme could be:

(1) list work elements (operations) in approximate order of starting
(2) insert resource information
(3) calculate operation times (durations)
(4) build up resource aggregations
(5) check use of shared resources (for example, crane)
(6) check financial aspects
(7) adjust plan as appropriate
(8) draw up programme
(9) make notes on thoughts about programme.

Although listed sequentially, the procedure outlined above will have elements of iteration and need not be done in the order indicated.

10.3 CONTROLLABLE MONEY

It was suggested (Section 10.1.2) that short-term planning should include financial considerations as far as possible. Not all money is controllable at site level so it is important to recognise what financial values are controllable at the level at which a plan is produced. In general, the money that can be controlled at site level is that related to labour and plant.

Site managers are able to determine (to some extent) the number of operatives and the types and numbers of plant employed and can influence their productivity. Changes in the numbers employed or the rates paid affect the expenditure of the project. Changes in the effectiveness of the use of the labour and plant affect the income of the project. A comparison between the earning and liability provides information for both pro-active and re-active control. Although the actual income and expenditure for a project is probably not known by the person carrying out the short-term programming, this should not stop finances being considered.

It is debatable whether materials are a controllable resource. Generally, materials need to be procured and made available, so in that sense they need to be controlled. They do not really offer the prospect of control in a financial sense unless bad practice is followed. This would usually be classified as waste, which may take many forms including wrong usage, damage, poor storage or degradation. These can be improved with better practice but in most cases are not as controllable through planning as are labour and plant.

The 'programmer' should be given an allowance in terms of labour money and plant money for each piece of work to be carried out. This can then be compared with an approximation to the liability calculated on the basis of average rates for hiring the resources. Any inaccuracies arising from this should be small enough to be ignored.

The terminology used to discuss finances has been introduced (Section 3.3). It is debatable as to whether the following discussion should be in terms of income and expenditure or in terms of earnings and liability. The approach adopted is to use earning and liability, as the analysis considers what will be the eventual situation and the terms are those commonly used.

10.3.1 Earning

Most estimates for projects are achieved using calculations based in some way on the labour and plant required. On the majority of projects, each type of work or bill item has been considered by the estimator and an estimate of the cost of carrying out the work has been produced. Usually this is broken down into cost figures which are estimates of the:

- labour
- plant
- subcontractors
- permanent materials
- temporary materials.

These figures form the basis on which a project is won and should therefore be fundamental to the control of the project. They are what have been allowed for in order to get the work done and should therefore be the allowance (or earning) for the purposes of short-term planning and control.

10.3.2 Liability

The liability on a piece of work is often not known until some time after the completion of the work, since it is affected by the amount of work completed in a given time (Section 3.3). Workers are paid some form of bonus, the actual value of which may not be ascertainable until the end of the payment period following the control period.

For the purposes of financial monitoring, the average cost rate for the various types of labour and plant can be used (if the actual cost rates are not available) to calculate both the planned and actual cost of the work. It is important not to delay the analysis either of the programme or of the control information in order to increase the accuracy of the information by a small percentage. Any delay means that the control action becomes considerably less effective than immediate action would have been.

10.3.3 Subcontractors

There is a tendency when planning to expect subcontractors to produce their own plans of work. This is done for several reasons including:

- subcontract labour is not specifically under the control of the overall project managers;
- information on the resource productivity is not readily available;
- numbers of operatives available are not known by the site management;
- numbers of operatives on site are not directly controllable by the site management;
- the payment to the subcontractors is often not controllable by the site management.

Despite these indisputable facts, it is suggested that the planning of subcontract work is often at least as important as planning the work of the contractor's own labour and plant and may sometimes be more important. This is because:

- The work of subcontractors impinges on the work of all other people on site and consequently the timely, or otherwise, completion of the work affects the work of many others.
- Realistic planning of the work enables targets to be set for the subcontractor.
- Realistic planning of the work provides a base, against which monitoring, pro-active control and re-active control can be carried out, to ensure timely completion of the work.
- Proper planning of the work helps to ensure that the correct payments are made.

On some work, such as building finishes and mechanical and electrical services installations, virtually all the work is often performed by subcontractors. The project cannot be planned properly unless the work of these subcontractors is made an integrated part of the overall plan. This means that such work must be subjected both to scheduling to suit the overall plan and to detailed control.

10.3.4 Comparison of allowance and liability

The comparison between the allowance and the liability should be done for both planned and actual situations. It is obviously possible to do this in absolute terms by comparing the earning with the liability and showing, for example, that a section of work has made a profit or loss of so many pounds. It is also possible to make the comparison at the planning stage by comparing the planned liability with the planned allowance.

This type of analysis has much in common with the use of variances by accountants to determine the health of a company. For short-term use it is advisable to use ratios rather than variances to compare the planned figures and the actual figures. These provide information for both pro-active and re-active control very quickly and hence offer much greater control capability than do variances. The theories for both methods are described in Chapter 12.

10.4 SHORT-TERM PROGRAMME AND USE OF RATIOS

An example is included here to show how a short-term programme can be produced, how it can meet the objectives (Section 10.2.1) and how it can be used comparatively for control purposes (Section 10.3.4). This example is based on the construction of a multiple span reinforced concrete viaduct. The output and cost rates are not intended to be realistic but the calculations could be attempted with what the reader considers to be realistic rates and the answers compared to those given here. Fuller, practical examples using the Tingham Tank Farm and the Tingham Hotel are provided in Chapters 15 and 16. Together these illustrate not only the method itself but its application to a wide range of types of construction.

10.4.1 Planned performance

Consider the programme of work shown in Figure 10.1. It is presented in the form of a resource bar chart which is introduced and discussed in Section 2.2.4. The quantities of work are shown below the bars on the chart. They are given in the relevant units for the work concerned. For example, the carpenters' gang are erecting or fabricating formwork and the units are therefore m^2, whereas the reinforcement gang are fixing reinforcement and the units of their work are therefore tonnes.

The appropriate cost rates and allowances are given in Table 10.1, where the allowances are given in £ per unit (for example, £ per square metre, £ per tonne) and the costs are given in £ per day.

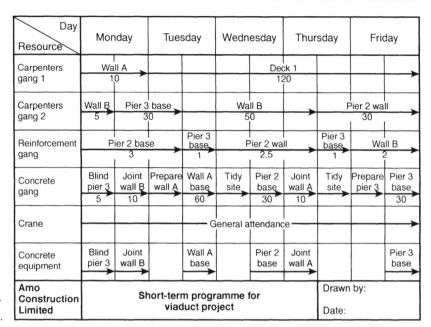

Day Resource	Monday			Tuesday	Wednesday		Thursday	Friday	
Carpenters gang 1	Wall A 10				Deck 1 120				
Carpenters gang 2	Wall B 5	Pier 3 base 30			Wall B 50			Pier 2 wall 30	
Reinforcement gang	Pier 2 base 3			Pier 3 base 1	Pier 2 wall 2.5		Pier 3 base 1	Wall B 2	
Concrete gang	Blind pier 3 5	Joint wall B 10	Prepare wall A	Wall A base 60	Tidy site	Pier 2 base 30	Joint wall A 10	Tidy site	Prepare pier 3	Pier 3 base 30
Crane				General attendance					
Concrete equipment	Blind pier 3	Joint wall B		Wall A base		Pier 2 base	Joint wall A		Pier 3 base
Amo Construction Limited	Short-term programme for viaduct project							Drawn by: Date:	

Figure 10.1 Short-term programme for viaduct project.

Table 10.1 Planned allowances and liabilities for viaduct project.

Allowances for work				Planned resource costs	
Work type	Unit	Labour allowance (£ per unit)	Plant allowance (£ per unit)	Resource	Planned cost (£ per day)
Wall formwork	m²	8.2	2.5	carpenters gang 1	180
Deck formwork	m²	7.6	2.5	carpenters gang 2	190
Base formwork	m²	5.6	0.5	reinforcement gang	180
Base reinforcement	t	80.0	10.0	concrete gang	150
Wall reinforcement	t	90.0	20.0	crane	300
Construction joint	m²	5.0	1.0	concrete equipment	50
Concrete base	m³	4.5	5.0		
Concrete wall	m³	5.1	7.5		

From this information, it is possible to calculate the planned performances as follows:

- planned labour liability

$$= \sum(\text{labour cost per day (Table 10.1)} \times \text{number of days the gang works})$$

$$= £3500$$

- planned labour earning

$$= \sum (\text{labour allowance per unit (Table 10.1)} \times \text{number of units})$$
(Figure 10.1))
$$= £3326.5$$

- planned labour performance

$$= \frac{\text{planned labour liability}}{\text{planned labour earning}}$$
$$= 1.05$$

- planned plant liability

$$= \sum (\text{plant cost per day (Table 10.1)} \times \text{number of days})$$
the plant works)
$$= £1750$$

- planned plant earning

$$= \sum (\text{plant allowance per unit (Table 10.1)} \times \text{number of units})$$
(Figure 10.1))
$$= £1337.5$$

- planned plant performance

$$= \frac{\text{planned plant liability}}{\text{planned plant earning}}$$
$$= 1.31$$

- planned overall liability

$$= \text{planned labour liability} + \text{planned plant liability}$$
$$= £5250$$

- planned overall earning

$$= \text{planned labour earning} + \text{planned plant earning}$$
$$= £4664$$

- planned overall performance

$$= \frac{\text{planned overall liability}}{\text{planned overall earning}}$$
$$= 1.13$$

From this it can be seen that the site is planning to work at 13% worse than the estimate overall and that the plant is going to be particularly badly utilised. At this stage, the manager should check the programme and agree to it only if no better way of performing the work can be found. The performances now become targets for the work.

The calculation of these values is independent of any allowance made in the estimate for price rises, profit and risk. It is of course possible to include such features but since they are not controllable at site level it is suggested that these are not included.

It is worth considering what should be a reasonable performance ratio at which to aim. If they are set unrealistically optimistic, and are unachievable, then they will quickly fall into disrepute. If they are too pessimistic and set too low a target then this of itself will tend to decrease the productivity of the project as workers are asked to achieve less. Since they are the targets for work, they should be slightly (5–10%) optimistic.

It is hoped that the potential power in these simple calculations can now be appreciated; they allow control action to be taken before work is carried out and hence when decisions can have greatest effect.

10.4.2 Actual performance

At the end of the week assume that the work done in the week is as shown in Figure 10.2 and the resources deployed are as shown in Table 10.2. These may be similar to the planned equivalent (Figure 10.1 and Table

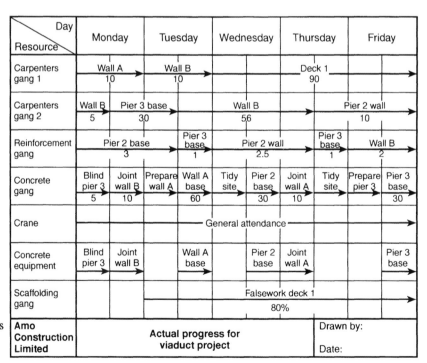

Figure 10.2 Actual progress for viaduct project.

Table 10.2 Actual allowances and liabilities for viaduct project.

Allowances for work				Actual resource costs	
Work type	Unit	Labour allowance	Plant allowance	Resource	Cost per day
Wall formwork	m²	8.2	2.5	carpenters gang 1	180
Deck formwork	m²	7.6	2.5	carpenters gang 2	200
Base formwork	m²	5.6	0.5	reinforcement gang	180
Base reinforcement	t	80.0	10.0	concrete gang	180
Wall reinforcement	t	90.0	20.0	crane	280
Construction joint	m²	5.0	1.0	concrete equipment	50
Concrete base	m³	4.5	5.0	scaffolding gang	280
Concrete wall	m³	5.1	7.5		
Falsework for deck	sum	1000.0	100.0		

10.1) or different depending on the actual progress during the control period. Here they look more similar than perhaps they are because of the presentation format used.

The allowances are the same as for the work that was in the original short-term programme, but they need to be augmented for the new work which was carried out but not originally planned. Such work could be brought about by variation orders issued during the week, by unforeseen work and by unpredicted changes in resource availability.

In this example, as in real life, some of this work has been tendered for and the estimators rates can be used. Some of the work may be extra work ordered by the client. If this is like work in the tender, the estimator's rates for similar work can be used; if it is completely different, a new estimate should be produced.

The costs for the resources would, at this stage, probably be the same as the planned costs with allowances made for the number of operatives that were available for work in each gang. The payment to the work force would not be known until after the control action has been taken.

Using these allowances and cost rates, the actual performances are calculated:

- actual labour liability

$$= \sum (\text{labour cost per day (Table 10.2)} \times \text{number of days the gang worked})$$
$$= \pounds 4820$$

- actual labour earning

$$= \sum (\text{labour allowance per unit (Table 10.2)} \times \text{number of units (Figure 10.2)})$$
$$= £3865.7$$

- actual labour performance

$$= \frac{\text{actual labour liability}}{\text{actual labour earning}}$$
$$= 1.25$$

- actual plant liability

$$= \sum (\text{plant cost per day (Table 10.2)} \times \text{number of days the plant works})$$
$$= £1650$$

- actual plant earning

$$= \sum (\text{plant allowance per unit (Table 10.2)} \times \text{number of units (Figure 10.2)})$$
$$= £1332.5$$

- actual plant performance

$$= \frac{\text{actual plant liability}}{\text{actual plant earning}}$$
$$= 1.24$$

- actual overall liability

$$= \text{actual labour liability} + \text{actual plant liability}$$
$$= £6470$$

- actual overall earning

$$= \text{actual labour earning} + \text{actual plant earning}$$
$$= £5198.2$$

- actual overall performance

$$= \frac{\text{actual overall liability}}{\text{actual overall earning}}$$
$$= 1.24$$

Once again, it can be seen that all aspects of the site were working at a worse output than in the estimate with the overall performance being 24% worse than estimated.

The efficiencies can be calculated using the planned and actual performances.

- labour efficiency

$$= \frac{\text{actual labour performance}}{\text{planned labour performance}}$$
$$= \frac{1.25}{1.05} = 1.19$$

- plant efficiency

$$= \frac{\text{actual plant performance}}{\text{planned plant performance}}$$
$$= \frac{1.24}{1.31} = 0.95$$

- overall efficiency

$$= \frac{\text{actual overall performance}}{\text{planned overall performance}}$$
$$= \frac{1.24}{1.13} = 1.10$$

These show that overall the site was working 10% worse than the plan. However, the plant was working 5% better than planned and the labour 19% worse. More detailed examination of the figures reveals that the labour efficiency is poor because of:

- extra work in the 'Wall B formwork' which was executed at a much lower rate than normal wall formwork;
- falsework to the deck being omitted from the programme.

It is reiterated that it is essential to have a realistic short-term programme which, although being achievable, is perhaps about 5–10% optimistic. This means that targets for the work force can be set which they are unlikely to achieve but which they see as within their reach. This is a good way to keep effort and, hopefully, productivity at the optimum level.

The information from the comparisons described above should be used to improve the programming and to make sure that the contractor is paid for all the work that is carried out. These are laudable and achievable objectives made possible by information that can readily be made available. It has been observed that short-term programmers soon become very proficient at producing these performance figures and acting on them as appropriate rather than just producing often wildly optimistic (or pessimistic) plans from which no lessons are learnt.

10.5   DETAIL FOR SHORT-TERM PROGRAMMING

Based on the defined principal objective of short-term programming (Section 10.2.1) much detail is required at this stage of the planning process. It has been suggested (Sections 2.2.4 and 10.2.4) that the resource bar chart is an excellent way of representing the short-term programme. This has been partially justified (Section 10.4) by high-lighting how easy it can be to assess a programme and the performance of a project using planned and actual comparisons of allowances and liability. The use of resource bar charts though is still uncommon in construction despite their many benefits. This section discusses the detail required for short-term programming and concludes with a discussion on the suitability of the resource bar chart as a planning tool and a means of representing the programme in the hope of persuading more to consider its use.

10.5.1   Detail of resources

As with all planning techniques it is important not only to identify all the resources required but also to be selective in the choice of resources for planning purposes (Sections 4.2 and 4.3). This is necessary because considering too many resources means that the planning process takes too long and becomes too complex; considering too few means that the programme loses its realism.

In the example (Section 10.4), the resources were grouped into gangs of workers. This is how they would normally operate and how they should be planned. It was not specified whether the gangs were the contractor's own labour or subcontractors. As stated at the start of the chapter this difference is more apparent than real for short-term programming.

None of the small items of plant and equipment such as compressors and scabbling hammers was considered in the example. It is assumed that there are enough of these available for the activities being planned. At some level of detail in the planning hierarchy it is essential to programme these things but this is likely to be at the ganger or worker or hourly programming levels and may well not be written down anywhere.

10.5.2   Detail of activities

The activities in the example represent typical levels of detail which would be expected in a weekly plan. They represent individual types of work in specific areas of the project. They usually have a deliverable for which an allowance has been or can be recognised. They are not necessarily single resource activities (Section 4.1) although they are often worked on by a single labour gang and a single plant resource at a time.

Activities such as the *general attendance* shown in the example for the crane resource are to be avoided if possible. Their use indicates that insufficient thought has been put into the planning with the consequential effect that there may be construction difficulties if the resource is unable to service the required activities at the required times. This is particularly important with cranes when they are required not only for very short duration activities such as lifting but also for long duration ones such as pouring concrete or holding things in place while they are fixed. The lack of a crane can stop all progress on work. This is discussed again in Section 10.5.4 and Chapter 15.

10.5.3 Time periods

Having defined a principal objective for short-term programming it is worth considering the time period a short-term programme can be produced for while still meeting the objective. For most site managers on the majority of projects, the short-term programme should be the detailed programme of work for one week with an overlap into the following week. At this sort of time-scale, the work should be programmed in half-day periods.

Longer periods than this do not provide enough detail to ensure that the resources are fully occupied. Smaller periods cause problems because there is not sufficient flexibility to enable variations in productivity and other problems to be worked around. In addition, the information may not be available in enough detail a week in advance to allow periods smaller than a week to be considered properly.

Although this is the general case, it is important to realise that some work may require much more detailed programming by the site management. Examples of this are the closure of a motorway to demolish a bridge; the disruption of sewage flow to enable a new connection to be made; and the withdrawal of power from an area in a hospital to enable a change of use to take place. In these cases, it is essential to plan everything in great detail and it would not be unreasonable to have time periods as small as half an hour. In some special cases such as switching flows on motorways, time periods of a few minutes may be required as the overall plan may only occupy an hour in total.

10.5.4 Resource bar charts

Resource bar charts have several potential benefits over other techniques used for short-term planning. In particular:

- it is easy to see if no work is available for a particular resource at any time;
- it is easy to see if more than one task is to be done by any resource at any time;
- it is excellent for communication to individual resources or gangs;
- it is easy to see that all resources are considered;
- if the quantities of work are included on the bar lines, it is simple to check the productivities of the resources.

No planning tool is ideal in all situations. Resource bar charts have the following potential drawbacks which may not be specific to the method:

- On a large project which is programmed in sections, it may be possible to miss out some of the project resources on the individual section programmes. This must be checked by the project management.
- On a large project which is programmed in sections, it may be possible to include the same resource on the programmes for two different sections. This must be checked by the project management.
- On a large, geographically dispersed project or one split into sections for control purposes, there is no way of checking that the movement of the resources and the time involved has been taken into account correctly. The project management must check this. Ideally, the movement of large pieces of plant, such as cranes, should be included on the plans but since this is not work related to any particular section or allowed for in the estimate, it is usually neglected.
- It is not immediately obvious which activities are being worked on at any one time since in the short-term programme they appear by the relevant resource but in the long- and medium-term programmes they appear as activities in their own right.

10.6 ORGANISATIONAL CONSIDERATIONS

Short-term planning is a continuous process. It impinges directly on more people than any other form of planning on a project and therefore it is perhaps the most important. In this section, it is assumed that the short-term programme is being drawn up by the contractor's organisation but similar points apply irrespective of the organisation concerned.

As well as being a continuous process for a project, short-term planning is also a part of the overall planning process in a company. Organisation for the overall planning process is considered in Chapter 11 but, for completeness, the rather more specific organisational requirements for short-term planning are included in this chapter.

It is suggested that:

- Short-term programmes are produced every week for the following 2 weeks.
- Short-term programmes are produced by the person directly responsible for the work, that is the engineer, section engineer, senior engineer or agent as the case may be.
- The person drawing up the programme should consult with all people possibly involved with the work before drawing up the plan.
- The person drawing up the programme should have access to the medium- and long-term plans for the work.

- The person drawing up the programme should know all the extra work which is required over and above the contract.
- The person drawing up the programme should know all the resources which will be available over the period of the plan.
- The person drawing up the programme should have time assigned when nothing else is to be done.
- When completed, the programme should be discussed with all people at the same or more senior levels on site to ensure that:
  - all the work is included;
  - the section integrates properly with the site;
  - all the resources on the site are utilised efficiently;
  - there is not too great a demand made on some resources by trying to use them on different sections at the same time.
- The programme discussion should be carried out at a formal meeting, the time for which must be allocated by management and should be fixed throughout the project. A meeting on Thursday for the following week is recommended.

These recommendations should be reviewed against the whole planning system in place within an organisation as, without information being made available between the parts of the process, the short-term programme might be trying to achieve the wrong project or company objectives. Further, the information required by the short-term planners might be hard to get and check without proper communication channels in place. It can be the case that what the planner could benefit from having is on the desk of a nearby office being used for an equally essential but different purpose. Such matters are discussed in the next chapter.

10.7 SUMMARY

- Short-term planning is taken to mean week-to-week and day-to-day programming.

- Short-term programming should be undertaken at least every week.

- The objective of short-term planning is to ensure that all the available resources are used in an efficient manner to achieve the objectives as set out in the medium- and long-term programmes and as modified by changes, errors and omissions to the project.

- Short-term planning should be carried out by people near the base of the management hierarchy.

- Short-term planning should be carried out by the people responsible for the assignment of the resources.

- Only controllable resources (labour and plant) should be programmed.

- Non-controllable resources (materials) should be planned for procurement and provision so that the controllable resources can carry out their work.

- The resources to be considered need to be selected and grouped appropriately.

- Subcontractors should be treated as labour and plant for the purposes of short-term planning since it is essential to control them to ensure smooth progress on the project.

- The resource bar chart is the best tool to use for short-term programming.

- It is essential to have a realistic short-term programme which, although being achievable, is perhaps about 5–10% optimistic, to set targets for the work force.

- The evaluation of financial ratios enables a programme to be controlled to the desired accuracy.

- The evaluation of financial ratios enables work progress to be controlled.

11

ORGANISATION FOR PLANNING

'Je n'ai fait celle-ci plus longue que parce que je n'ai
pas eu le loisir de la faire plus courte.'

BLAISE PASCAL
(Lettres Provinciales)

In English: 'I have made this letter longer than usual only because I
have not had the time to make it shorter.' For 'letter' please read
'project'. If this wonderful statement reads true with the word project
then it could be because not enough time was spent planning it or
because the planning effort was not itself planned or efficiently
organised. We have to find enough time for planning and we have to
plan and organise the planning. Please note that shorter is not
necessarily better – planning must still meet all the objectives of good
planning but be organised better to complete it more efficiently. *Vous
comprenez?*

11.1 INTRODUCTION

Planning is expensive. People have to spend time thinking about it, doing
it, talking about it and checking it. *Not* planning is even more expensive.
Ordering materials for delivery at the wrong time, ill-considered
sequencing of work, large amounts of waste, wrong numbers of key
resources, a large amount of managerial time spent trouble-shooting, and
a lack of information are all potentially very expensive. They are also
signs of poor planning, or even a complete lack of it.

It is essential to recognise the costs and people involved in planning at the outset of a project and ensure that the required resources are devoted to the planning task throughout the project. To a large extent this recognition is a function of higher management; however, all managers irrespective of their position in the hierarchy must provide themselves and their subordinates with the necessary resources for planning and allow the required time for the task.

Organising planning should allow for what is desirable (in a planning sense) and what is obligatory (in a contractual sense). With so many different procurement systems available for construction projects the obligations of the parties should be understood. An introduction to this topic is provided.

This chapter discusses the organisation of the planning function. It contains sections on the planning processes, including the hierarchy of planning and control that exists on a project; planning the planning function; the engineers' and architects' roles in planning and control; the clients' role in planning and control; organisation for planning and control; and planning in contracts.

The chapter as such provides a link between the preceding chapters which are concerned principally with addressing issues pertaining to planning, and the following chapters which deal with the related functions of monitoring, control, evaluation and contractual matters. The preceding chapters covered mainly planning tools and techniques, but discussed them in practical terms which necessarily related them to the planning process and the role of management. Chapter 1 used an introduction to the planning and control cycle of management to explain the contents of the book. That introduction is discussed further here in the context of a planning and control hierarchy in a company. Chapter 10 went beyond techniques and included specific points on the unique art of short-term planning and how to organise for it.

This chapter extends coverage of organisation for planning to introduce general considerations (applicable to all the preceding chapters on planning) on the various types of planning that a company or project might encounter. These different types of planning (strategic, tender, long-term) may benefit from using any of the techniques discussed in earlier chapters.

11.2 HIERARCHY OF PLANNING AND CONTROL

Many plans exist for a project. However, the work gets done only once (hopefully) and these plans must somehow be integrated. Figure 1.2 shows the general multilevel site planning and control cycle for a project from the perspective of a contractor's organisation. A similar structure exists for the client's representative, although the work covered in the plans differs from that included in the contractor's plans. In addition, the plans produced by the clients' representative take information from the contractor's organisation and feed information into the contractor's organisation in the form of external constraints. Figure 1.2 shows how the different levels of personnel interact and have an involvement in planning and other related functions. The interaction is now discussed for the different functions in the planning and control cycle.

11.2.1   Planning levels

It can be seen in Figure 1.2 that the structure for planning is quite complex and that, although several people produce plans, only one piece of work that gets paid for is done. The people with the most direct influence over the work done are the people who produce the short-term site plan of work. This short-term plan is drawn up with reference to:

- the project manager's plan;
- the control action that the project manager takes;
- external influences such as constraints or variations imposed by the client;
- the most up-to-date information regarding the state of the project, the resources and their productivity.

The project manager's plan is more remote from the actual. It is drawn up with reference to:

- the objectives for the project set in the head office;
- longer term variations;
- the project manager's view of the current status of the project and its resources.

It seeks to influence the site plan rather than the actual work carried out and, in doing so, it will naturally have less effect on the work than the detailed plan.

The head office plan is even more remote from the work. It is drawn up to achieve the objectives for the project as set out in the company policy. It is based on relatively little information (which may be out-of-date). It can only influence the work done by influencing the project manager's plan which in turn influences the site plan which directly affects the work carried out.

Each level of planning should be confined to the area over which the planner has direct control. At section level on site, labour and plant are the two main considerations. At main site level, the subcontractors, some minor materials and some overheads are included. At head office level, all things are included although those controlled at lower levels will not be included in great detail.

11.2.2   Planning detail

The general methods which can be used for planning and control are the same at all levels of the hierarchy but the contents of the plans and the levels of detail vary. For example, it is not normally necessary for the company board of directors or partners to know in detail what construction process is to be used. Conversely, it is not usual for a section agent to know the overall financial forecasts of the site. Indeed there are some things which cannot and should not be included in some plans at some levels.

At all levels of planning, the projects are broken down into smaller work packages (often called operations or activities). A project may be divided in many different ways depending on the level within the hierarchy for which the plan is intended (Section 4.2). For the board of directors, a project may be one activity, yet for a section agent in charge of one bridge on a motorway contract, that bridge alone may have tens or even hundreds of operations. Decisions also have to be made on what technique to use in producing the plan. The techniques used for planning and their advantages and disadvantages at different levels of this hierarchy are discussed in Chapter 13.

11.2.3 Control

Within the planning and control cycle (Section 1.2.2) the control which can be exerted effectively varies in a similar manner to the planning (Section 11.2.1).

At the lowest level in the hierarchy, re-active control action (Section 1.2.2) can be taken to affect the work directly (since the people involved are close to the work and their plan related directly to the work carried out). Pro-active control (Section 1.2.2) at this level seeks to ensure that future plans are realistic and achieve the targets set by higher levels of management.

The project manager's pro-active control seeks to improve the project manager's own plan by comparing the performance with the original plan. Since the project manager rarely does the collection of the information from the project, the control action is necessarily rather limited in scope and effectiveness. Reactive control at this level does not affect the work directly, but rather seeks to affect the work by directly affecting the site plan. In so far as it affects a plan, it might be considered to be pro-active.

Similarly, the head office control is even more remote from the work and can have the least effect of all the control actions. Its pro-active control seeks to change the head office plan to be more realistic in the light of the most recent, though obviously flawed, information whereas the re-active control seeks to change the project manager's plan and therefore is rather pro-active in nature.

Control methods and mechanisms, which can be used at all these levels, are described and discussed in Chapter 12.

11.2.4 Planning, control and estimating

It is important to mention at this stage the interaction between planning and control and the estimating process (Figure 1.2). Estimating forms the basis of the original head office plan and therefore is fundamental to the success of the project (and sometimes the survival of the company). The accuracy of the estimate depends on many things but the most important, and the most difficult to ensure, is an up-to-date accurate quantification of the performance of the resources on different types of work under different site conditions.

This information is most readily made available by correlating the information collected at site level during the planning and control cycle. This must therefore be fed back to the estimators in enough detail to allow them to interpret it and use it in future projects.

If the estimates are inaccurate, or the site staff think that they are being unfairly treated in the control cycle, it is often because of their own reluctance to feed back this information.

11.3   PLANNING FUNCTION

The process of planning can be considered to be a project in its own right, vital to the successful completion of the works. When viewed in this manner, it is obvious that the planning should be planned if only to ensure that:

- the objectives of the planning are known;
- the right number of correctly skilled people are assigned to the planning task;
- all the required information is available or can be made available.

11.3.1   Stages of planning

In terms of a general procedure to be followed for good planning, it is suggested that the following steps be carried out:

Understand the project to be planned, the plans which are required for it and the major constraints which are imposed upon it or upon the planning process itself   This should mean that the whole of the hierarchy of plans (Figure 1.2) is defined in this step. In practice it more often means that the higher levels of the hierarchy are assumed to produce their plans in isolation initially and that this stage is concerned with the level of project manager's plan and below. At this stage the size of the planning problem, the amount of detail and the time scale in which the various plans must be produced begins to become apparent.

Break the work down into the constituent parts, each of which can be tackled independently   The long- and short-term plans (for example) should be separated at this stage because of the different objectives, time-scales and people involved. On large projects, it may also be necessary to break the production of the master project programme into sections either because of the size of the task, the complexity of the work or the time available.

Assign people to carry out the various functions   It is essential to ensure that the people are suitably qualified, correctly briefed, have all the necessary information available (or know where to obtain it) and have all the necessary tools (such as computers and computer packages) at their disposal. It is also necessary to ensure that they are given enough time to produce the plans properly and that planning is high enough on their list of priorities to ensure that it gets done properly. On large projects where the planning has had to be split into sections, the briefing of the planners is important in order to ensure that everything is considered and in the same manner.

Put the individual section plans together to form a project plan   This step is only necessary on large projects. It is essential at some stage to have a programme for the whole project rather than relying on separate plans for individual sections. This step may take considerable time, effort and skill, and should not be underestimated. Typically, it may be necessary to assign a person or a team permanently to this task on a large complex project. Without this step, there is likely to be an imbalance in the sections, leading to problems on the project overall.

Check the plans produced   This is a step which is often ignored but which is as essential as the others. The plan should be checked for correctness and 'goodness'. Chapter 13 describes how plans can be evaluated.

11.3.2   A planning flow chart

As an example of how part of the planning function can be represented on a flow chart, Figure 11.1 shows a simplified process for resource planning. It encompasses the considerations appropriate to planning resources as covered in Chapter 7, but is also relevant to other planning techniques (see Chapters 3 to 8) for which, at approximately the same time, information is required by the planner(s) about the project, the activities and the resources. It can be seen that the process is by no means linear and can involve many iterations, as the project is broken down into activities which have resources assigned to them, and these activities are then re-combined in the project model to see the collective resource implications.

The whole process can be seen to be potentially very confused and disorganised. It is the task of management to organise the planning

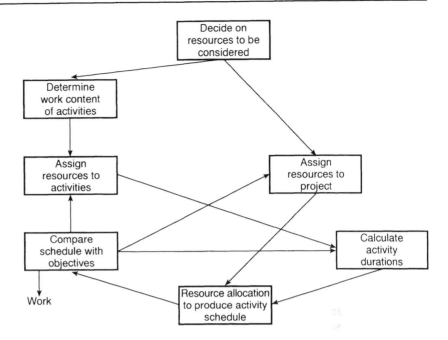

Figure 11.1 Simplified resource planning process.

(Section 11.6), so that planners do not spend time waiting for others to provide data or keeping others waiting.

11.3.3 Planning tips for better plans

Throughout the planning process, several points, if kept in mind, can help the planner to produce better-than-average plans. These include the following.

Patterns One of the main ways that project performance can be improved is by recognising a pattern in the construction process or the working sequence, and then using the planning process to take advantage of it. If a plan is based on a particular pattern of working, the plan may be very efficient but the work may need to be strictly controlled in order to ensure the pattern is adhered to. Changes in the work that may change or remove the pattern can have effects which are much more wide-ranging than might be expected.

Repetition Repetitive elements of a project can make the difference between a successful and an unsuccessful project and the same is true in planning. It is important to look for repetition and form generic plans for these sub-elements. For some types of work, for example precast bridges, floors of buildings, plumbing installations and building finishes, aspects of the construction are likely to be repeated several times. Planning on the basis of cyclic working not only leads to a better plan being produced with less effort, but also enables the work to be done more efficiently as the work method is honed in the repeating cycle.

Standardisation Standardisation aids communication and understanding. Standard terms, standard colours for bars on bar charts and standard output formats should be used whenever possible. It is important that standardisation should not be used in such a way as to stifle initiative, since this would tend to increase cost. Standardisation arises naturally through proper training at company level.

Detail Sometimes it is the detail of a plan which is important. It is possible to hide things by taking too coarse an approach. This can be an argument for considerable detail in quite high level plans and one reason for having large numbers of activities in master programmes. The detail may of course be advantageously hidden in tender plans (in order to win work), but this can cause problems later.

Targets A programme is meant, among other things, to be a target. It is important to make a programme realistic in all aspects. Because it is a target, it should be between 5 and 10% optimistic. In order to achieve this, it is essential to have good, reliable measures of the plans. Chapter 10 discusses how this measure can be obtained and used for short-term programmes. Similar methods can be used with longer-term programmes, and these not only help to ensure the plans are realistic, but also assist in integrating the various levels of planning.

Uncertainty avoidance Uncertainty should not be allowed to be an excuse for not planning. Uncertainty is a fact of life in construction. It cannot be avoided. The act of planning helps to reduce the uncertainty in many respects and therefore must be carried out. When using standard planning techniques to plan an uncertain process, it is essential to record the assumptions made and communicate these to the recipients of the programmes.

Information technology Computers can help in the planning process but they will not (yet) do the planning for you, and should not be allowed to make any decisions, such as splitting activities (Sections 4.7 and 5.6). It is essential to ensure that the planners have been properly trained in the use of the tools which they are provided with, and that they understand the way the tools work. See Chapters 4 to 8 for indications of the effects of different assumptions.

Communication It is important not to lose the thinking which goes into the plan but to communicate it to all the parties which may be affected by it. Networks, despite all their problems, are useful to assist in some types of thinking and in communicating it to users. Other techniques are useful for communicating other ideas to other people. Bar charts are particularly good in many instances. Good communication can help solve a lot of problems and give rise to a lot of improvements as, by the simple act of communicating, problems are identified and solutions debated.

11.4   ROLES OF CLIENTS AND THEIR REPRESENTATIVES

11.4.1   Planning by clients

Planning by clients may involve many of the techniques illustrated elsewhere. Clients are particularly interested, when considering the finished work, that:

- it should provide the required facilities;
- it should present the desired image;
- it should meet the maintenance requirements;
- it should be operational by a specified date;
- its cost should be within the allowed budget.

Clients are also interested in the appraisal and design stages of a project in order that they can:

- determine whether they can afford the desired facility, or what they can afford;
- appreciate the environmental and social impact of the facility;
- identify the risks in proceeding;
- estimate the costs involved;
- estimate the project duration;
- work out the probable cash flow;
- prepare for raising investment funding;
- identify critical dates;
- understand the uncertainty in considerations of time and money;
- evaluate the (proposed) contractor's work programme.

During the construction stage clients need to:

- have funds available to pay the contractor at the appropriate times;
- be aware of the responsibility of contract variations and claims for extra time and money;
- be represented in monitoring the progress and quality of the work;
- be able to assess the value of claims on a knowledgeable basis.

These considerations are similar to those of contractors, but from a different viewpoint. Clients are less concerned about, for example, resource allocation provided the work gets done on time and within budget, and they have the facts and figures for payment and claims' purposes.

The evaluation of contractors' plans is important as it helps with the identification of:

- potential problems in construction;
- areas of high-risk;
- possible claims;
- unclear areas of understanding;
- levels of experience;
- current productivity rates and prices;
- the suitability of the plans in meeting the client's requirements.

In summary, clients should be conversant with planning techniques in order to generate their own plans, monitor progress, evaluate contractors' plans and understand what the limitations of the plans are. This is a lot to expect of a client so, as with engineering design, they are allowed help in the form of professional advisers, usually architects or consultant engineers.

11.4.2 Role of clients' representatives

Different sectors of the construction industry, even within a single country, operate in different ways and use different contract conditions which specify the roles of the parties differently. In this section, for reasons of brevity, engineers, architects and other clients' representatives are referred to collectively as clients' representatives.

Clients' representatives have traditionally not been directly involved with the planning of construction projects, preferring to leave the task to the appointed contractors. This is a narrow view of the duties of the clients' representative and it does not provide clients with the service for which they are paying.

A number of procurement methods (or contract systems) which place a far greater emphasis on collaborative planning by all parties are now frequently used. In many places contract arrangements known as *management contracting* are frequently used; the essential feature of these is that the main contractor subcontracts all the work to trade contractors. The role of main contractors is thus to coordinate the work which in essence means that they plan and control the work. Since main contractors tend to operate under a payment scheme in which they share the risk with their client, they act in many senses as the client's representative. A contract system known as *construction management* in which the traditional client's representative is replaced by a construction manager is also found in many places. This arrangement includes mechanisms for encouraging pro-active planning and control by the construction manager. Attention has also recently been directed to systems in which construction management is specifically defined so that the client's representative can become significantly involved in all aspects of planning and control.

The clients' representatives should, it is strongly suggested, undertake planning on all projects for the following reasons:

- The plans produced can be used as targets for the contractor's work. It is important to set realistic targets.
- The plans allow budgets to be produced rapidly and accurately. They should be provided for the client as early as possible in as accurate a manner as possible. They are useful when checking contractors' submitted plans.
- The availability of a detailed programme of work enables variations to be evaluated at the design stage in a sensible manner.
- It is necessary to plan their own work in order to determine the resource requirements in terms of supervision for the contractor and the production of their own, usually design, work.
- The construction phase is often only a part of the client's overall project. It is essential to plan the construction phase very early in order to ensure that it fits in well with the rest of the project.

Clients' representatives can be instrumental in the planning that a contractor carries out. They help to put together the contract documents which contain a specification of the planning that should be done and communicated to the client. There appears to be general agreement that the specification of the planning should cover:

- the client's requirements in terms of the level of detail to be achieved. This is very difficult since the level of detail can be specified in many ways, which may be incompatible.
- the frequency of production and/or updating of the programme.

In addition the specification may cover:

Format of the programme to be submitted   At one extreme, this can mean (for example) specifying that the contractor must submit a project network, whereas at the other extreme it could mean the specification of the type of network, the number of activities, the information to be provided for each activity, and the detailed layout of reports. Such specifications may be useful if the project is part of a larger overall project for the client, since it can be made to fit in with the overall plan with the least effort for the client. However, it may be detrimental to the management of the construction project if contractors use a different system for their own control purposes or if contractors have no expertise in the use of the specified system. Both of these latter situations occur remarkably frequently.

Computer package to be used   This has the same benefits and drawbacks as the specification of the format (above).

Resource demands of the activities and the overall programme   This enables the client to make much more informed decisions regarding extra work, extensions of time, and so on, and also allow the client much better control. The unwillingness of many contractors to provide this is perhaps a measure of how little detailed planning they do.

Financial implications of the activities in the programme This is common in some forms of contract and can even be used as part of the payment mechanism. Under many traditional contract systems, contractors are usually reluctant to provide financial information since they believe that it gives away some of their commercial edge. The financial implications of a plan are obviously of benefit to the contractor since they form the basis of the budgets and cash flow forecasts. These are helpful to the client for the same reason. In addition they are helpful to both parties in checking the plans and monitoring the goodness of the plans (see, for example, Chapter 10 on the use of financial information for monitoring short-term planning). Since it can help all parties in the planning and control process, it is recommended that requesting more financial information is considered when drawing up the specification.

Frequency of planning updates In most projects, monthly updates are ideal but on some complex projects, or ones which progress very rapidly, it may be that regular weekly updates are required. Obviously, if a particularly difficult situation is foreseen, the client's representative should specify much more frequent updates during the crucial period.

When a programme is submitted, a client's representative should check that:

- all the work is included;
- the resource demands are reasonable;
- all the imposed constraints are met;
- the logic of the programme is reasonable;
- the meaning of the logic is clear;
- all the specification clauses are complied with.

Moving the work forward without agreement on a plan will in general lead to considerable problems later in the project and should be avoided by all parties.

When monitoring and looking to control the work of a contractor, the client's representative should consider the following at the end of each control period:

- The progress of each activity – there is no simple single measure of progress on an activity and one should therefore include the time elapsed, the time remaining, the cost incurred and the cost remaining in this process.
- Deviation from the plan in terms of time, resources and money.
- Reasons for the deviation from the plan in terms of its possible effect on finances, resource demands, overall completion and attainment of any intermediate completion dates – it is suggested that the responsibilities for these deviations be allocated to the parties at this stage rather than at the end.
- The goodness of the future plan.

There is clearly much for clients and their professional advisers to get involved with and much that they should be interested in – it is their money (and time) that is being spent as well as the client's and the contractor's.

11.5 OTHER TYPES OF PLANNING

11.5.1 Pre-contract planning

It is important for all organisations involved in a project to plan the work of the project continuously. This should start from an organisation's first involvement with the project and continue until its last involvement or until contract completion.

For contractors, the first involvement is usually at tender stage. Planning at this stage enables a company to:

- find the best way of doing the work;
- evaluate the finances of the chosen method and compare them with those of other methods;
- demonstrate to the client that the contractor is capable of completing the work in the chosen manner.

For clients' representatives, the first involvement is often before construction tender stage since they may well have been involved in the design. In the case where the client's representative was involved with the design, they should have had a plan from their first involvement and have produced an outline plan for the construction before sending out tender documents. This would be in order to:

- set targets for the contractor as to the time period and likely cost of construction;
- inform the client as to the likely cost and time of construction and likely cash flow pattern in order that the client can sanction the project.

Tender planning uses the same techniques as planning at other phases of the project. However, the results obtained can be expected to be different for the following reasons:

- there is very little time available for planning at the tender stage;
- there is often very little detail available on which to base the plans;
- the objectives of the programme (as set out above) are different to those of programmes produced at other stages.

Despite these differences, when planning at the tender stage it is important to follow the steps set out in Section 11.3.1 for the general planning process. Assumptions are necessary at each stage, and decisions are made following investigation by the planning team. It is important that the thought put into planning at this stage is not lost or wasted, so it is essential to record all the decisions, the reasons for them, and the assumptions made. This should be in as standard a form as possible in order to maximise the benefit. The records of the process should be communicated to the client's representatives and to the construction team, should the project be won.

11.5.2 Strategic planning

Strategic planning is undertaken at company level and sets out the main company objectives. It is often only produced in financial terms. Strategic planning is not covered in this book and readers interested in this should consult general management or accountancy texts.

11.6 ORGANISATION FOR PLANNING AND CONTROL

There is no best way to organise all projects. Organisation depends on, amongst other things:

- project complexity
- project size
- type of work
- project duration.

Planning is, or should be, part of the quality assurance (QA) procedures or total quality management (TQM) system operated by the companies involved with the project. The required organisation, timing and responsibilities for the execution of the procedures should thus be written down and made available for inspection. As with all quality procedures, there is a danger that the procedure itself becomes more important than the quality of the product and it is one of management's functions to ensure that this does not happen. The quality documents help to ensure consistency; the quality itself comes from the commitment of the staff.

With such a wide range of types and sizes of construction projects and the variety of different contractual arrangements used, it is obviously impossible to suggest organisational models for the planning and control process that could be operated on all projects. The following suggestions for organisation are for large projects. Smaller projects may be organised like a single section of a large project.

11.6.1 Strategic programmes

These plans set the objectives for all the company personnel and are very important. They should:

- be drawn up to reflect an organisation's objectives;
- not be restrictive as to their means of achievement;
- be reviewed at least 6 monthly or more frequently if the need is recognised from the control action which becomes necessary.

11.6.2 Long-term programmes

The long term is defined as a period extending more than 2–3 months into the future, the whole period of which should receive similar detail in the planning process. The programmes produced for the long-term requirements of a project should:

- be for the whole of a project or for the remaining part of it;
- be reviewed at 3 monthly intervals;
- be drawn up by the contract manager or equivalent person or by a planner;
- be in more detail for the first 6 months than for the rest of the project to reflect the degree of uncertainty about future work;
- provide enough information to allow 'long lead time' materials and subcomponents to be ordered for specific delivery dates;
- provide enough information to allow key resources to be available for specific dates;
- be in enough detail to allow project budgets to be calculated and their accuracy assessed;
- be compared with the strategic plan to ensure its targets are achieved or to feed information upwards in the management hierarchy to allow revision of the strategic plan;
- be checked against all programmes for the project produced by organisations earlier in the command chain to ensure that the targets are being met;
- be checked by people lower in the command chain for realism.

11.6.3 Medium-term programmes

The medium term is defined as a period which extends up to 1, 2 or 3 months into the future, the whole period of which should receive similar detail in the planning process. It is recommended that the programmes devised for this sort of period should:

- cover 3 months of a project;
- be updated monthly;
- be drawn up by the agent or section engineer responsible for the work;
- allow the resources and materials to be available on time;
- contain enough financial information on controllable money to allow re-active and pro-active control to be taken;
- contain all work known about at the time including any extra to the contract;
- be drawn up to meet the targets set out in the long-term plans as amended by any variations;
- be evaluated against the long-term targets;
- be evaluated both physically and financially to allow pro-active control to be taken by alteration of either the medium- or long-term plans or both.

These points clearly differ from the list for long-term planning and serve to illustrate the need for careful organisation of the planning process.

11.6.4 Short-term programmes

Short-term planning is a continuous process. It impinges on more people directly than any other form of planning on a project. It is therefore perhaps the most important form of planning carried out on a project. Comparing the above Sections (11.6.2 and 11.6.3) with the similar list for short-term planning (Section 10.6) the differences in what should be done are apparent. It is assumed that the short-term programme is being drawn up by the contractor's organisation but similar points apply irrespective of the organisation concerned.

11.7 PLANNING AND CONTRACTS

Up to this point planning has been considered as part of a general construction management strategy. This section focuses on the specific interaction between the planning function and contractual obligations. Whatever obligations exist, it should only be necessary to modify slightly the usual procedures used for project planning and control.

There are many procurement systems that can be used on a construction project so the discussion here is kept as general as possible. Where specific examples are needed a conventional contract is assumed and civil engineering terminogy used; thus the employer (client) engages a contractor to execute the works with the contract being administered by an engineer. The words builder and architect (rather than contractor and engineer) would be the equivalent parties in many building contracts in which there would be reference, as appropriate, to planning obligations.

11.7.1 Time for completion

The vast majority of construction work is performed under contract. In English law, as in most legal systems, the general obligation in respect of time is for the contractor to complete the work within a reasonable time.

Where a contract contains express provisions as to the start and completion dates, those provisions prevail. It is rare, particularly for large projects, for the start and completion times to be unspecified. The time for completion remains as set out in the contract until some event occurs for which the employer is at risk. Thus, for instance, if by the terms of the contract the employer bears the risk of bad ground conditions and such conditions eventuate, then the contractor is entitled to an extension of time if the groundworks will take longer than anticipated. In this way, the original time for completion would be extended forward in time. The precise mechanism of extension depends on the terms in the contract.

Many construction contracts have an extensions of time clause under the powers of which an extension may be computed. If there is no such clause, a reasonable time will be substituted for the time originally agreed. In the latter case it is sometimes said that time is put 'at large', but in practice this simply means that the original time is modified to produce a new reasonable time in all the circumstances.

Where an extension of time clause exists, this clause may specify the way in which the extension is to be computed. It is far more common, however, for this question to be left open and thus the means by which extensions may be computed need to be examined in detail. This is discussed in Chapter 14.

11.7.2 Timing and sequence of work within the construction period

In English law (and in most other legal systems) contractors may in general plan their work as they like providing they complete the work by the contract completion date. Where a working sequence is specified in the contract, or intermediate completion dates are set for elements of the work, these constrain contractors considerably.

It is rare for plans (or other programmes or method statements produced by a contractor) to be directly incorporated into a contract. Certainly, amongst the reported English cases where this happens, it is usually a result of the inadvertant inclusion of the plan into the contract documents. Where the plan is a contract document, the contractor is entitled to perform the works in accordance with the plan and if the plan becomes impossible to achieve for reasons at the employer's risk, the contractor can claim damages for breach of contract. Usually such a result is not intended so plans, method statements, resource schedules and so on are usually not incorporated into the contract.

11.7.3 Plans/programmes/method statements called for under a contract

Although plans, programmes and method statements are rarely made into terms of a contract, employers usually require some indication of the means by which contractors plan to undertake the work. Once contractors are appointed, most standard conditions of contract oblige them to submit a plan of work which shows how they intend to undertake the work and meet the completion date. This is valuable to the employer in many ways:

- It enables the employer to refuse approval to means of construction which conflict with the employer's other objectives. Where such disapproval is unreasonable, the employer must compensate the contractor if alternative methods are more expensive. This approval mechanism nevertheless enables the employer to exercise a degree of pro-active control.

- Where design information, free issue materials, access, possession of areas of the site and so on are to be provided by the employer under the contract, the contractor's plan enables the employer to plan for the provision of such information at a time which does not delay the contractor.
- At progress meetings, the contractor's plan provides a useful basis for discussing progress and future work.
- In the event that the contractor is unable to keep up with the plan in such a way that the contract completion date is jeopardised, the employer can take whatever action the contract allows to require the contractor to accelerate.
- In the event that delays occur which are at the employer's risk, the plan provides a convenient starting point for discussions of extensions of time.

In the British civil engineering industry, plans and programmes provided by the contractor after the award of the contract are known generically as Clause 14 programmes. This name is derived from the clause under the family of contracts known as the Institution of Civil Engineers contracts by which they are required. These programmes are not part of the contract itself. They are the result of a mechanism in the contract. Thus a failure to meet a date in the Clause 14 programme is not, of itself, a breach of contract.

11.7.4 Risks

The contract sets out which events are at the contractor's risk and which are at the employer's risk. If the contract is silent in respect of an event which has occurred, a legal inference is drawn as to which party is to bear the risk. If the contract is silent on the point, the law supplies an answer to which party is at risk; for instance, where the contract is silent as to the responsibility for unforeseen ground conditions, it is usually the case that the contractor bears the risk (for an English case on this point see Bottoms versus Mayor of York (1892)).

This book does not consider how any particular contract allocates risks. Events which delay or expedite the progress of the works are considered simply to be (a) the contractor's risk or (b) the employer's risk. All occurrences may be so divided. Thus the terms of a contract should influence the parties in deciding how to model the project so that risks and their consequences can be planned for, monitored, controlled and handled in a way which avoids contractual dispute situations.

11.7.5 Resources

The question of resources is also a matter of contractual concern. Contractors who have tendered for work will have done so on the basis of an assumed resource allocation. It is frequently possible to accelerate the works by employing additional resources so it may be argued by an employer that an event at the employer's risk has not caused a delay because the contractor may employ additional resources (that is, accelerate) to recover the programme. The validity of such an argument depends on any terms in the contract which specify the level of resources which the contractor must have available or terms which oblige the contractor to accelerate. Generally speaking, in the absence of such express requirements, the contractor is not obliged to employ resources at a level in excess of those which were reasonably anticipated at the time the agreement was made.

It is normally implicit in civil engineering contracts that the works may vary in scope or type within reasonable limits; the contractor must allow for this in the tender, together with the possibility that different resources may be required in order to execute the works. This is very different from requiring the Contractor to increase the resource employment in order to accelerate the works. This is explored further in Chapter 14.

11.8 SUMMARY

- Many plans are produced for a project.

- The nearer the plan is to the work the greater the influence it can have.

- The planning process itself needs to be planned.

- Planning exists in a hierarchy in an organisation and in a project.

- Planning detail differs in different types of plan.

- Re-active control by people high in the hierarchy can only affect the plans of the lower levels and not the work directly.

- Planning and control information should be fed into the estimating function.

- Good planning should follow procedures and learn from experience, but a linear planning process cannot be defined.

- Clients should be actively involved in all stages of project planning.

- Clients should set targets for contractors.

- Clients' representatives should know about project planning to understand the contractor's plan, link it with their own and the client's plan, and set targets for the contractor.

- Planning requirements should be specified by the consultant in the project specification.

- Information generated at the tender planning stage should be communicated to the project team.

- Organisational considerations are different for long-, medium- and short-term planning.

- Contractual obligations should be borne in mind when organising the planning function.

- Even the very best plan will go wrong if:
  - the site manager chooses to diverge from it for no particular reason;
  - insufficient resources are provided at specific times;
  - sustained bad weather intervenes.

12

MONITORING AND CONTROL

'I only ask for information.'

CHARLES DICKENS
(David Copperfield)

A simple line. An apparently simple request. Gathering information should be easy to do. The difficulty is in gathering the right information and representing it in a way which is useful to somebody. People in senior management are often heard to utter this statement in exasperation as they attempt to control projects or, worse still, as they prepare (or evaluate) claims' submissions often long after the time when the information should have been collected.

12.1 INTRODUCTION

During any construction project the three inter-related factors of time, money and quality need to be controlled. An improvement in one can be achieved at the expense of one or both of the others. Good planning and control should mean that all three could be improved simultaneously. Managers on a project must decide on acceptable targets for each of these factors and take action to ensure they are achieved.

The importance of quality has been recognised in recent decades and a considerable amount of literature is devoted to it. Indeed some management theories are based to a large extent on quality concepts. This chapter concentrates on the time and money (physical and financial) aspects of project control.

Before control action can be taken it is usual to have some evidence on which to act. This might take the form of such things as:

- spotting mistakes
- recognising lack of progress
- identifying areas of poor quality.

It is generally negative observations that give rise to the need for control action, although positive observations give good guidance on how to do things efficiently and hence allow a degree of pro-active control. There is much that can be observed, measured and recorded and much that actually is. Sometimes the recorded information is in a form which can help to justify taking certain control action; at other times the information is in a form which appears useless for control. This chapter looks at sources of information and techniques for monitoring work and their usefulness for control.

In the short term, monitoring enables the progress of the resource gangs deployed on the works to be examined and controlled; and the supply and use of materials to be monitored. It is also possible to use financial information for control in the short term, provided that the allowed costs can be determined from the project information (Section 10.4). It should be remembered that an important aim of short-term planning is to meet the objectives of the medium- and long-term plans (Section 10.2). Information about time and money is needed for controlling to the master programme and for appropriate resource planning in both the long and short term.

Financial information is useful and can be determined quite quickly if finances have been considered in the short-term planning. The financial information is often not considered until valuations have been completed and the appropriate figures plotted against the predicted financial profile for the project. The financial profile could indicate such things as cash flow, income and expenditure, and with appropriate software can be plotted over a master bar chart programme. This information is very useful to senior management in making major decisions on future planning for a project (for example, whether to accelerate or not).

Control of both the financial and physical aspects of a project is essential for all parties, be they the client, designer, main contractor, subcontractor or supplier. They are all able to exert control at different levels, on different things and at different times. This chapter examines the control process with a particular emphasis on techniques that would be used by main contractors. The techniques are, however, directly applicable to control by others.

There are many possible financial models of a project. A model found in many places is a bill of quantities or a schedule of rates. A bill of quantities is used in examples as appropriate in this chapter.

12.2 PHYSICAL AND FINANCIAL CONTROL

Control action can be conveniently divided into two types, re-active and pro-active. These words have arisen earlier (Chapters 1, 10 and 11) but are discussed and defined more fully here.

12.2.1 Re-active control

Re-active (or feedback) control is the type of control often thought of in construction. It consists of setting targets (usually in planning), performing the work, collecting information regarding progress, comparing progress with the target and finally taking direct action to make the work come closer to attaining the targets. This form of control action is shown diagrammatically in Figure 1.1 by the arrow connecting *control* to *work*.

Re-active control requires a system for collecting information about what is actually produced. This is also the fundamental source of productivity information for the estimating function on future projects. There is necessarily a lag between the work being carried out and any control action being taken. The collection of information, its comparison with the targets and the decision as to what action to take, are operations which all take time. Furthermore, the uncertain nature of construction must be considered together with the need to avoid precipitate action based on too little information. It may therefore be a long time after an event occurs before control action is possible. In rapidly moving projects, this may mean that the work is finished before the management team are sufficiently well informed to take the necessary steps.

12.2.2 Pro-active control

Pro-active (or feedforward) control consists of controlling the targets which are set. When a plan is made (or a target set), it is checked independently against an agreed set of criteria which can change as more information about the project and working methods becomes available. Where discrepancies are found, control action is taken. This adjusts the plan by, for example, altering the level of resources on the project. Pro-active control is exercised before any work is carried out. It relies on the existence of two independent assessments of work and productivity. This form of control action is shown diagrammatically in Figure 1.1 by the arrow connecting *control* to *plan work*.

12.2.3 Physical and financial models

Both the financial and physical aspects of work must be controlled and the control applied to each must be coordinated. In most construction situations, it is just as unacceptable for a project to finish at the proper time – but excessively over budget – as it is to have it finish within budget – but very late. In order to control both aspects in an integrated manner, it is important to understand the relationship between the two. In construction it is common to monitor and measure them using different people carrying out completely different functions.

The financial aspects can be modelled using:

- A bill of quantities or schedule of rates which define the elements of work to be done in terms of the different types of work (for example, square metres of formwork, cubic metres of concrete and linear metres of hand rails).
- A break-down of the work into cost heads showing the amount of labour, plant, materials and so on that are required.
- The cost of the labour, plant and materials used on the project.

The physical aspects of the work can be modelled using the activities from the programme which define the elements of work to be done in terms of different structural components (for example, the abutment of a bridge, the columns on a floor of a building or the car park to a supermarket).

In many cases managers do not make use of any relationship between a plan of work and a financial model. In order to exercise control it is beneficial to develop such a relationship. For a bill of quantities or schedule of rates several methods have been suggested including the creation of a contractor's measure bill or a bill split. There is also strong support for the suggestion that a bill of quantities should reflect the working methods better, and operational bills have been developed although they are not widely used.

Note that bill splitting is a process whereby each work item in a bill of quantities is divided into percentages which are then, if appropriate, allocated to different activities comprising a plan of work. Similarly, each activity on a programme of work can be linked to all the appropriate bill items. The process can be carried out either for all items and activities or for those which it is most important to be able to control. It can take a long time to do this and is sensibly aided by using integrated computer software. In such cases the preparation time for a bill split is reduced to a few hours for a single person on most projects of reasonable size. The technique becomes very powerful in terms of project control when a system for the accurate monitoring of the work is in place to provide progress information in a form that can be compared with the bill split.

12.3 COLLECTING AND USING INFORMATION

12.3.1 Monitoring and record keeping

Monitoring and record keeping are tasks for the present and the short term. They are useful if carried out properly in that they provide vital information for project control purposes and for good office management of tasks such as invoicing, calculating bonus payments and preparing valuations. Their comprehensiveness (or lack of it) often only becomes apparent in the long term when possible contractual entitlements are being evaluated.

Monitoring and record keeping are often seen as chores and are not carried out as systematically or as thoroughly as they should be. Monitoring without record keeping means that record keeping is consigned to the memory (which is not wholly reliable!). Record keeping without monitoring means a vivid imagination!

The importance of doing both tasks properly cannot be over-emphasised. To go back in search of missing information is very frustrating for all concerned. Perhaps the information was recorded but cannot now be found; perhaps the information was never recorded in which case the search is futile. Although clairvoyancy is not in the job description of those involved with making and keeping the principal records (and therefore some detective work is inevitable), much time is wasted, often by very senior (and hence very costly) people, trying to make good a deficiency in these relatively simple, sometimes tedious, tasks.

For those in middle and junior management who are usually charged with the tasks of monitoring and record keeping it is worth remembering the possible uses of the collected information. To those in senior management it must be worth pointing out the importance of the tasks and the possible uses of the information. Possible uses are listed below:

- making managerial decisions (taking short-term control action);
- assessing the effects of variations (preparing valuations and possible replanning);
- updating project resource output listings (for planning the next phase);
- updating company resource output listings (for estimating and future planning);
- calculating financial positions;
- evaluating contractual entitlements (preparing or assessing claims).

12.3.2 Sources of records

Some of the diverse sources of records on a project are:

- diaries;
- field books;
- time/bonus sheets;
- quantity surveying (QS) valuations;
- orders/invoices/delivery notes/receipts;
- job card and coding systems;
- general correspondence (letters and memoranda);
- requests for information (RFI), confirmations of verbal instruction (CVI), architect's instruction (AI), engineer's instruction (EI), variation (or change) orders, measurement memos, daywork sheets, charge notes;
- drawing issue register;
- minutes of meetings;
- progress reports;

- progress charts;
- progress (or record) drawings;
- photographs;
- video tapes;
- memory.

It is strongly advised that reliance on the last of the sources above is minimised. Memories are notoriously unreliable and may not be available when required due to the break-up of project teams and departures from the company.

Because many of the above sources essentially offer lists of discrete data there is a need for sources offering better continuity of reporting. Diaries, progress reports, record drawings (Section 12.3.3) and progress charts (Section 12.3.5) are particularly useful in this respect.

Computers are excellent for storing and manipulating data. Their use in monitoring and record keeping is increasing to the point where all information should be recorded on them. This can be by keyboard, by electronic data transfer, or by scanning documents. With all information stored on (and able to be interpreted by) computers, the scope for improved management of both project and company is great.

12.3.3 Progress drawings

Progress drawings are produced specifically for monitoring progress on a project. Two forms of progress drawing are traditionally used:

- annotated (coloured up) layouts;
- annotated (coloured up) bar charts.

These are essentially graphical techniques which can provide only limited information but are very useful, especially in conjunction with diaries and progress reports.

It is desirable to show:

- the location of the work done,
- the amount of work done at a given time (often percentage complete),
- the work on-going at a given time.

The location of the work can be shown shaded on a progress drawing with supplementary information about the date and quantity of work involved.

12.3.4 Proportion of work done and work on-going

The proportion of work done on an activity is usually shown by shading a proportion of a bar on a bar chart and identifying the appropriate date(s) (Figure 12.1). Adjacent time periods can use different colours to distinguish them or, especially if recorded on computer, the charts can be saved on disk or printed out each week, month or whenever.

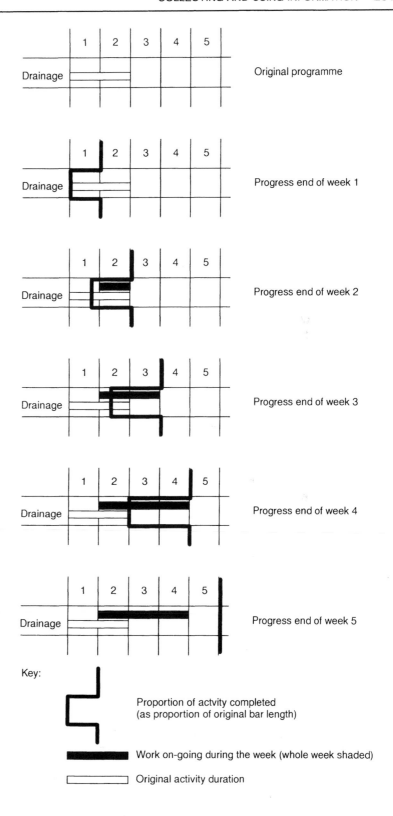

Figure 12.1 Marking up a
progress bar chart.

The work on-going at a given time can be shown by shading the time period on a bar chart during which the work was on-going. This is essentially producing an as-built bar chart. This can give a good picture of a project and it can also show, if annotated accordingly, other important information such as the rates of working at particular times and the actual quantity of work completed.

For projects on which the original quantities do not change (!), marked-up bar charts are easy to produce and maintain. They can be difficult to update when variations occur or when quantities are remeasured, especially when the percentage of completed work is depicted. This is because when the quantity changes, the effect could be accommodated in the activity by changing the duration, changing the resources, or changing both. This makes a nonsense of trying to divide up the original activity bar according to what proportion of the work has been done. What was 50% complete at a particular time would be only 25% complete at the end of the next week if, in the meantime, the original quantity doubled and no work was done; and doubling the resources allocated to the activity might enable it to be completed within the original duration.

There is no easy solution to this except to acknowledge that other sources of records may be required when the records are subsequently consulted for whatever reason. It is an argument for acknowledging that the information on marked-up bar charts can be unreliable and/or incomplete and they should be kept as simple communication tools. When computers are used, there is more scope for making the record drawing more comprehensive, holding information linked to the diagram but not shown on it. The problem of allowing for changes to a project remains.

A marking-up sequence is shown in Figure 12.1. For a typical activity not subjected to variations, the sequence shows how the proportion of work done at a given time can be denoted by a vertical line which intersects the duration (the original bar) at an appropriate point. 'Progress end of week 1' shows zero work completed at that stage. The sequence also shows work on-going during a time interval by shading a bar during that time interval. 'Progress end of week 3' shows that work was going on in weeks 2 and 3 and that two-thirds of the drainage work was completed by then. In summary, the sequence shows that work was delayed by 1 week and then took 3 weeks to complete, in equal amounts each week, rather than the 2 weeks planned.

This simple example shows the usefulness of such a simple system (remembering that it becomes fairly meaningless when significant variations are issued) in conveying and recording information. Sometimes however, detailed progress information is lost due to the marking-up methods used. For example, when a bar chart is shaded in just one colour to indicate percentage complete, the amount for each time period ultimately becomes lost, as shown in Figure 12.2 for the same situation as Figure 12.1.

The *end of project* bar chart in Figure 12.2, remembering that this final bar chart might be the only version available to an observer, shows

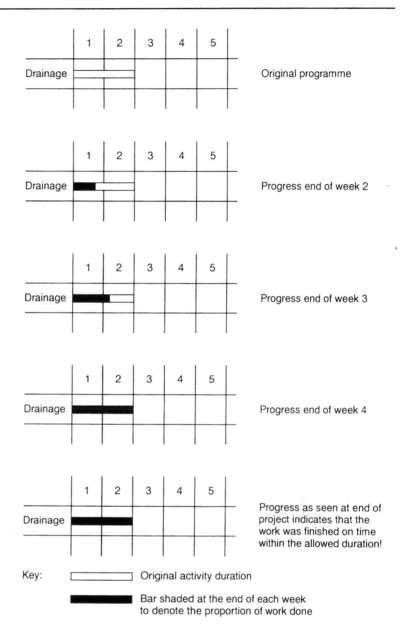

Figure 12.2 Poorly marked-up progress bar chart.

no information about how much work was done when; indeed, one interpretation could be that all the work was completed on time and within the specified duration.

12.3.5 Progress charts

The methods described above give a measure of the physical work done on a project. The methods described in Section 12.5 give a measure of the financial situation on a project. If both these measures were employed on a project, it would be rare to have them indicate the same situation. This is often attributed (by people who should know better) to such factors as learning effects (Section 4.7.2) and different prices of materials from phase to phase in the project. In reality the situations differ because too little thought has been given to the planning process and too little detail has been included in the final plans which are used for monitoring. If enough detail is included in these plans, several very useful progress charts can be developed and used for control purposes.

Probably the most useful chart which can be produced is a special case of the cumulative resource profile (Figure 2.23) where each activity is assigned one abstract resource (Section 3.2.8) for each day (or week or any other convenient time period). This abstract resource is called an *activity day* (or *activity week* etc.), the cumulative use of which can then be calculated by the aggregation process described in Section 7.3. If all the activities are approximately the same size (in value or cost terms) and are sufficiently detailed to assume (realistically) constant productivity throughout their durations, then the activity day chart is an excellent measure of the progress of work throughout a project in both physical and financial terms.

It can be used as a control chart by plotting the number of activity days completed at any time on the same axes as the planned line. This allows both the absolute deviation from the plan and also trends to be monitored and control action taken where necessary. Obviously if alterations to the original project occur, then it becomes necessary to adjust either the original plan or the actual figures or both. This technique is so useful and robust that, if the plan is designed with the aim of control in mind, and the variations are expected to be relatively small, the chart could be used as a payment mechanism.

An example of an *activity week* chart is shown in Figure 12.3 and includes both a planned and an as-built progress line.

Alternative resource-based examples of the *activity day* chart are commonly used as quick monitoring and control charts, although they are not without problems. Actual key resources are used instead of the activity day resource. Examples are square metres of brickwork, linear metres of kerbs and volume of concrete placed. Usually one key resource is chosen as the leading parameter for the project and its cumulative production graph is calculated by aggregation. This is then used in a similar manner to the activity day plot for monitoring and control.

Problems can occur when this latter technique is applied to a project which has no obvious leading parameter. In most construction projects the leading parameter changes throughout the duration as the work changes from foundations to superstructure to finishes. This requires different leading parameters to be considered at the different phases (cubic metres of excavation at the start, cubic metres of concrete during the superstructure phase and linear metres of pipework during the finishing phase for example). The technique is most likely to be useful if the phases are distinct. If they are concurrent (as is quite common with fast track systems), the technique has problems.

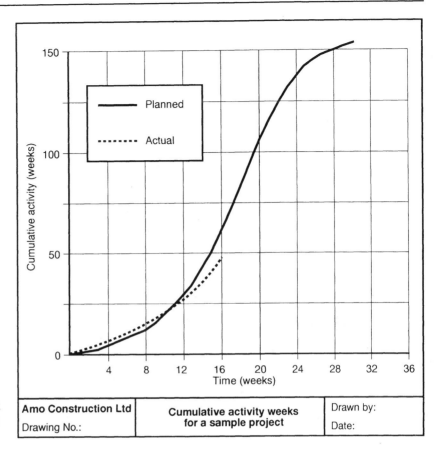

Figure 12.3 shows the cumulative activity (weeks) plotted against Time (weeks).

Legend: Planned, Actual.

Amo Construction Ltd Drawing No.:	**Cumulative activity weeks for a sample project**	Drawn by: Date:

Figure 12.3 Progress chart using abstract resource.

Figure 12.3 shows the planned and actual cumulative activity weeks for a sample project. In fact this is the activity weeks chart for the Tingham Tank Farm long-term programme (Figure 15.12). It can be seen that the planned progress on the project is far from linear and is in fact close to the traditional S-curve. The actual progress (activity weeks completed) shows that after a good start when the project was ahead of schedule, it fell behind at Week 11. The planned progress rate was not achieved between Weeks 8 and 15 and only in the most recent week has progress improved significantly. The chart highlights these points but does not indicate why progress was not as planned; other techniques described in this chapter are needed in order to address this problem.

12.3.6 Job card and coding systems

Job card and coding systems are used in manufacturing industry more than in construction but they do offer systems for monitoring and control which have advantages and drawbacks.

For a job card system, a project is split into tasks (which may be the planned activities) and a job card made up for each. This card contains information about the task including some or all of the following:

- the task name
- the quantities of the major materials required
- the resources assigned to the task
- the allowance for the work
- the planned start and finish dates of the task
- the actual start and finish dates of the task
- the engineer/manager responsible for the task
- details of any variations affecting the task.

A job card is set up ahead of work on a task and is issued to a ganger or foreman responsible for the task when the work is carried out. When work is carried out on the task, information is added to the card providing a continuous record of the work on it. The main problem with the job card system is the difficulty in ensuring that a card is completed in a timely and accurate manner by the ganger.

Similar problems surround the use of coding systems in construction. Coding systems are used to group similar types of work together in order to enable a contractor to gather information about the productivity of the work force on a particular project, to compare it with company records and to update these records for use in estimating, planning and controlling future projects.

The work of the project is grouped and coded. As an example, all concreting work could be given the code C, concreting columns CC, concreting walls CW, concreting columns on the first floor of a building could be CC1. This coding requires the equivalent of a standard method of measurement (a document to assist in the preparation and use of a bill of quantities) to be developed by the contractor, although it would be to suit the contractor's own work rather than general construction. Gangers and foremen are required to fill in, on a labour allocation sheet, the code of the work which each member of the gang has been working on and the time they spent on it. The amount of work done in each grouping or coding is obtained from site records. From this information, it is a simple matter to calculate the productivity.

Problems encountered in this technique include the difficulty of ensuring the accuracy of the information collected on the labour allocation sheets. Experience has indicated that more than two characters of code can yield unreliable information but two-character codes are usually felt to be insufficient for the purposes of reasonable control in general construction.

12.4 FINANCIAL CONTROL

Financial control involves measuring or estimating the financial parameters of a project, comparing them with target figures for such parameters and, if there is a disparity, taking action to improve matters in the future. Since much financial planning is based on inaccurate estimates, the information which accumulates over time can tempt managers to revise the financial plan as well as to initiate control action. Such revision of the financial targets should only be made with the agreement of senior management (for example, area managers or directors). If cost targets were allowed to rise without report, the viability of the project might be in danger and a timely opportunity of renegotiating the flow of revenue might be lost.

12.4.1 Re-active and pro-active financial control

In a well-managed project there are two principal mechanisms by which financial control may be exercised. The first of these is termed re-active or feedback control (Section 12.2.1) which is useful to maintain the overall targets of a project and for processes extending over significant periods of a project, such as large earthmoving and concrete placing. There is the opportunity to learn from experience (which is the basic meaning of feedback control). Many operations, however, have a duration commensurate with the reporting interval. By the time that control action can take place the damage has been done, the losses have been made and the disaster is complete.

What is termed here pro-active or feedforward control, attempts to solve this problem by ensuring that the operational plans and the current financial plan are coincident and substantially attainable. The control action on nonconformity is either to call for a revision in the operational plan of work to bring it into line with the financial plan or to recognise deficiencies in the financial plan and modify it. As the construction period for a piece of work approaches, there is usually an increase in awareness of the practical problems associated with it. One of the effects of pro-active control may be to cause a revision of a project's financial plans when cost targets, previously estimated and agreed, cannot be met.

12.4.2 Effects of delay

The time taken to assemble and report information has an important bearing on the effectiveness of control action. The time lag between control measurement and control action can make the effect of the action counter-productive. This is an extremely important factor in the control of a construction project particularly since formal reporting can often take weeks.

Using a simplified mathematical model of a control system for a project with weekly reporting, it can be shown that if there is no delay between measurement and control action, the minimum period of disturbance over which it is possible to exert some beneficial control is about 3 weeks; if there is a 1 week delay between measurement and action, the minimum period of disturbance which may be attenuated rises to about 9 weeks. Several days are normally required to assemble and process site measurements, so disturbances with a period of under 6 weeks are most probably not controlled by this mechanism, even under ideal conditions.

For day-to-day control action, considering such things as variability in morale, availability of work or labour and the need to ensure good working practices, a project manager must rely on enthusiastic and well informed site supervisors. The control system that they exercise should minimise the need for a lot of formal control action but their knowledge of the direction of and need for informal control action often comes from the formal control mechanism.

12.4.3 Pro-active financial control

The objective of pro-active financial control (Section 12.2.2) is to produce a timely and practical plan (for carrying out each activity) which conforms to the overall project plans and cost estimates and current knowledge of the project. The procedure for pro-active control is simple. Well in advance of construction the site manager submits a proposed method of work for carrying out an activity. This proposal should contain (at least) the duration of the activity and the number of hours of each resource that are estimated for completion of the activity.

Cost information can be calculated directly for the proposed resources. It is convenient at this stage to consider how costs might be saved on each activity and the cost implications of reducing the duration. For example, the cost of some concreting work using a concrete pump might, under certain conditions, be reduced to the cost associated with employing crane and skip at the expense of increasing the time to do the work.

The latest agreed plan is taken as the standard against which the proposed plan of work is assessed and against which agreement should be sought on all physical and financial parameters. The assessment must take into account the current knowledge of the project (which should have improved as construction proceeds). For example, if the original agreed plan was drawn up on the assumption that the water table was 10 metres below ground level and it has subsequently been found to be 2 metres below ground level, this could cause significant changes to the work method and the cost plan for activities involving work below ground. In some cases, pro-active control would involve designing a new work method for the activity; in other cases it would involve coming to a new agreement on the cost plan and duration for the activity.

Two points about pro-active control are often discussed:

The optimum performance of a project is usually not the sum of the activities all carried out in individually optimum ways It is important to utilise the resources on site fully and effectively and it may therefore be advantageous to perform some activities in a sub-optimum manner in order to optimise the whole project. This balancing between the optima for the individual activities and the project optimum is a complex task which should be undertaken by senior project personnel. It is suggested that this be considered at the regular project progress meetings (Section 10.6).

Using the latest agreed financial plan against which to evaluate the current work plan can be a problem as it may be different from the original estimate   This is good for project control as it sets the most realistic targets but, from a company viewpoint, it can hide variations from the original estimate which could give rise to wrong or insufficient control action.

12.4.4   Re-active financial control

The objective of re-active financial control is to determine the deviations of past progress from the project financial plans and to take direct action in the event of adverse deviations. The final objective is to achieve or improve upon the financial performance planned for the project. The components required for re-active control are:

- a financial plan (Sections 7.5 and 7.6) expressed either as the financial curves or as elemental parameters;
- a measure of project completion;
- measures of site cost;
- agreed methods of comparing performance with plan;
- methods to identify the areas of adverse performance;
- methods of improving future performance.

The last five points are discussed in the following section.

12.5   ASPECTS OF RE-ACTIVE FINANCIAL CONTROL

The aspects of financial control discussed below require some understanding of project finances and their planning. These topics are covered in Chapters 3 and 7. Types of financial control are covered in Sections 12.2 and 12.4.

12.5.1   Measures of completion

Because projects are complex, any single measure of a project by itself is liable to be misleading in some respects. For example, people may be measured by their height with great precision. This information is useful in many ways. If trying to buy clothes perhaps height is the single most useful dimension for general application but there would be limited confidence in the ability to choose a well-fitted suit using only this dimension. Several measures, not just height, are required for a comprehensive picture, and the same is true of a construction project.

The basic measurement quantity for a construction project is the amount of work completed. It is possible to make a global percentage estimate of this but it is likely to be fairly inaccurate and, hence, of little use for diagnosis and control action. Another measure of the state of completion of a project can be achieved by producing a detailed (measured) interim bill of quantities. This gives a more accurate picture but the cost of producing it and the time taken to produce it significantly reduces its usefulness as a routine measurement system for control.

A more practical measure of project completion can be produced by dividing a project into elements through which the physical and financial planning is carried out. The chosen elements may be types of work or activities.

Elements based on types of work have the advantage that they can be expressed in physical units. For example, if a type of work was *concrete placed*, the natural unit would be cubic metres and all concrete of high and low quality, mass and structural, would be lumped together into that category. This method produces the illusion of physical measurement.

Elements based on activities must be expressed purely in terms of percentage complete, estimated by someone on site. Estimates can be based on how much of the activity duration has been completed or how much of the activity value has been completed. For use in physical planning it is preferable that the percentage completion estimate is based on the duration of the activity.

12.5.2 Measures of construction cost

Some aspects of site cost are often beyond the control of site or project management. For example, when a subcontract has been agreed for erecting the fencing to a site, the cost of that work is fixed. Agreed prices for the supply of materials fix the unit costs, but project management must protect the project against the effects of waste, pilfering and non-delivery. In contrast, site managers can exercise relatively strict control of internal labour and plant costs. As they commonly comprise 40% of the total costs, the quality of control of these latter factors is crucial for control of the financial viability of a project.

Most companies operate a system that ensures that external expenditures arising from a project are attributed directly to that project. In other words, the total payments for labour, materials, subcontractors and external plant are known. The figures are often compiled monthly for the submission of invoices or for payment. They tend to be accurate but also a month or more in arrears so they are only of use in the control of very low frequency movements.

Records of the allocation of labour and plant can be made on a half-daily basis by site supervisors. With some training and explanation, a supervisor may make this allocation to the elements of work comprising a work plan. The workers' trade gives an indication of the broad classification of work on which they have been employed within an element; alternatively the supervisor may be trained to indicate the classification of work. Plant may be treated in a similar manner. If there are problems in collecting the allocation information, perhaps from literacy difficulties, a travelling cost clerk may be used to visit the supervisors and make the records. A cost clerk may improve the accuracy of the records by asking questions but the responsibility must remain with the site supervisor. The classification of costs is often done using a company-wide system consisting of detailed cost codes but this can be difficult to get right (Section 12.3.6).

After collecting site records, the number of worker- and plant-hours for each classification of each activity can be made available. This can be readily assigned to the appropriate heading and aggregated for each week or other control period. The total daily worker-hours for the project or section should be compared with the daily timekeeper's records of the workers on site, to ensure that the labour costs are covered and allocated as well as possible. A similar check on the total available plant hours is required.

The result of following these procedures is that a breakdown of the labour and plant costs attributable to each activity, split between main types of work, is produced. Although a degree of estimation is involved in producing the individual daily figures, there is overall conformity with the actual site costs and, with honest reporting, the figures aggregated over an activity and over a period are a more reliable estimate of the actual costs.

12.5.3   Comparing financial performance with plan

There are two main ways of assessing the financial performance of a project. The first and most important way uses the most recent revised, agreed expenditure pattern which is known as the budget. It takes into account all current project knowledge and the best available forecasts. The second way uses the expenditure pattern agreed at the commencement of the project which, although out-dated, is closely related to the recovery of money by way of interim accounts. Whichever method is used the expenditure pattern forms the baseline for the comparison.

The comparison of actual performance with the planned performance requires the definition of the term *budget* as an agreed, planned cost. At tender stage or upon the award of the contract, budgets may be prepared for the overall project and for individual sections; these are referred to as the original budgets. As construction proceeds, circumstances usually cause some revision of these planned expenditures, which are known as revised budgets. The revised budgets provide a realistic basis for comparison taking into account improved knowledge of the project and the original budgets provide a fixed basis of comparison which is strictly related to revenue for a fixed price contract.

VARIANCES

Departures from a baseline are known as *variances*. This is unfortunate since the term has a well-established definition in statistics and is often used to undertake studies of the uncertainties inherent in construction projects. Nonetheless it also has an established meaning in the financial context and it is used here in accordance with the definitions expressed in Figure 12.4.

The comparisons made to evaluate performance can be designed to suit the company and the project. Those discussed here are frequently used in practice and have proved valuable. Original figures carry the subscript o and revised figures carry the subscript 1.

Imagine that at some stage in the project the incurred costs are c, the latest revised budget was to incur costs b_l and the original budget for costs at this stage was b_o. The value of the work actually completed to date in terms of the latest revised budget is v_l and in terms of the original budget is v_o. As a result of experience to date the best estimate for the

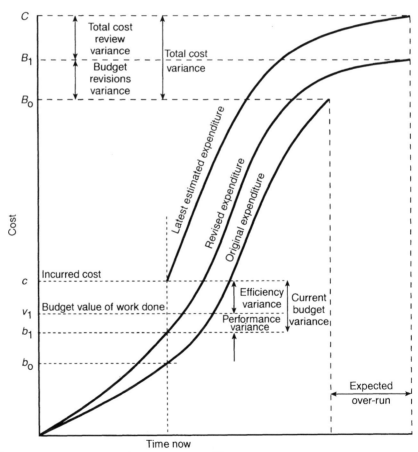

Figure 12.4 Project variances.

total cost of the works is C which may be compared with the latest budget cost B_l and the original budget cost B_o.

Three variances describing the current position of the project arise directly from these figures:

(1) current budget variance $= c - b_l$
(2) performance variance $= v_l - b_l$
(3) efficiency variance $= c - v_l$

Three more variances may be used to describe the present view of the state of the project at completion:

(4) total cost variance $= C - B_o$
(5) budget revision variance $= B_l - B_o$
(6) total cost review variance $= C - B_l$

A further variance, which is an obvious combination of the total cost review and the efficiency variances, serves to isolate the work that is still to be done and this is:

(7) future budget variance $= C - B_l + v_l - c$

These figures can be prepared for the overall project, for individual sections or types of work, or for different classifications of cost such as labour and plant if the financial planning calculations described earlier have been carried out and forecasts are prepared. When budgets have been prepared so that it is difficult to calculate the budget value of the work done to date, then an approximate estimate of this quantity may be formed, assuming that all costs have altered uniformly, as

$$\text{budget value of work done} = \text{incurred cost}$$
$$\times \frac{\text{previous estimate of total cost}}{\text{revised estimate of total cost}}$$

Where performance is reviewed at regular intervals, variances (1) to (3) may usefully be calculated for each period in addition to the cumulative form.

Variances as described for project control are very closely related to the use of variances as defined by management accountants for company control. They provide a wide view of a project and may be used at many levels. For example, they should be used by the owner, by the main contractor for long- and medium-term control (at intervals of 1–3 months) and by major subcontractors at similar intervals. Their use for short-term control, however, is problematical and since this is the most effective timescale for control action, contractors must use another method.

RATIOS

The ratio method for financial control should be used at all levels of the project organisation. It is straightforward to operate and provides information for both re-active and pro-active control. Its use is most valuable for short-term planning and control as demonstrated through an example in Section 10.4.

The following ratios are defined:

$$\text{planned performance} = \frac{\text{planned liability}}{\text{planned earnings}}$$

$$\text{actual performance} = \frac{\text{actual liability}}{\text{actual earnings}}$$

$$\text{efficiency} = \frac{\text{actual performance}}{\text{planned performance}}$$

These can be calculated either for total money or independently for individual cost heads such as labour and plant. If the latter is done, it can indicate that a different method of working has been employed to the one envisaged at tender stage.

It might be assumed, in an ideal world, that all performances and efficiencies would be 1. However, the real world is far from ideal and it is sometimes advantageous to have values other than unity.

By calculating the ratios and keeping a record of them over time, it is possible to see:

Effectiveness of the planner It is important to set realistic but slightly optimistic targets for the workers. If a planner sets too high a target, the efficiency is very low; if the planner sets too low a target, the efficiency is high (or at least 1 since workers tend to have a habit of not working beyond expectation).

General performance compared to the estimate This is of use to middle and higher management in predicting what the outcome of a project will eventually be.

Planned performance should be approximately 1 If it is greater than 1, the work is planned to cost more than the estimate; if it is less than 1, the work is planned to cost less than the estimate. Since it must be assumed that the estimators have considered the work and their suggested cost is realistic, any deviation from this is an indication of a need for pro-active control. This means checking the plan to see if it is correct. If the plan is accepted by all concerned as being correct, then it is against the planned performance that the site should be judged and not against the tender or estimated figures.

Actual performance should be approximately 1 The same reasons as indicated for the planned performance apply. A project is won on the understanding that the estimator's rates are correct; a value of performance greater than 1 indicates that the project will lose money. This is perhaps an indication that re-active control is required.

Efficiency is a measure of the workers against the plan If the plan has been accepted by all levels in the organisation, the efficiency indicates the need for re-active control since it shows the deviation from the planned performance.

These figures are very robust and can be used if the work changes during the period for which the plan operates either because of extra work or because less work is done. It is important to know that the correct amount of work has been done so, in addition to the ratios, it is necessary to monitor the amount of liability or earnings.

The ratios should be seen as indicators and their interpretation left to the project management. Suggesting that the ratios should be approximately 1 is debatable. A project manager may want the ratio for performance to be 0.5, that is, the project has actually spent only half as much as it has earned.

The ratios could also be expressed the other way up (earnings/liability or planned/actual). They could be interpreted just as easily as above and the ratio for performance would increase when the site performance increased.

12.5.4 Identifying areas of poor performance

Although it is necessary to ensure the overall financial health of a project, it is also desirable to improve a good performance so managers should be interested in rather more than the overall performance parameters of their project.

When a set of variances shows that a section of work can be improved there is a need for more detailed information so that a problem area may be located more precisely. If the data collection systems on the project are precise enough and a computer system is being used, it is possible that cost information at the next level of detail (under a work classification for the section) may be available on an exception basis. With or without this information, it is useful to bring together the responsible managers and supervisors to study the problem. Experienced project managers often pay attention to problem areas during their site inspections and are able to diagnose the source of the trouble. Detailed inspections are, however, undertaken somewhat infrequently and misleading conclusions can be drawn. Personal observation together with the contributed knowledge of subordinates appears to offer the best chance of defining the problem and finding a solution.

12.5.5 Improving future performance

In order to improve work performance the working methods frequently need to be redefined. Working practices will have evolved over time and it may require senior management authority to get these practices changed. A change in method requiring increased exertion or reduced payment could be expected to meet resistance.

In such circumstances there is some pressure on the project manager to ensure that the revised methods do in fact solve the problem; if it does not, the authority of the managers may be undermined. The techniques of method study and work analysis, together with the proper use of limited trials, may be used to ensure that the proposed new work method offers enhanced efficiency.

12.6 UNCERTAINTY AND RELIABILITY

12.6.1 Uncertainty of information

Since the cost of gathering information is high, some form of estimation is usually necessary. Estimation leads to a loss of accuracy which results in uncertainty. The uncertainty of any individual financial measurement on a construction project depends on the methods used to produce it, but experience typically points to a range of uncertainty of 10% when applied to an activity by good staff.

If an estimate equals the target figure then, given a symmetrical distribution of error, the outcome is just as likely to be over the target figure as under it. Assuming that the distribution of error is normal, the *standard error* is approximately 4% of the mean for a ±10% range, so if the estimate is 4% above the mean there is a 16% probability that the target is being met and exceeded despite the measurement. There would be a 1 in 6 chance that any control action was inappropriate and a decision between the costs of inappropriate action and the costs of inaction would be necessary.

On a construction project with tight margins, a real 4% cost slippage could be crucial and methods of improving the reliability of the data are required. The method used most often is analogous to the repeated observations made in land surveying to improve the reliability of an observation, but here the quantity is split into n sub-elements which are measured rather than making the same observation n times. This increases precision but is really a reflection of the compensating errors in the small estimates.

If the allocation of labour to an activity on a foreman's daily report sheet is inaccurate (which it certainly can be) it is still reasonable to expect the total allocation to an activity over a week to be rather more reliable. This assumes that the error is random rather than being a specific misrepresentation for other reasons (such as optimistic guessing).

Similarly, the earnings on an activity are subject to the errors in the estimates of completion of work and defects in the earnings model. Despite this the estimates of earnings over all on-going activities can be expected to be more reliable than the estimate of a single activity.

12.6.2 Reliability of information

The uncertainty inherent in information that is sincerely recorded is discussed above. One matter which is considered less often and is more difficult to assess is the possibility that those transmitting the information have reasons for supplying doctored figures. This may be done without any malicious intentions: for instance, if a project manager has had a run of good luck with weather and other factors and has made a good profit on the early phases of the project, but foresees a number of difficulties later on, it may seem convenient to apply the early profits to cover the later losses so that the project is reported as one making a uniform profit – even during difficult times. The manager's reasons for doing this are not laudable, but they are in one sense not dishonest as nothing is being taken away from the company in the long term. The reasoning is, however, based on a perception of the way in which company senior managers perceive the performance of site managers. Early successes may be overlooked when a subsequent problem develops.

In order for accurate reporting to be ensured, senior managers must have a policy – and must communicate it – that success is viewed on a global basis and that misreporting is a more serious breach of company rules than making a loss. Without reliable records, senior managers cannot exercise proper financial control.

12.7 CONTROL AND SUBCONTRACTORS

Subcontractors are widely used throughout the construction industry. It is considered that their use is beneficial for several reasons including:

- it removes the requirement to maintain a large work force in a rapidly changing industry;
- it consequently removes the requirement to operate a large human resources office with its sizeable overheads;
- it removes the requirement to maintain a work force skilled in a range of specialisations;
- it provides the opportunity to employ personnel who are skilled in one particular aspect of construction (such as bricklaying);
- it provides an opportunity for the contractor to pay when paid and hence improve the cash flow for the main contractor and reduce the risk on a project.

For these reasons some construction contractors operate almost exclusively through subcontractors. However, employing subcontractors to carry out work has many drawbacks including:

- Subcontractors may have little or no commitment either to the project or to the main contractor and will readily change their allegiance.

- The line of control between operatives and management may be longer than using own labour and could consequently be less efficient.
- The use of subcontractors can increase the management problems on a project because it creates interfaces (a major cause of problems) between organisations.
- The use of many subcontractors can cause many such interfaces both between the main contractor and the subcontractors but also, and more importantly, between the subcontractors themselves.
- Problems caused by one subcontractor can cause large consequential problems for other subcontractors and the main contractor, who will almost certainly not be able to claim the costs of these problems from small subcontractors.
- The skilled operatives of subcontractors may be more expensive to employ on a time basis.
- The subcontractor has to have its own management and this must be paid for.
- The subcontractor's management may not be skilled in planning and controlling their work.

With such persuasive arguments for and against the use of sub-contractors, main contractors will continue to make their own decisions based on how they balance the arguments for and against. If subcontractors are used, it is essential for the main contractor to plan their work to integrate with all the other work on the project. This requires at least as much detail to be included as for the planning and control of the main contractor's own work.

The simple rule for the main contractor is:

Irrespective of who is actually doing the work, plan and control it as though it is your own work.

12.8 UPDATING PLANS

As work progresses on a construction project, it is inevitable that there will be deviations from the original plans of work. Progress will not be as expected, alterations will be required either by the contractor or the client and more information will become available. As the deviations get larger and the information becomes more certain, the manager must consider the usefulness of the original plan and ask whether the plan needs to be changed.

12.8.1 When to update

There are no rules regarding when a plan should be updated but the quality statement for a project should state when a plan is to be reviewed. The suggestions in Table 12.1 may be useful.

Table 12.1 When to update plans.

Type of programme	When to update
Short term	weekly on a rolling programme basis reviewed daily
Medium term	monthly on a rolling programme basis reviewed weekly
Long term	(1) 3 monthly (2) when the deviations from the existing plan are such that the existing plan is meaningless (3) when the client requests an update

Table 12.2 Updates and responsibilities.

Element to update	Cause	Responsibility
Activity list	extra work required by the client	client
	extra work caused by the contractor	contractor
	work missing from original plan	contractor
	more detail available	contractor/client
	more detail required	contractor/client
Activity duration	extra work required by the client	client
	extra work caused by the contractor	contractor
	work missing from original plan	contractor
	more detail available	contractor/client
	more detail required	contractor/client
Fixed dates	restriction imposed by client	client
	restriction imposed by contractor	contractor
	restriction imposed by outside authority	client/contractor
Logic connections	restriction imposed by client	client
	restriction imposed by contractor	contractor
	restriction imposed by outside authority	client/contractor
	omission from original plan	contractor
	error in original plan	contractor
Resources	extra work required by the client	client
	extra work caused by the contractor	contractor
	work missing from original plan	contractor
	more detail available	contractor/client
	more detail required	contractor/client
	better productivity information available	usually contractor but could be client
	better availability information available	usually contractor but could be client
Finances	extra work required by the client	client
	extra work caused by the contractor	contractor
	work missing from original plan	contractor
	more detail available	contractor/client
	more detail required	contractor/client

12.8.2   What to update

There are various elements of a plan, all of which may need updating. Table 12.2 indicates the need for alterations to the elements of the plan and who may be responsible.

Updated plans should be used for all the tasks that original plans are used for, namely:

- to provide timing information,
- to muster resources,
- to set targets,
- to provide control information,
- for pro-active control,
- to produce budgets,
- to avoid problems,
- to communicate up and down the management hierarchy,
- to communicate with other organisations.

Invariably there are differences between two generations of plans and these differences are important. The original plans are the ones on which the project was won and the prices based. Deviations from it must be recognised and explained. If at all possible, differences must be quantified in terms of time and money and the relevant parties charged.

12.9   SUMMARY

- Monitoring and record keeping are important functions for project control and should utilise all means available in order to be more likely to yield the right information when it is required.

- Although carried out in the short term, records are required for various purposes throughout and after a project.

- It is inadvisable to rely too much on memory for record keeping.

- Diaries and record drawings are important for providing continuity of progress reporting.

- Inappropriately marked-up bar charts are useless records for future use.

- Graphical progress charts can be good for control.

- Progress charts can be drawn for any key resource, including financial and abstract resources.

- Activity days or weeks can be particularly useful measures of progress.

- Re-active control seeks to change the performance of the resources on the project to meet the plan better.

- Pro-active control seeks to change the plan to be more realistic in the first place.

- Re-active and pro-active control should exist together on a project.

- The time-scale involved in its operation means that re-active control can only really be of use for work of long duration.

- Control of any type requires reliable information to be collected.

- The information used for control can be used for many other purposes (for example, claims analysis and updating estimating information) if it is in the correct form.

- The optimum performance of a project is often more than the optimum performance of the normally recognised individual parts of the project.

- Variances are good for overall control especially by clients.

- Variances can be developed for categories of work or sections of a project.

- Variances are really only useful for medium- and long-term control.

- Ratios provide a measure of plans and lead to better planning.

- Ratios enable rapid control action to be taken.

- Ratios can be applied for long-, medium- and short-term control.

- Variances and ratios should both be applied pro-actively and re-actively.

- Not all information for control is reliable.

- Controlling subcontractors is more difficult than controlling one's own resources.

- When updating plans it is important to use the right plan for the right purpose.

13

CHOICE AND EVALUATION

'You pays your money and you takes your choice.'

ANON.

This should really be the other way round. In construction planning you choose your technique and produce a plan before a huge commitment to expenditure when you execute the work. A bit more time spent choosing the best way to plan, then checking and evaluating the plan would perhaps cost a little more in the planning but might lead to considerable cost savings in construction. A good plan would leave a site keen to execute the work, not the planner.

13.1   INTRODUCTION

This chapter covers two very important aspects of practical planning. The first of these is the selection of planning technique. With modern computer packages concentrating almost exclusively on a small range of network-based methods, it would appear that the choice is now becoming a choice between planning packages rather than techniques. This chapter shows that a positive choice of technique should be made, bearing in mind the objectives of planning at various stages in a project, the applicability of the various methods, and their strengths and weaknesses. A large number of planning techniques are covered but, since new variants are continually being produced, the list of those included is not exhaustive.

The second aspect covered in this chapter is the evaluation of plans produced by any method. Plans are often produced and then used without enough thought being given as to whether they are correct or how good they are. In Section 13.3 the need for the evaluation of plans is discussed and methods for considering both the correctness and the goodness of a plan are described.

13.2 CHOICE OF PLANNING TECHNIQUE

Considerable research has been done on the techniques available for project planning and control. Some of the techniques are so well established as to have standard documents for their definitions and use. Others are less generally accepted and consequently less widely written about. This section contains tables describing some of the techniques, their application, their benefits and their drawbacks. It also suggests some points which should be considered when choosing the method(s) to apply in any situation.

13.2.1 Types of planning

Planning for a project is carried out by many people at many stages of the project in order to achieve many different objectives. The objectives and those responsible (for a contractor) are summarised in Table 13.1.

13.2.2 Planning techniques and their application

Table 13.2 gives an outline of the main features of the more common project planning techniques used in construction. It also provides an indication as to when the techniques might be applied. Many of these techniques are described and discussed further in Chapters 2, 5, 7 and 8.

There are 24 techniques listed in Table 13.2. Several could be chosen in many situations. More than one is usually used on any project. To help choose a technique their strengths and weaknesses are now discussed.

13.2.3 Strengths and weaknesses

Table 13.3 provides a summary of the advantages and disadvantages of the main techniques.

Table 13.1 Objectives of types of planning

Type of planning	Planner/responsible person	Objectives
Pre-tender	head office specialist estimator	determine the phasing of the work to be done and the major temporary works requirements allow pricing of the work more accurately than by the application of standard rates predict cash flow
Contract long term	specialist planner, contract manager/agent	set overall work method for the project ensure that final and intermediate dates are met allow manager to muster resources predict cash flow allow pro-active control of time, resources and finance
Medium term phase	section agent, section engineer, agent	provide more detail and update the long-term programme
Short term	section engineer, section agent	make use of available resources to achieve the objectives of the long- and medium-term plans
Post-completion	contract manager, agent quantity surveyor	evaluate the entitlement in terms of time and money

Table 13.2 Application of planning techniques

Planning technique	Main features	When used
Activity list	the basis of most other planning techniques. A list of all the activities to be carried out. Might also contain times for carrying out the work and resources required	for simple projects or parts of projects, at very outline stage or for great detail
Bar chart (Gantt chart)	chart shows activities against time at which they will be done, can show the float	pre-tender planning, tender planning, contract planning, section planning, phase planning, simple projects, to show results of other more complicated techniques (such as networks)
Linked bar chart (linked Gantt chart)	chart shows activities against time at which they will be done together with some of the logical connections between the activities	tender planning, contract planning, long/medium-term planning, section planning, simple projects
Cascade chart	a partially linked Gantt chart in which the activities have been put in such an order that earlier ones are shown at the top and later ones are shown at the bottom	tender planning, contract planning, long/medium-term planning, section planning, simple projects, to show results of other more complicated techniques (such as networks)

(Continued)

Table 13.2 (*Continued*)

Planning technique	Main features	When used
Resource bar chart	chart showing the activities which each resource or gang of resources will be doing at any time	short-term planning, all types of project
Arrow diagram	a network showing activities on arrows between nodes which are events representing instants in time. Used to model the fact that one activity cannot start until other(s) have finished	contract planning, long/medium-term planning, complex projects
Ladder diagram	an extension of arrow diagrams that allows the modelling of overlapping activities by use of non-zero duration dummy activities	contract planning, long/medium-term planning, complex projects, projects with overlapping activities, repetitive activities (linear projects)
Activity-on-node diagram	a network that shows activities in boxes (or nodes) linked together with arrows representing the logical relationships between the activities	contract planning, long/medium-term planning, complex projects
Precedence network	an extension to activity-on-node networks using four different types of connection between the starts and finishes of the activities	contract planning, long/medium-term planning, complex projects, projects with overlapping activities but not very suitable for building finishes, repetitive activities (linear projects)
Line of balance	a target graph showing rate of completion of a series of similar units and a bar chart for the completion of an individual unit. Combined to give targets for completion of activities of the units throughout the project. Considering resources allows the planner to balance the speed of completion of the different activities	repetitive projects, linear projects, long/medium-term planning
Time-chainage chart	combination of line of balance and Gantt chart with the activities plotted on a chart showing time along one axis and distance along a road (chainage) along the other	road projects, linear projects, long/medium-term planning, contract programmes, tender programmes
Progress chart	graph showing one measure of completion (for example, height of an earth fill dam) against time	simple projects, projects with one major work type
Resource aggregation	summing of the resource requirements of all the activities for each time period	all projects with Gantt chart, network or time-chainage chart

(*Continued*)

Table 13.2 (*Continued*)

Planning technique	Main features	When used
Resource levelling	evening out the use of resources by moving the activities in time	all projects with Gantt chart, network or time-chainage chart
Resource smoothing	evening out the use of resources by moving the activities within their float time	all projects with Gantt chart, network or time-chainage chart
Resource-oriented scheduling	models the construction process by concentrating on the use of resources rather than on activities	used in addition to network techniques when resources are considered to be the main planning restriction over part of the project
PERT	network-based technique that allows the consideration of activities with uncertain duration. Gives probability of completing project by any date	long term planning, project evaluation, risk analysis
GERT	network-based technique that allows modelling of choices within the network logic	long term planning, project evaluation, risk analysis
Burman's doubtful outcome analysis	alteration to network model that allows modelling of activities which have uncertain outcome or uncertain duration	long term planning, project evaluation, risk analysis
Roy's method	a network-based technique that allows overlapping and tied activities	as standard network techniques
Method statements	description of how each piece of work is to be carried out including timing, resource and spatial considerations	throughout all projects
Simulation	computer modelling of project (network techniques are specific versions of simulations)	ill-defined projects
Monte Carlo simulation	usually network-based techniques in which the duration, cost and resources of an activity are defined as distributions. The project model is analysed many times to provide a statistical view of the project outcome	project evaluation, risk analysis
Mass haul diagrams	diagram showing the movement of earth in volume and distance terms for a project	road projects, linear earth-moving projects

Table 13.3 Strengths and weaknesses of planning techniques

Technique	Strengths	Weaknesses
Activity list	the fundamental planning tool	no relationships between activities shown; no durations
Bar chart (Gantt chart)	well understood, simple, good for communication	does not show relationships between activities; cannot represent uneven rates of progress on an activity
Linked bar chart (linked Gantt chart)	quite well understood, simple, good for communication	confusing when many connections are included; cannot represent uneven rates of progress on an activity; confusing when float shown
Cascade chart	good for communication	confusing when many connections are included; cannot represent uneven rates of progress on an activity; cannot be easily produced for simple projects; confusing when float shown
Resource bar chart	ensures that all the resources are fully utilised	difficult to ensure that the programme achieves the objectives of the long/medium-term programme
Arrow diagram	easy to computerise, can produce Gantt chart, shows simple relationships between activities	training difficult; only allows FS connections; use of resource connections both solves problems and causes them; cannot represent uneven rates of progress on an activity; dummies difficult to explain to some levels of staff; cannot model process type activities
Ladder diagram	easy to computerise, can produce Gantt chart, appears to allow complex relationships between activities to be modelled	training difficult; difficulty of interpretation; use for monitoring; cannot represent uneven rates of progress on an activity; not good for communicating timing of activities and resource demands
Activity-on-node diagram	easy to computerise, can produce Gantt chart	training difficult; only allows finish–start connections; cannot represent uneven rates of progress on an activity; use of resource connections both solves problems and causes them
Precedence network	easy to computerise can produce Gantt chart, appears to allow complex relationships between activities to be modelled	training difficult; difficulty of interpretation; use for monitoring; cannot represent uneven rates of progress on an activity; not good for communicating timing of activities and resource demands
Line of balance	good for projects which contain many identical units; provides completion targets for all activities	training difficult; variations in individual units causes problems in the analysis; float in the programme for the individual units causes problems with the analysis; projects which are part repetitive and part not cannot be modelled easily; learning effects are neglected; not very commonly computerised

(Continued)

Table 13.3 (*Continued*)

Techniques	Strengths	Weaknesses
Time-chainage chart	good for communicating; can be shown against a longitudinal section of the project; shows areas of conflict very clearly	difficult to read for nonexperts; not usually computerised; cannot represent activity float; variable rates of work are difficult to model; cannot represent intersections easily
Progress chart	very simple and clear	rarely one and only one major measure of a project; difficulty arises in change-over from one leading parameter to another
Resource aggregation	shows resource demands	
Resource levelling	produces a plan of work within specified constraints	does not produce best answer, only better than average; very time consuming; balance between different resources difficult to achieve; variable resource demand activities difficult to model
Resource smoothing	produces a plan of work within specified constraints	does not produce best answer, only better than average; very time consuming; balance between different resources difficult to achieve; variable resource demand activities difficult to model
Resource-oriented scheduling	models what is really happening with resources on sites	not well recognised, not well developed
PERT	easily computerised; provides an estimate of an uncertain project end date	does not provide start and end dates for activities except as in ordinary networks; assumptions on activity uncertainty have little or no basis in reality; floats in uncertain networks have little or no meaning; assumes activities are statistically independent
GERT	easily computerised; allows modelling of uncertain process	does not provide start and finish times for individual activities; difficult to explain; not well established
Burman's doubtful outcome analysis	models an uncertain process in a simple manner; when used in conjunction with ordinary networks it may help assign realistic durations to activities	not well established
Roy's method	allows modelling of tied activities in a sensible manner	not well established, no software available
Method statements	draws together information for all aspects of the work; should ensure that all planning has been done	
Simulation		difficult and expensive to set up

(*Continued*)

Table 13.3 (*Continued*)

Techniques	Strengths	Weaknesses
Monte Carlo simulation	good for evaluating projects	requires many runs (hundreds of thousands for projects with several uncertainties); does not provide start and finish times for activities
Mass-haul diagrams	provides a first attempt at earthworks phasing and hence at construction phasing	does not take into account the differing haul conditions along a site; does not take into account the differing haul conditions for full and empty lorries; difficult to take into account the effects of haul on construction phasing

13.2.4 Choice of technique

Plans of work for construction projects appear to fail for many reasons. People blame many things including:

- the general uncertainty inherent in a project;
- changes induced by the client;
- the poor performance of a major subcontractor;
- the ground conditions which were not as predicted.

They even sometimes admit that the plan was not very good in the first place. One thing that rarely gets blamed but in fact is often the root cause of the problems is the choice of planning technique.

When deciding on the techniques to be used in the planning and control of a construction project the choice varies from company to company and from project to project. There is rarely one and only one correct choice. The list below contains points which should be considered before a choice is made. The order is not meant to imply any order of importance as this may change from time to time and from project to project. A manager should at least have considered these points and be able to justify the final decision:

- the background and experience of the planner;
- the background and experience of the client;
- the wishes of the client;
- the availability of computer software;
- the company standard methods;
- the type of project;
- the size of the project;

- the duration of the project;
- the complexity of the project;
- the time available for planning;
- the objectives of the plan;
- the site organisation.

13.3 PLAN EVALUATION

Project planning is a simulation process in which various planning techniques are used to create a model of a project and to analyse how the model behaves under certain conditions that might arise. In standard simulation procedures, the most important step to be carried out is a check that the model provides a sensible representation for the range of anticipated conditions. In the case of project planning this checking procedure is termed plan evaluation.

13.3.1 Evaluation considerations

Project plan evaluation generally occurs in two situations:

- assessing the nature of the proposed plan;
- reviewing and updating existing plans.

In reviewing existing plans, the evaluation process (Figure 13.1) should be an essential part of the planning process in the planning and control cycle (Figure 1.1). Figure 13.1 shows that as work progresses on site a series of monitoring procedures records the progress. The nature of the monitoring procedures differs from site to site and from project to project. The information recorded is used to update the plan to the current position. If any deviation has occurred from the initial plan a new plan will effectively have been created. The updated plan should be evaluated to check that it is satisfactory. If it is not, a re-plan should be considered. This, in turn, must be evaluated. This form of evaluation compares the differences between the current progress of the work (an as-built plan) with the estimated progress as given by the current plan.

The assessment of proposed plans (pro-active control) is shown on Figure 13.1 for the pre-construction stage of the planning and control cycle. This form of evaluation may be either comparative, that is, comparing the characteristics of two proposed plans, or

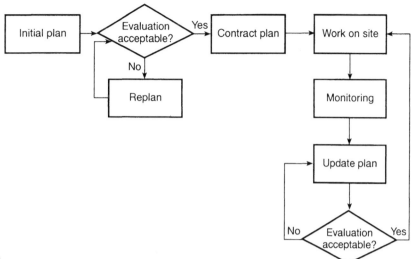

Figure 13.1 Plan evaluation

quantitative, where the merits of a single plan are examined against objective criteria.

Evaluation in the pre-construction phase of a project may be used to illustrate where the sensitivities lie in the plan. This in general creates an iterative process in which the initial plan is evaluated and then altered, if necessary, to produce a more satisfactory plan. This replanning cycle is unlikely to lead to the best solution, even if the nature of the best solution could be defined.

Such evaluation has great value in the post-completion phase of the project where the basis of contract claims (Chapter 14) can be illustrated by the analysis of the plan under various scenarios.

13.3.2   The need for plan evaluation

As trends in the construction industry develop towards larger and/or more complex projects, regulated under a wide range of different forms of contract, a greater awareness of the need for project plan evaluation has emerged. If good control over complex projects is to be achieved, a sound understanding of the nature of the project and its susceptibility to likely external factors is invaluable. This may be achieved through evaluation.

These trends in the construction industry combined with powerful computing technology has led to a large research effort towards the development of automated, intelligent planning packages. These packages often employ artificial intelligence (AI) computing techniques, and are more commonly termed *expert systems*. They are designed either to replace or aid the construction planner in the planning process. The introduction of such technology may remove some of the planner's knowledge of the project, thus making errors and omissions more likely. The formalised checking procedures of project plan evaluation may highlight these errors.

With many possible project procurement systems, and thereby different forms of contract, a cautious attitude has developed to the impact of particular contract conditions. In, for example, the fast track form of contract, construction work commences before all the detailing of the design has been completed. Although this leads to a shortened period between the project inception and the final handover to the client, the scope for errors and omissions is greater than in a more traditional form of contract. Good procedures for checking and evaluating plans are highly desirable in such circumstances.

13.3.3   Criteria for evaluation

All project plans are different, even for the same project. In general, it would be expected that a plan would contain a programme of the expected progress of activities on site and a financial model of the project. As the evaluation of a plan should, in theory, cover every possible factor that could conceivably affect the project, a large number of criteria for evaluation present themselves.

For example, the time analysis of individual activities (under the restrictions imposed by the logical connections) can be used to produce an overall expected duration for a project. This estimated duration may be altered by an almost endless list of possible factors:

- any unexpected delays in the project;
- poor activity duration estimates;
- changes in the sequencing of the activities;
- the expected mood of the project manager on the day.

Any project has different susceptibilities to changes in the different factors. Such susceptibilities are highlighted if a full evaluation of a plan is performed. In a more practical environment, it is only possible to consider a limited number of criteria. Some of these criteria are listed below, with typical questions that demonstrate their use.

Time considerations

Start and finish times for the project	– Does the project fit within the allowed time?
Activity durations	– Are the activity durations realistic for the tasks being performed?
Fixed periods/fixed dates	– Are any fixed periods or fixed dates being violated?
	– Are earthmoving activities in the winter shutdown period?
	– Is there any violation of contract periods?

Money considerations

Earnings and income	– What are the expected and actual rates of income and earnings?
	– How do they compare?
	– What is the lag between the two?
	– Is the project on schedule?
Liability and expenditure	– What are the expected and actual rates of liability and expenditure?
	– How do they compare?
	– What is the lag between the two?
	– Is the project on schedule?
Cash flow	– Is there a positive cash flow?
	– How do the expected and actual cash flows compare?
Profitability	– Is the project making a profit?
	– Will it make a profit?
	– Will the profit be as large as expected?
	– What is the percentage profit versus turnover for (a) the project or (b) the company?
Delays in payment	– Are the delays in payment as expected?
	– Will the profit increase if payment is delayed longer?
	– Will the project suffer if payments are received later than expected?
Inflation and interest rates	– What will be the effect of increased/ decreased inflation or interest rates?
Front-end loading of earnings	– To what extent has the project been front-end loaded?
	– Will a significant change in the quantities in a bill of quantities have a benefit?

Materials and resource considerations

Continuity of use for resources reusable materials	– Is there a good continuity of plant and usage?
Multiple use of plant (using plant for more than the task they were acquired for)	– Can the types of plant be reduced by dual usage?
Smoothing of resource usage	– Is the resource usage smoothed? – Could it be done more effectively?
Actual level and comparison to the planned level	– Are the resource levels as expected, comparing to both the planned resource schedule and the implied plant from the bill of quantities?
Implied productivities	– What are the implied actual productivities? – How do these compare with the expected/planned productivities? – What effect will a small improvement or deterioration have?

Logic considerations

General logic	– Is the plan logical? – Can the project be built this way using the technology available?
Types of logic being employed	– Is it all true logic or is there resource logic which may constrain the plan?
Bottlenecks in the plan	– Are there any highly important activities?

Completeness of a plan

– Is the plan complete?
– Does it contain all the necessary work?
– What level of detail does it contain?
– Are there any areas with significantly more or less detail than average?

Safety considerations

– Is the proposed method of work safe for all the operatives involved?
– Is the proposed method of work safe for the public?
– What safety equipment is necessary?
– Is the necessary equipment available?

Quality considerations	– Can the required quality of workmanship be achieved?
	– Are the right working methods being used?
	– Is the sequencing right?
	– Are any protective measures necessary for finished work?
Environmental considerations	– Are the working methods sympathetic to the environment?
	– Are extra protective measures necessary?
	– Can care for the environment be demonstrated?
	– Are disruptions minimised?

From this list, which is far from exhaustive, it is apparent that there are a large number of criteria that will affect the project if they are altered. The degree of this effect depends on the response of the project to the alteration. The effect may be unnoticeable or conversely the resultant problems could be virtually insurmountable.

The benefits that can be gained from the evaluation of project plans are considerable yet traditionally, within practical planning exercises, the level of manually performed evaluation is limited. This reluctance may be due to:

- the large volumes of data that have to be processed for large projects;
- a lack of standard methods for evaluating projects;
- human nature – few people enjoy undergoing or undertaking large-scale rigorous checking procedures;
- a lack of understanding of where problems may exist.

As automated planning systems are introduced, the desire for similar plan evaluation systems will increase. This in turn will generate many more standard methods for evaluation of the various criteria.

With the growing emphasis on quality assurance and better computing power available, there is likely to be a rapid growth in computer-intensive, automated plan evaluation.

13.4 SUMMARY

- No single planning method is the best in all situations.

- Different techniques have different strengths and weaknesses.

- A plan should be drawn up to achieve some objectives.

- Evaluating an individual plan of work is difficult.

- Plans should be evaluated for correctness and goodness.

- It is possible to evaluate a plan for correctness by ensuring that the necessary construction logic is obeyed and that the objectives are achieved.

- The goodness of a plan is more than checking for the achievement of the objectives and is a mixture of many things, some of them objective and some of them subjective.

- When evaluating plans one should take into account:
  - time considerations
  - financial considerations
  - material and resource considerations
  - logic considerations
  - the completeness of the plan
  - safety
  - quality
  - environmental considerations.

14

EVALUATING EXTENSIONS OF TIME

'Felix qui potuit rerum cognoscere causas.'

VIRGIL
(*Georgics*)

In English: 'Lucky is he who has been able to understand the cause of things'. Even luckier is he or she who has been able to find a way to record all appropriate information, quantify realistically the effects (delays) and put forward a quantified case for or against a claim for an extension of time. It would be nice if luck were not needed and all parties could agree on applicable methods for assessing the knock-on effects of the cause of things.

14.1 INTRODUCTION

Extensions of the time allowed for carrying out a project are important considerations because of the nature of the contract (Section 11.6) which forms the basis of so many construction projects. Planning techniques are frequently used on a project for the evaluation of possible extensions of time caused by changes to the project. There are many problems inherent in this. Some problems arise because the techniques were not developed for retrospective analysis, other problems arise because the techniques are wrongly applied. This chapter describes, with illustrative examples, some methods which can be applied to evaluate extensions, the problems which might be encountered and possible solutions to them. It should be remembered that the calculated extensions of time should be used as negotiating positions rather than statements of extensions. The information contained in the chapter is equally valid for clients and contractors.

The objective of using any technique to evaluate a possible extension of time is to produce a model of the project which, when it has alterations imposed on it, behaves like a project managed by an experienced contractor.

14.2 NETWORK-BASED TECHNIQUES

Network techniques (Chapter 5) are often used for evaluating extensions of time because they appear to give a scientific basis to the process and are generally accepted as providing a reasonable model of reality on a range of (but not all) construction projects. There are, however, several significant problems which must be taken into account in their application. These are discussed in the remainder of this section before specific techniques are introduced.

Basic network planning theory assumes an unlimited number of resources are available when constructing a network and calculating the critical path. Most networks for construction are drawn assuming a limited number of resources and therefore must include resource constraints along with the technological constraints between the activities (Section 5.8.1). Under these circumstances, the apparent critical path is resource-constrained and the meaning of a critical activity, as commonly used, does not signify that the project is necessarily lengthened if the activity is delayed or extended – it normally means that the resource assignments must be changed if the activity is delayed or extended. This is usually a much easier management task than speeding up a project that shows a delay on a theoretically correct non-resource-constrained network.

Networks rely on the activity models and the links between them. These links must be sufficiently good models of the reality of the construction process to allow changes that are made to the activities and the links to be used for predicting the effects of changes on the real project. For many situations this is reasonable but, in some situations, such as when there are large numbers of interacting resources on a project or where there is insufficient or too much detail in the network model, the reliance on the model is misplaced.

In some of the more complex network models which use precedence connections and/or multiple calendar effects, the assumptions regarding the nature of the activities and the links between them determine the answers obtained in any calculations carried out. The behaviour of the model is not necessarily what would be expected by the experienced construction professional.

Three basic techniques have been proposed for using networks to evaluate extensions of time. These are briefly described and discussed below. All other methods are variations of these. Whichever technique is used, it is important to point out that any technique is exceedingly unlikely to give the correct answer, if indeed there is a correct answer. What can be expected is that the chosen method gives a reasonable approximation to reality.

14.2.1 Addition method

The addition method is a commonly employed method of analysis, although it is rarely written about in detail. The method assumes that the critical path duration is the project duration. Such an assumption is, in general, not realistic in non-resource-constrained networks and the inclusion of resource constraints causes other problems as discussed below. The steps involved in applying the addition method are now described.

1 PRODUCTION OF NETWORK EQUIVALENT OF THE MASTER OR CONTRACT PROGRAMME

The contract programme is used as a basis from which to calculate an extension of time, although any other programme which indicated the original concept for the project could be used. Even on quite large and complex projects, it is common to have contract programmes in non-network form and in relatively little detail. This is claimed by contractors to be because of the lack of detail available and shortage of time at the start of the project. It may also be that contractors put too little effort into initial planning or judge that there is more to be gained in claims by the submission of an outline programme.

In order to apply alterations to a programme and to evaluate their effect, a network-based programme must be produced. This programme must incorporate the thinking of the initial programme and not just give the same answer that the contract gives for the key dates of the project. It should incorporate the timing, resources and finances of both the individual activities and the programme as a whole and it should reflect the restrictions originally envisaged in the activity connections.

The network equivalent of the contract programme must contain sufficient detail to allow all the recognised variations to be imposed in a realistic manner.

2 IMPOSITION OF VARIATIONS TO NETWORK EQUIVALENT CONTRACT PROGRAMME

The variations which have occurred on the project should be analysed both individually and collectively to determine their effect on the activities, the project resource demand and the network. When the analysis has been done, the results can be imposed on the network in any or all of the following ways:

Alteration of activity duration This is perhaps the most common alteration. It can be used to model the situation of a change in the amount of work in the activity under a constant activity resource demand (Section 4.4). The duration used for an activity should not necessarily be the actual duration of the activity since this would imply that the change in duration was all accounted for by the variation being considered. This is not always the case. For example, the productivity of the resources implied in the original programme might have been in error or the resource productivity might have been affected by many small variations. A change in activity duration may result in the need to change the network connections.

Alteration of network connections This models a change in work method which may, for example, be brought about by changes imposed by the architect or engineer. If the network is correct and contains only technological constraints, the need for changing connections to model changes in method will not happen as frequently as when resource constraints are included. If activity durations are changed and the lags (Section 5.4) in the network are used to model work on activities rather than the delay between activities, it is often necessary to change the durations on the lags.

Alteration of activity resource requirements This models a change in work content of an activity or a method of carrying out an activity. Care must be taken to ensure that changes in duration and resource content are compatible.

Alteration of fixed dates This models the change of availability of such things as areas of the site and information or other imposed restraints.

Alteration of project resources This models the management action taken by a contractor to achieve the objectives of the contract.

3 ANALYSIS OF UPDATED NETWORK

A network analysis is performed to provide a new project duration, resource demands, financial budgets etc. The critical path is deemed to define the contract duration.

PROBLEMS WITH AND POSSIBLE SOLUTIONS FOR THE ADDITION METHOD

Some of the problems with the addition method are now described:

- The results of the analysis of the updated network do not necessarily bear any resemblance to what happened on the project. Supporters of the addition method argue that it is intended to show an entitlement not an actual extension.
- Many of the variations do not have the desired (or indeed predictable) effects if mitigating action is not included.

- Resource constraints in a network give wrong answers for the extensions of time and may cause projects to be held up for obviously insignificant problems.
- Having no resource constraints in a network gives a ridiculously short duration for the contract network's critical path, unless other scheduling procedures are employed.
- The level of detail in the contract network determines the alterations which can be imposed in a sensible manner.

Bearing in mind the above points, an analysis of the addition method for reasonableness should consider the following points:

- A reconciliation between the original and updated contract programmes at activity and resource level.
- A logic listing for the network of the original and updated contract programmes including an explanation of the types of links (technology, resource).
- A list of the fixed dates used in the original and updated contract programmes with reasons behind them.
- The resource outputs assumed in the original and updated contract programmes.
- The resource demands of the activities in the original and updated contract programmes.
- The project resource profiles for the original and updated contract programmes.
- A list of assumptions in the network analysis used for example, splittable or non-splittable activities, work or delay lags and whether SS and FF lags together imply progress constraint throughout the activities (Chapters 4 and 5).
- An analysis showing how the original and updated contract programmes actually took space and access into account. This could be in the form of profiles or histograms for space resources.

14.2.2 Subtraction method

This method is an attempt to work backwards from an as-built programme, removing the identifiable effects of variations caused by one party and leaving a programme which represents the original contract programme and the effects of the variations caused by the other parties to the contract.

The steps in the subtraction method can be summarised as follows.

1 PRODUCE AN AS-BUILT PROGRAMME FROM SITE RECORDS

This would normally mean interpreting the project records and constructing a bar chart but there is no reason why this format should be used. Other representations such as time-chainage charts would be equally valid. The level of detail of this programme would be determined by site records.

2   PRODUCE AN AS-BUILT NETWORK FROM THE AS-BUILT PROGRAMME

This step requires the conversion of the as-built programme into a network. If representation techniques other than a simple bar chart have been used in the first step, this step will need considerable care. It is essential that all the logic links are included. No indication was originally offered with this method as to whether or not all types of connections (technology, resource etc.) should be treated equally. It is suggested that the different types of connection should be treated differently.

3   REMOVE FROM THE AS-BUILT NETWORK THE EFFECTS OF THE VARIATIONS CAUSED BY ONE PARTY

This can be done in one of four ways:

- alteration of activity durations
- alteration of activity resources
- alteration of project resources
- alteration of fixed dates.

PROBLEMS WITH SUBTRACTION METHOD

There are several potential problems with this method but the main ones are:

- Production of the as-built programme in enough detail from the project records is often remarkably difficult.
- Production of the as-built network is a great problem as it is necessary to decide what connections to put in. Resource constraints are required in order to ensure the as-built programme is achieved, but these give problems when removing variation effects.
- Parallel delays can occur and not be recognised by the analysis.
- Resource effects are not taken into account in any reasonable manner.

14.2.3   Stage addition method

The staged addition method is based on the addition method but contains alterations to try to overcome many of the potential problems of that method. The objective is to model what happened in reality, including the management decisions and mitigating action. The method is the addition method applied at control points throughout a project with the project replanned from those points to the end. This mimics the idealised running of a project in which the project management assesses the status of the project at regular intervals (for example, monthly meetings) and takes corrective action.

The steps in the staged addition method are:

1 PRODUCE A CONTRACT NETWORK WITH NO RESOURCE CONSTRAINTS

The guidelines set out for the addition method (Section 14.2.1) should be followed.

2 CARRY OUT A RESOURCE SCHEDULE TO GIVE A CONTRACT PROGRAMME INCLUDING RESOURCES

The resulting programme should be checked against the original to ensure that it is the same. If it is not, alterations should be made to the scheduling procedure, the fixed dates or the priorities used in the scheduling procedure to make it agree in all significant aspects.

3 DECIDE ON A CONTROL PERIOD

This should be about 1 month and should mirror the control periods that were actually used when carrying out the project since this increases the similarity of the results.

4 CARRY OUT THE FOLLOWING STEPS FOR EACH CONTROL PERIOD

- List all the variations known about in that period and whether they affect that control period or some other.
- Impose the variations on the contract network as for the addition method.
- Carry out a resource schedule using the resources used to obtain the contract schedule. This gives a notional extension of time caused by the variations in that period.
- Impose any mitigating action taken and reschedule. A comparison of this with the programme produced at the previous control point provides a reasonable estimate of the extension of time at that stage of the project. Mitigating action may indicate that the client does not have to pay for an extension of time but may have to pay for extra resources, resource inefficiency or decreases in management efficiency.

5   REPEAT ALL STEPS FOR EACH CONTROL PERIOD

PROBLEMS WITH STAGED ADDITION METHOD

The main problems with the staged addition method are:

● The method relies on the ability of the resource scheduler. Many of the commercially available planning packages offer a resource scheduling facility but few provide the user with much control over its method of operation. What is required is a technique which provides a schedule of work similar to what might be expected to be produced by a planner.

● The method takes a considerably longer time and usually more thought than either the addition or subtraction methods. This is both because of the repetitive nature of the technique and also because of the need to consider carefully the mitigating action taken throughout the project.

14.3   EXAMPLES WITH NETWORK-BASED TECHNIQUES

This section contains illustrative examples of the use of networks for evaluating extensions of time. It is included to illustrate problems which can occur and which should be guarded against. The term *resource constraint* is used to refer to all constraints which are not true technology constraints (Section 5.8.1). The commentary refers principally to the network-based methods of addition, subtraction and staged addition (Section 14.2). The staged addition method has similarities to another technique – the modified time slice method – so a passing mention is made in this section to that technique. The time slice method is not specifically a technique for evaluating extensions of time but it can be useful to that end. Its use and application in evaluating extensions of time is covered in more detail in Section 14.5 with an example in Section 14.6.

14.3.1   Example 1

Consider the simplified network shown in Figure 14.1. The activity durations are shown in days and the activity names are deliberately non-specific. It represents a worst case scenario for extensions of time with no parallel paths to allow for adjustments.

This network would be used for the construction of all five storeys of a five storey building. In this case, the logic connections would be technology constraints representing, for example, that floor C cannot be constructed until floor B has been completed.

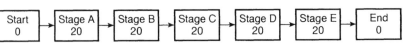

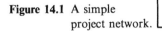
Figure 14.1 A simple project network.

The same network might also be used to represent the construction of five separate housing units by a single resource gang. In this situation, the logic connections are all resource constraints representing, for example, that the resource will move from unit B to unit C.

The question arises as to what the effect of a problem on one of the activities would be in each situation. The effects are discussed by considering two scenarios.

FIRST SCENARIO

Work has progressed as programmed through to day 15 when it is realised that there is a problem with the design of stage B which means that no work can be started on it until day 30.

This would be imposed on a network by setting a fixed earliest start time for the activity *stage B* at 30 giving a finish date for the project of 110 and an extension of 10 days.

For the multi-storey building project, this analysis gives the 'correct' answer.

For the housing project, although the technique shows the same answer, any reasonable contractor would change the order of carrying out the work and the project would be completed in the original 100 days.

Both the addition and the subtraction method would show a delay of 10 for both projects, as would the time slice method (Section 14.5) applied as originally intended. This is because resource connections are treated as technological connections.

The stage addition (and the modified time slice) method would both show the correct differences between the two projects.

If such a thing happens once, the site or project management could be expected to recognise the change of order. If there were hundreds of such cases, it might be expected that site management would not get all decisions correct and an allowance should be made for this disruption. It is difficult to quantify this effect but considerable extra time might be necessary.

It should be noted in all these examples that:

- the later a variation is known, the more likely it is to cause a problem
- a variation should be imposed when it becomes known about not when the effect occurs.

If it is considered reasonable to change the duration of an activity by adjusting the resource levels on the activity, the situation is very similar and the client would only be expected to pay extra costs rather than allow an extension of time.

SECOND SCENARIO

Work progresses on programme to day 15 and then extra work is recognised as being necessary in the activity *stage B*. The extra work is to be carried out by the same type of resource as all the other work and will take 10 days.

This would be imposed on the network by increasing the duration of the activity to 30 days. An analysis of the network shows that the new duration of the project would be 110 days indicating an extension of time of 10 days.

Assuming in the first instance that it is not possible to speed up the activity *stage B* by increasing the number of resources on the activity – a reasonable assumption for many construction activities and one which is commonly made in analysis – the multi-storey building project would indeed take 110 days and this duration could not be reduced.

The housing project would also take 110 days assuming the same resources on site. However, it may also be possible to change the resource levels on site and complete in 90 days by employing a second gang.

In this case, a check needs to be made on the availability of service resources (such as cranes) and an allowance needs to be made for management efficiency and the productivity of resources. Since the second gang is not required until day 70 (or day 80 to finish the project by day 100) and the problem is known about at day 15, it would seem that a reasonable contractor could finish on time by claiming extra cost from the client.

The staged addition method (and the modified time slice method) would show this while the other methods would not.

The later the problem is known about, the greater the number of problems which would be caused and the more a client would be expected to pay.

14.3.2   Example 2

Consider now the project shown in Figure 14.2. This is an extension of the project in Figure 14.1, showing a path of work parallel to the critical path.

The way that the network is drawn implies that there will be two resource gangs on site, assuming all activities are to be carried out by the same type of resource. The effects of variations are discussed by considering the same two scenarios used above.

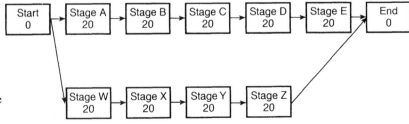

Figure 14.2 A second simple network.

FIRST SCENARIO

Work has progressed as programmed through to day 15 when it is realised that there is a problem with the design of stage B which means that no work can be started on it until day 30.

This would be imposed on a network by setting a fixed earliest start time for the activity *stage B* at 30, giving a finish date for the project of 110 and an extension of 10 days.

The discussion for this scenario is identical to that for the single line network (Figure 14.1).

For the multi-storey building project, the simple critical path analysis on the updated network gives the correct answer.

For the housing project, although the technique shows the same answer, any reasonable contractor would change the order of carrying out the work and the project would be completed in the original 100 days.

Both the addition and the subtraction method would show a delay of 10 for both projects as would the time slice method applied as originally intended. The stage addition and the modified time slice method would both show the correct differences between the two projects.

The later a variation is known, the more likely it is to cause a problem.

If it is considered reasonable to change the duration of an activity by adjusting the resource levels on the activity, the situation is very similar and the client would only be expected to pay extra costs rather than allow an extension of time.

SECOND SCENARIO

Work progresses on programme to day 15 and then extra work is recognised as being necessary in the activity *stage B*. The extra work is to be carried out by the same type of resource as all the other work and will take 10 days.

This would be imposed on the network by increasing the duration of the activity to 30 days. An analysis of the network shows that the new duration of the project would be 110 days, indicating an extension of time of 10 days.

The situation here is rather more complex than in the original example.

Assuming in the first instance that it is not possible to speed up the activity *stage B* by increasing the number of resources on the activity – a reasonable assumption for many construction activities and one which is commonly made in analysis – the multi-storey building project would indeed take 110 days and this duration could not be reduced.

The duration of the housing project would depend on the progress of the second, non-critical path. If the contractor had started this at its earliest time, the work could be completed within 100 days with no change of resource levels but with a change in the order of activity performance. The changes would use up the float (Section 5.2) on the activities making the task of the contractor more difficult and it should be asked if this is unreasonable. It has long been held that the float belongs to contractors to allow them to organise the work more effectively. However, the float here is not true float because of the inclusion of resource constraints. If all the resource constraints were removed, the changes would only require use of less than 2% of the total float in the project as opposed to the 100% implied by the inclusion of the resource constraints.

As in Example 1, a check needs to be made on the availability of service resources such as cranes, and an allowance needs to be made for management efficiency and the productivity of resources. Since the second gang is required over most of the duration of the project, it would seem that a reasonable contractor could finish on time by claiming extra cost from the client.

The staged addition method and the modified time slice method would show this whereas the other methods would not.

The later the problem is known about, the greater the number of problems which would be caused and the more a client would be expected to pay.

14.4 PRACTICAL CONSIDERATIONS FOR THE ADDITION METHOD

The addition method attempts to show the impact on the contract programme of the various key or important variations by imposing them all at once on a programme which models the original intention of the contractor. In general this is a potentially flawed approach, as indicated both in the analyses of the example project (Section 14.3) and in the initial discussion (Section 14.2).

The rest of this section covers action which should be undertaken if the method were to be used despite its problems.

14.4.1 New programme

The original contract programme is commonly submitted and accepted in a bar chart format but contains too little detail to be of use in the addition process. A fundamental step in the process is therefore the creation of a new contract programme (the expanded contract programme) which contains more detail, in terms of the defined activities, than the original. This new programme needs to be in a network format showing the connections that were originally thought to exist between the activities.

The creation of such a new programme is difficult and open to mistakes in the translation; it may also sometimes be based on subjective judgement. It is therefore essential that the new programme is compared in detail with the original. This comparison should prove that the expanded and original contract programmes are the same in all aspects including overall duration, activity connections, activity durations, resources and productivities.

The concept that the important variations can be recognised without including them in the analysis is sometimes stated but in general is misguided. It is essential that all variations are included since the interaction of the variations is usually complex. This must include changes, errors and omissions caused by all parties.

14.4.2 Recommendations

It is suggested that the following points are considered:

A reconciliation between the original and new contract programmes should be provided It is essential to the success of the method to be able to demonstrate that the two programmes are the same at least in logic terms and overall time terms. Even this is not sufficient for the addition method and the contractor should also show the similarity at activity level. A combined programme can often be produced from a computer package by printing the original and the updated contract bar chart in terms of summary or hammock activities (Section 4.2) on the same output.

The contractor should provide a logic listing and drawing of the network on which any argument for an extension is to be based This is fundamental to the addition method and small errors or inconsistencies can cause large variations in final results.

The contractor should provide an explanation of the logic including which are technological constraints, which are resource constraints and which are other constraints The reasons for this are discussed in Chapter 5. The addition method relies on the these logic links and some of them may just not be able to carry the onus of supporting a project.

The contractor should provide a list of assumptions made in the network drawing and in the analysis packages Such assumptions affect the correctness of the network model used (Chapter 5) and the answers obtained by the calculation of extensions of time. Examples of assumptions which are required are:

- splittable or non-splittable activities;
- work or delay lags;
- whether SS and FF lags operate only on the starts and finishes of activities or whether they control progress throughout linked activities.

The contractor should provide an idea of the resources in the new contract programme and the outputs which they were expecting Even if not required in the original contract they should be asked for at this stage. This should be in both activity and aggregated terms since it is important to know what the original intention is before variations can be evaluated.

Resource demands and productivities are essential to allow the validation of individual and combined delays.

It is often informative to see a financial picture of a project by linking the financial model (such as a bill of quantities) to the activities and aggregating over time This provides an estimate of expected and actual, spend and earning rates for the contractor and client. It goes some way to checking the resource loadings but would also serve as a check on actual production.

It is important to take account of any mitigating action necessarily taken by the contractor Without resource constraints in the network, the mitigating action could be simple, as changes in the order of the activities and the consequent delay in the project would tend to be reduced.

Allowance should be made for the change of productivity caused by many small variations.

The whole of the analysis of the addition method relies on the length of the critical path being an indicator of the project length. This must mean that resource constraints are contained in the network and that they are treated in the same way as other constraints. An analysis of a simple example project (Section 14.2.1.) is given including and excluding resource constraints. In this it was seen that the different types of constraint must be treated differently. The engineer/manager has to answer the question

> *Should a project be held up or appear to be held up because resources are allocated in the wrong place or because a minor resource is not available?*

This is a difficult question which cannot have a general answer, but more often than not the answer is in the negative.

It is unrealistic not to consider errors, omissions and alterations caused by the contractor in the analysis since the effects of the combination of variations cannot be predicted.

Many of the problems of an incorrectly applied addition method could be overcome by employing the following steps which could be done quite quickly:

- Validate the expanded contract programme in terms of overall duration, activities, connections, resource productivities and resource demands.
- Evaluate the effects on the individual activities of all the changes.
- Add the changes into the network a month at a time. The changes should be added in when they are known about, not when they occur. The result should be examined to determine if the critical path has any resource constraints in it or any activities which are stretched. The connections concerned with these situations should be examined to determine if they are still valid and the network changed as necessary.

14.5   TIME SLICE METHOD

This method was developed for the client and contractor to carry out joint monitoring of live projects and as such can be a valuable tool in project control. It can be used for evaluating extensions of time and it has similarities with other methods introduced above. This section contains a brief outline of the method in the context of a discussion of its use in evaluating extensions of time.

Fundamentally, it is an adaptation of the staged addition method and seeks to model the performance of the project. It relies on a network existing for the project and this may also be one of its problems if it makes no distinction between the different possible meanings of the connections.

14.5.1   Time slice method applied retrospectively

In the time slice method applied to live projects, the parties reschedule all future work from defined control points and use the results of each re-schedule from the control point to the end of the project to determine if the project is ahead or behind and by how much.

Rescheduling means not only re-calculating the critical path but also providing a detailed review of the network, its activities and connections before the calculation is done.

The application of the method retrospectively can be described by the following steps:

(1)   Prove that the original contract programme is credible.
(2)   Move forward in time to the first selected stage and record on the programme the progress achieved to date.
(3)   Reschedule the outstanding work, maintaining the same sequence and logical connections as were present originally.
(4)   If delay has occurred, identify the possible causes and apportion time according to contractual ownership of the delay-causing event(s), taking due account of concurrency. Generally these delays are in the activity at the head of the critical path after rescheduling. Since only the net effects of delays within the time slice are calculated by the rescheduling, then concurrency of delay is automatically taken into account.
(5)   Correct the rescheduled programme to account for any changes in the planned sequence and then repeat stages 2–4 until project completion is achieved.
(6)   The total entitlement is the sum of the individual entitlements calculated at each time slice. It should be noted that if better progress than planned is achieved then rescheduling would bring the finish date forward in time. This does not mean that any previously calculated entitlement is lost to the contractor.

Several points in this procedure need discussion. First, the original contract programme needs to be in a network format in order for the various scheduling and rescheduling steps to be carried out. If resource constraints are included in this network, the original programme might be credible but the rescheduled ones would almost certainly not be. If technological constraints only were included then it would be necessary to apply a resource scheduling procedure to it to obtain the original programme.

As mentioned earlier, when applying variations in the time slice or any other method, they should be applied when they are known about, not when their effects actually occur.

14.5.2 Some common mistakes and their effects

It is best to discuss the method and illustrate some pitfalls against a suitably defined objective. For the time slice method the basic objective (Section 14.1) can be extended as follows:

> *To model a project in such a manner that, given variations to it, the model is able to behave like the project. In this behaviour it should, as closely as possible, model the changes in duration of the project which would be predicted at the time at which the variations became known, allowing for reasonable actions being taken by the contractor but otherwise following the proposed method of work. If possible, it should also model the changes in resources and costs brought about by the variations.*

One common problem when methods are developed is that people seek to reschedule the work from the start of the project each time a delay is recognised. This is contrary to the objective:

> ... *it should, as closely as possible, model the changes in duration of the project* which would be predicted at the time at which the variations became known ...

A model which reschedules from the beginning is effectively modelling the opportunity to rebuild a part of the project which had already been constructed and should not be allowed.

Networks are commonly drawn containing both technological constraints and resource constraints and the critical path calculated using both. The different types of constraints are treated identically throughout the analysis and this causes many problems. The difference between a resource-constrained critical path and one which is not resource constrained should be examined.

The true critical path so often referred to is a concept which in reality is meaningless because of the different types of connection in the network.

It is necessary to treat different types of connections differently and to use resource scheduling techniques or similar ideas rather than just critical path analysis.

14.6   EXAMPLE WITH TIME SLICE METHOD

14.6.1   Simple project

Consider the example below. The scenario is quite complex and some time should be spent understanding what is happening and why. The connections which represent the technology of the method are represented in Figure 14.3. The connection between the *brickwork* activities has not been included because it only represents the contractor's preferred method of work showing the intended progress of resources.

Consider the following variations to the project:

- The contractor failed to start building 1 until week 3 for reasons which are not in dispute and for which the responsibility is accepted as the contractor's;
- In week 2 building 2 suffers a delay which is caused by the client.

14.6.2   Effect of variations

The variations that arise can now be applied at the times at which they became apparent. There is no clarity in the example as to when this was and therefore it is suggested that a worst case scenario should be adopted in which it is assumed that the delays are known about only just before they occur.

In this case, the failure to start building 1 would be known about at the start of the project and the situation would be as shown in Figure 14.4. It is important to note that the delay to the building would not have extended the project because an experienced contractor would have been expected to take mitigating action which in this case would have meant increasing the number of bricklaying resources to enable the two buildings to be worked on together. Whoever caused the delay might be expected to pay for the extra resources required but would not be expected to pay for an extension to the project. It is a matter of

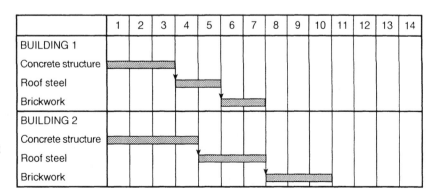

Figure 14.3 Original bar chart showing technological logic.

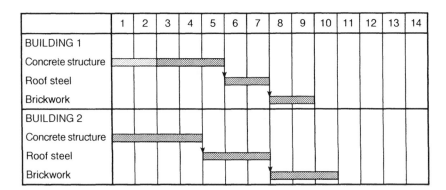

Figure 14.4 Project with variation to building 1.

judgement as to whether or not the provision of extra resources constitutes reasonable mitigating action; following the example it is assumed to be so here.

The second variation affected only building 2. It took effect in week 2 and, for this exercise, it is assumed that it became known about at the start of week 2. This is illustrated in Figure 14.5 which shows the state as known about at this time with a reschedule done from start of week 2. The work actually carried out is shown in black.

It can be seen that this variation is quite important to the project. It extends the project duration to 11 weeks and reduces the amount of time for which the doubling of resources is required. This is brought about because the path which has been extended contains only technological constraints. If this delay were caused by the client then it is suggested that the client should pay for an extension to the project but should not pay for extra bricklaying resources.

Figure 14.6 shows the effects of the next stage of the delays. In this, there are concurrent delays to building 1 and building 2 . They occur at the start of week 4. The actual work done on the project is shown in black and it can be seen that one-third of building 1 *concrete structure* is complete while half of the original building 2 *concrete structure* and all the previously known delay to it is complete.

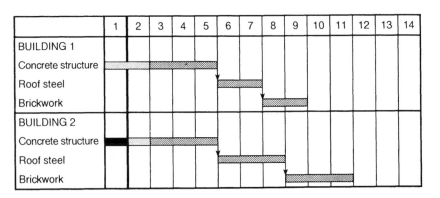

Figure 14.5 Project with variation to building 2.

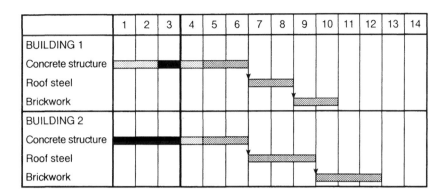

Figure 14.6 Project with concurrent variations to the two buildings.

Rescheduling from the start of week 4 shows once again that the project is delayed by a week and now is forecast to take 12 weeks. The variation which caused this delay can clearly be seen to be the one which affected building 2. The variation to building 1 had no effect at all on the project duration but did have an effect on the timing of the resource demands and, if it was extra work rather than delay, would affect the project costs.

It would be important if this variation to building 1 occurred without the concurrent delay to building 2. In this case, the reschedule would be as shown in Figure 14.7 and the project would not have been delayed but the resource demands would have to be altered in the future to enable the two brickwork activities to be carried out simultaneously for 2 weeks.

All these stages have been carried out in a manner which models the behaviour of a reasonable contractor. That is, as the variations have become known, the work has been rescheduled assuming the same technological logic and wherever possible maintaining the concept of the work method which in this case meant using a single bricklaying gang. The ownership of the delays to the project and the extra cost are apparent from the analysis.

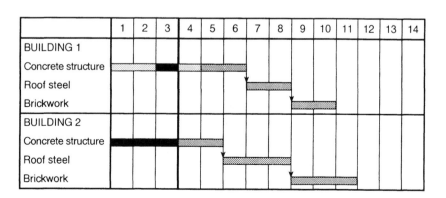

Figure 14.7 Project with variation to building 1 only.

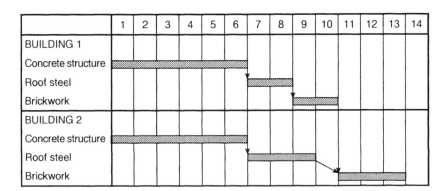

Figure 14.8 Project evaluated by the addition method.

On a more complex project, the same method should be followed and the rescheduling process would be the same in principle but rather more time-consuming.

14.6.3 Comparison with addition method

The example serves to illustrate a potential fundamental flaw in the addition method. In this method, all the constraints for the example project, including the resource constraint between the two brickwork activities, are treated equally, the variations imposed and the critical path recalculated from the start of the project. The project is then as shown in Figure 14.8.

It can be seen that the project duration is now 13 weeks giving an extension of 3 weeks which is obviously an unrealistic situation. The simple addition method is therefore, in this instance and indeed in general, an unsatisfactory model of reality to use for the evaluation of extensions of time caused by variations. It always overestimates the extensions of time caused by variations and provides no idea as to the ownership or apportionment of the delays.

If the resource constraint is removed and the addition method employed, the result for this project would be as shown in Figure 14.6 but it would be impossible to allocate responsibility for the delays and extra costs due to the various variations.

14.6.4 Conclusion

This section has discussed the application of the time slice method to the evaluation of extensions of time. It has been shown to be a good model of the process actually carried out on a project. As applied at present, it suffers from two main problems:

- the time at which variations are added into the project;
- the need to consider future resource connections at each time slice.

If these problems are tackled, it is suggested that the method is an excellent technique for evaluating extensions of time.

14.7 SUMMARY

- Planning techniques are often used to evaluate extensions of time.

- Networks appear to give a more scientific approach than other techniques.

- There are many possible problems with the use of networks for evaluating extensions of time.

- All methods which can be used for evaluating an extension of time caused by a set of variations rely on the fundamental information which is available.

- A large number of variations affects the productivity of both resources and management and should be considered in any analysis.

- There needs to be agreement on the original contract programme from which extensions are calculated.

- The main problem with network-based techniques is the recognition and treatment of the different types of constraints.

- The critical path calculated using a resource-constrained network is unlikely to be the project duration.

- All methods for evaluating extensions of time must differentiate properly between the different types of constraints in networks.

- The addition method applied on a network containing resource constraints tends to give the largest possible answer for the extension of time although the amount by which the delay is overestimated is impossible to determine.

- For the addition method to be analysed for reasonableness, considerable information over and above the network and the variations is required.

- The time slice method applied on a network containing resource constraints tends to give too small an answer for the extension of time due to the contractor, although the amount by which the delay is underestimated is impossible to determine.

15

TINGHAM TANK FARM PROJECT

15.1 INTRODUCTION

This chapter contains an extended example. It discusses the planning of the Tingham Tank Farm, which is introduced briefly in Appendix A with the other Tingham Development projects. Section 15.2 contains all the available information for the project including drawings and quantities. It is hoped that the information provided is sufficient for the project to be realistic and that the simplifications enable the discussion to concentrate on the planning aspects. Section 15.3 contains a short description of the choice of planning techniques, the data required for these techniques and examples of the application of the techniques to the project.

15.2 TINGHAM TANK FARM

15.2.1 Project description

The project consists of the construction of six underground reinforced concrete tanks, a valve house, a control and administration building, roadways, car park and pipework. Pipework is needed to connect the tanks to the valve house and to connect the new construction to an existing pumping station approximately 1150 m from the valve house. The layout is shown in Figure 15.1.

The control building has been let as a specialist subcontract and the main contractor has to provide access and facilities for the subcontractor.

The site is very congested due mainly to the large amount of earthworks which is to be undertaken. This means that the valve house cannot be constructed until the backfill to the tanks is complete.

The site is situated near the top of an escarpment which falls away sharply to the south east of the site. The ground conditions on site are moderately stiff clay under 300 mm of topsoil.

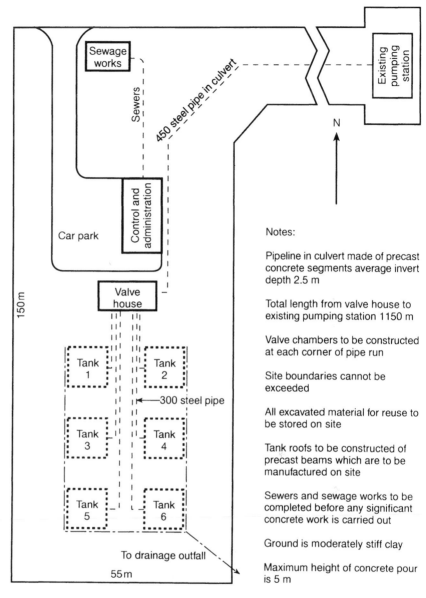

Figure 15.1 Tank Farm site — general arrangement (dimensions in millimetres).

Notes:

Pipeline in culvert made of precast concrete segments average invert depth 2.5 m

Total length from valve house to existing pumping station 1150 m

Valve chambers to be constructed at each corner of pipe run

Site boundaries cannot be exceeded

All excavated material for reuse to be stored on site

Tank roofs to be constructed of precast beams which are to be manufactured on site

Sewers and sewage works to be completed before any significant concrete work is carried out

Ground is moderately stiff clay

Maximum height of concrete pour is 5 m

15.2.2 Tanks

The tanks are approximately $10\,\mathrm{m} \times 10\,\mathrm{m} \times 10\,\mathrm{m}$ reinforced concrete cubes with 1 m thick base and walls. The roofs are each made of 10 precast concrete beams which must be constructed on site.

The tanks are to be lined with a chemical proof lining which will be installed by a specialist subcontractor. The company nominated for this work will require 8 working days on site for each tank after completion of the walls and before installation of the roof beams. Access to the inside of

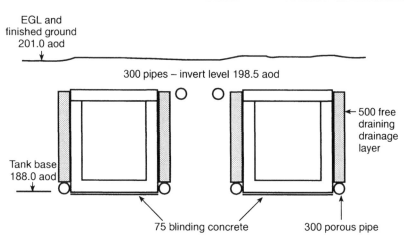

EGL and
finished ground
201.0 aod

300 pipes – invert level 198.5 aod

500 free
draining
drainage
layer

Tank base
188.0 aod

75 blinding concrete   300 porous pipe

Figure 15.2 Typical cross-section through tanks (dimensions in millimetres).

Notes:
(1) Horizontal concrete surfaces to be covered with bituthene sheet waterproof membrane
(2) All vertical concrete surfaces to be treated with bituminous paint
(3) All backfill to be with suitable material
(4) Tanks not to be backfilled until concrete has attained 28 day strength

the tanks for the contractor's labour must be provided and 20% use of a crane must be allowed.

The tank roofs will be finished with a waterproof membrane.

The sides of the tanks are to be coated with a waterproof bituminous paint and a 500 mm thick drainage layer is to be provided. The pipes under the drainage medium are to be connected together and discharge into a stream flowing off the hill 200 m from the south east corner of the site.

15.2.3 Pipework

The pipework from the tanks to the valve house is to be 300 mm diameter steel pipes.

The pipework from the valve house to the existing pumping station is to be 450 mm diameter steel pipes. These are to be installed in a precast concrete culvert for which the average depth to invert level is 2.5 m.

All the precast units for the culvert are to be made on site.

15.2.4 Drawings

The drawings in Figures 15.1 to 15.6 illustrate the general arrangement of the site and provide details of the major construction work.

15.2.5 Basic quantities

The quantities shown in Tables 15.1 to 15.5 are the major quantities for the measured work of the main contractor. No information is provided for the work to be carried out by the specialist or nominated subcontractors.

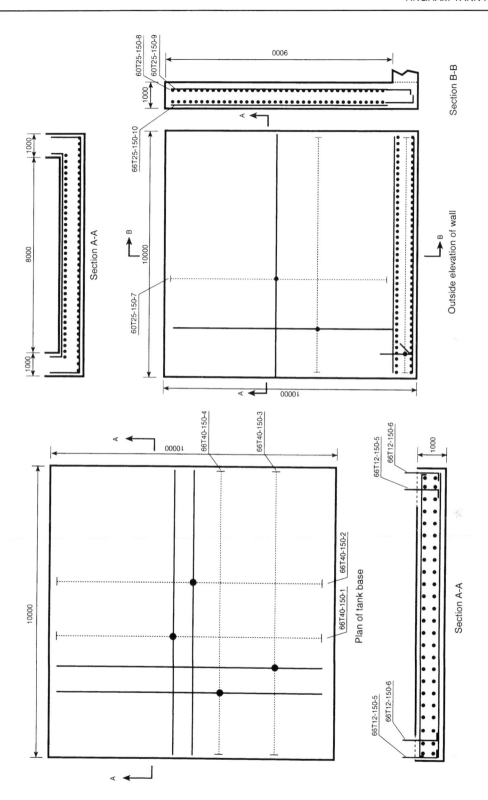

Figure 15.4 Tank wall reinforcement detail (dimensions in millimetres).

Figure 15.3 Tank base reinforcement detail (dimensions in millimetres).

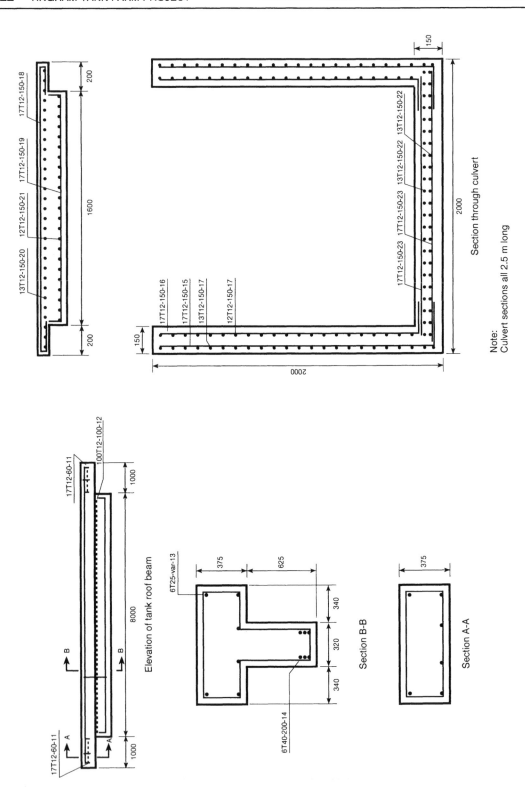

Figure 15.6 Precast culvert details (dimensions in millimetres).

Figure 15.5 Precast beam details (dimensions in millimetres).

Table 15.1 Base reinforcement per tank.

Diameter (mm)	Weight (tonnes)
40	26.04
12	0.36

Table 15.2 Wall reinforcement per tank.

Diameter (mm)	Weight (tonnes)
25	38.76

Table 15.3 Precast beam reinforcement per tank.

Diameter (mm)	Weight (tonnes)
40	6.51
25	2.54
12	4.38

Table 15.4 General quantities per tank.

Item	Quantity per tank
Base concrete	$100\,m^3$
Wall concrete	$342\,m^3$
Base formwork	$40\,m^2$
Wall formwork	$585\,m^2$
Beam formwork	$200\,m^2$
300 mm porous pipe	$40\,m$
Drainage material	$200\,m^3$
Bituminous paint	$400\,m^2$
Waterproof membrane	$100\,m^2$
Specialist lining	$324\,m^2$
Falsework	$1200\,m^3$
Construction joint	$40\,m^2$

Table 15.5 General site quantities.

Item	Quantity on site
450 mm steel pipe	$1\,150\,m$
300 mm steel pipe	$240\,m$
Bulk excavation	$24\,000\,m^3$
Car park	$900\,m^2$
Excavation for 450 mm pipe	$2\,875\,m^3$
Excavation for drainage pipe	$200\,m^3$
Excavation for 300 mm pipe	$450\,m^3$

15.3 PLANNING THE TINGHAM TANK FARM

This section contains examples of both good and bad planning practice, with the good and bad points brought out as appropriate in the continuing discussion.

15.3.1 Method statement

A method statement (Section 2.9.1) is an essential part of planning this project from a contractor's point of view. It also enables the owner to ensure that the contractor has fully appreciated the problems inherent in the work and has solutions to them. Method statements contain different things depending on both the stage of the project at which they are produced and the target audience for the statement. In this example it is assumed that the method statement is being produced by the main contractor for their own use at the start of the project.

The method statement for this project should therefore include (or refer to) the following components, discussed further in Section 15.3.2.

Sketches showing the layout of the site during construction If, as usual, the site layout is planned to change during the project, a series of phase diagrams should be used. In this project, only one major phase is predicted and this is shown in Figures 15.7 and 15.8.

Descriptions of any special construction methods to be adopted A sketch or description of, for example, the tank construction sequence should be provided and could be added quite easily to Figure 15.8.

List of the allowances in the contractor's tender for the various types of work This is shown in Table 15.6. These allowances should be in terms of the controllable money for the level of management for which the plan is intended. Most commonly, as in this example, this will be in terms of labour and plant money for contractors rather than the total value to the contractor. It should also be noted that these allowances should be for the work to be done, not necessarily for the work which will actually be paid for. Thus there is an allowance for the scaffolding to the tanks and for the formation of the construction joint between concrete pours for the tank even though in some cases these may not be paid for directly. The allowances should be available from the estimating documents for the project. If they are not, it is perhaps an indication that the estimate has not considered the work in enough detail.

List of major resources to be used and the outputs assumed This can only be deduced here from Tables 5.7 and 5.8.

A long-term programme of work for the project This should show activities, timing and usage of major resources and include all necessary supporting information (Tables 15.7–15.9 and Figures 15.9–15.11) as well as the main representation of the plan, a Gantt chart (Figure 15.12).

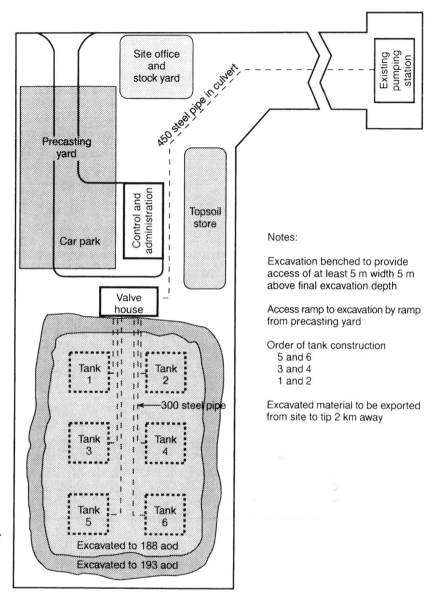

Figure 15.7 General arrangement of site showing excavation limits.

Within the figure:

Site office and stock yard

450 steel pipe in culvert

Existing pumping station

Precasting yard

Control and administration

Topsoil store

Car park

Valve house

Tank 1

Tank 2

Tank 3

Tank 4

Tank 5

Tank 6

←300 steel pipe

Excavated to 188 aod

Excavated to 193 aod

Notes:

Excavation benched to provide access of at least 5 m width 5 m above final excavation depth

Access ramp to excavation by ramp from precasting yard

Order of tank construction
5 and 6
3 and 4
1 and 2

Excavated material to be exported from site to tip 2 km away

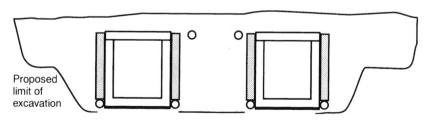

Figure 15.8 Typical cross-section through tanks during construction.

Proposed limit of excavation

Table 15.6 Contractor's allowances for work.

Item	Unit	Labour (£ per unit)	Plant (£ per unit)
Base concrete	m³	7.00	2.10
Wall concrete	m³	9.40	2.90
Base formwork	m²	12.00	2.00
Wall formwork	m²	12.00	2.20
Strip base formwork	m²	2.00	0.20
Strip wall formwork	m²	3.00	1.00
Beam formwork	m²	12.00	0.30
300 mm porous pipe	m	1.50	0.30
Drainage material	m³	6.50	0.50
Bituminous paint	m²	4.50	0.05
Waterproof membrane	m²	16.50	1.00
Falsework	m³	2.00	0.20
Construction joint	m	8.50	0.50
450 mm steel pipe	m	5.40	4.25
300 mm steel pipe	m	5.00	4.00
Excavation for 450 mm pipe	m³	1.60	1.50
Excavation for drainage pipe	m³	1.60	1.50
Excavation for 300 mm pipe	m³	1.60	1.50
Base reinforcement	tonne	160.00	10.00
Wall reinforcement	tonne	165.00	10.00

Resource implications for the project   Extra resource information is given later in Figures 15.13 to 15.19.

Financial appraisal of the programme   An example is given in Figure 15.19 and is discussed in 'finances' in Section 15.3.2.

15.3.2   Choice of technique and activities for long-term planning

TECHNIQUE

The tank farm project is a complex civil engineering project with several different types of work to be carried out in a restricted area. There are highly repetitive elements in the precasting work; linear elements in the 450 diameter pipeline; ordinary construction work with some repetition in the tank construction; building work in the two buildings; and mechanical and electrical (M&E) work in the pipelines and control equipment.

For long-term programmes, there appears to be no single ideal planning technique which would adequately model all the work to be undertaken. In addition, because the site is congested, the position of the work on the site should be taken into account when planning and when presenting the plan.

If a network were chosen, a simple activity-on-node or flowchart network (Section 2.5.2) would suffice as a basis for most long-term planning. Such a network (Table 15.7 and Figure 15.9) illustrates the manner in which linear and repetitive elements of projects can be represented in simple networks. The repetitive elements which are independent (precasting of roofs and culverts) are split into separate manageable activities and shown as possibly being worked on together. Repetitive elements which are inter-related (all the work on the tanks) are split into convenient manageable activities which are linked together to represent the construction method. The elements are divided into thirds, each third representing the work on a pair of tanks.

The choice of planning method is largely determined by the construction method. Linear work (the culvert and pipeline) is split into activities based on the amount of work and the level of detailed control required. The activities are linked together according to the construction method and provide an overlapping activity model.

A precedence diagram (Section 2.5.3) could be used for either the contractor's or the client's planning, although the activities could vary. Examples of networks for both purposes are shown in Figure 15.10 (with Tables 15.8 and 15.9) and Figure 15.11 and are discussed below.

Both types of network would be unsuitable for much of the communication required of a planning technique and a bar chart should be considered for a long-term programme. A bar chart (Section 2.2) is shown in Figure 15.12. It should be noted that the four models use very different activities.

The choice of representation technique is rather more complex than just the choice between different types of network or between networks and bar charts. The networks produced for the project (Figures 15.9, 15.10 and 15.11) depend on the activity selection. The choice of activities limits the type of network which can be used. It would, for example, be impossible to produce a meaningful activity-on-node network from the activities used for the two precedence networks. However, neither type of network is sufficient for planning purposes and a bar chart is also necessary. Although it would be possible to produce a bar chart using the activities in any of the network models, here a different set of activities has been used for illustration purposes.

The bar chart, although indicating when activities will be done, is not ideal for repetitive or linear work. In this project, there is considerable repetitive work in the precasting of approximately 600 units. This must be carried out on site and should therefore be planned in some detail. There is also linear working in the actual construction of the culvert and fixing of the pipe within it.

For the repetitive element of the work, it may be worth considering using the line-of-balance technique (Section 2.4), whereas for the linear work a time-chainage approach (Section 2.3) may be useful. The usefulness of the different approaches should be balanced with the

Table 15.7 Durations and resources for network shown in Figure 15.9.

Activity	Duration (days)	Labour	Carpenters	Steel fixers
Start	10			
Set up site	4	5		
Install batching plant	15	5	3	
Construct precast yard	15	5	2	1
Precast roofs 1/3	70	4	3	2
Precast roofs 2/3	70	4	3	2
Precast roofs 3/3	70	4	3	2
Precast culverts 1/3	70	4	3	2
Precast culverts 2/3	70	4	3	2
Precast culverts 3/3	70	4	3	2
Excavate tanks 1/3	5	3		
Pipe connections 1/3	12			
Blinding and bases 1/3	10	4	3	2
Tank lining 1/3	20			
Tank walls 1/3	25	4	3	2
Backfill tanks 1/3	5	2		
Excavate tanks 2/3	5	3		
Blinding and bases 2/3	10	4	3	2
Tank lining 2/3	20			
Tank walls 2/3	25	4	3	2
Backfill tanks 2/3	5	2		
Excavate tanks 3/3	5	2		
Blinding and bases 3/3	10	4	3	2
Tank lining 3/3	20			
Tank walls 3/3	25	4	3	2
Backfill tanks 3/3	5	1		
Excavate pipe track 1/3	20	2		
Construct culvert 1/3	40	4	1	1
Culvert pipework 1/3	20			
Excavate pipe track 2/3	20	2		
Construct culvert 2/3	40	4	1	1
Culvert pipework 2/3	20			
Excavate pipe track 3/3	20	2		
Construct culvert 3/3	40	4	1	1
Culvert pipework 3/3	20			
Pipe connections 2/3	12			
Pipe connections 3/3	15			
Tank roofs 1/3	10	5	1	
Tank roofs 2/3	10	5	1	
Tank roofs 3/3	12	5	1	
Offices and roads topsoil strip	4	2		
Excavate roadworks	4	2		
Excavate office site	4	1		
Site drainage	40	3		
Construct roads	20			
Construct prefabricated office	30			
Sewers	25	4		
Sewage works	20			
Landscaping	20			
Finishes	10	4		

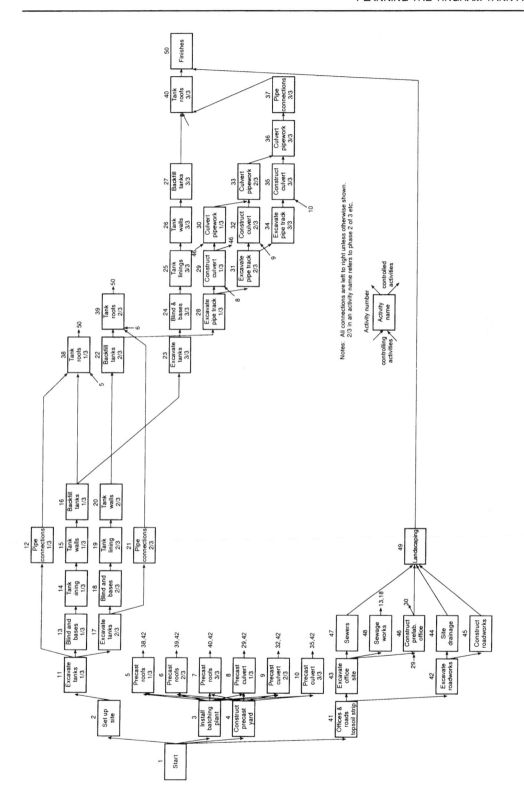

Figure 15.9 Activity-on-node network.

Table 15.8 Durations and resources for network shown in Figure 15.10.

Activity	Duration (days)	Labour	Carpenters	Steel fixers
Start	10			
Set up site	4	5		
Install batching plant	15	5	3	
Construct precast yard	15	5	2	1
Precast roofs	205	4	3	2
Precast culverts	180	4	3	2
Excavate tanks	30	3		
Blinding and bases	40	4	3	2
Tank lining	60			
Tank walls	70	4	3	2
Backfill tanks	55	2		
Excavate pipe track	60	2		
Construct culvert	100	4	1	1
Culvert pipework	60			
Pipe connections	30			
Tank roofs	32	5	1	
Offices and roads topsoil strip	4	2		
Excavate roadworks	4	2		
Excavate office site	4	1		
Site drainage	40	3		
Construct roads	20			
Construct prefabricated office	30			
Sewers	25	4		
Sewage works	20			
Landscaping	20			
Finishes	10	4		

difficulty of their integration with the plans for the whole project. In this project, it was not felt to be necessary since the network could be drawn to provide enough detail without the use of too many activities. The bar chart used could be improved by including some annotation to clarify certain points without destroying the bar chart as a communication tool.

It should be noted that the level of detail available in the initial information limits the amount of detail that can be included in the plan. There is no detail for the buildings and the planner therefore has to make assumptions. In the representations shown in Figures 15.9 to 15.12, the assumption is that there is enough information available to set the duration of the relevant activities. Here it can be seen that the plan is to be used for setting the target start and finish times for the subcontractors to carry out the required work. With no other detail available, it is impossible to tell whether the target is reasonable. It will also be difficult to use the activities for monitoring and control since any measure of completion would be too uncertain and open to interpretation. The choice of activities is discussed further in the next section.

Table 15.9 Durations of logic links for network shown in Figure 15.10.

From Activity	To Activity	Lag type	Duration (days)
Start	set up site	FS	
	install batching plant	FS	
	construct precast yard	FS	
	offices and roads topsoil strip		
Set up site	excavate tanks	FS	
Install batching plant	precast roofs	FS	
	precast culverts	FS	
Construct precast yard	precast roofs	FS	
	precast culverts	FS	
Precast roofs	tank roofs	SS	70
	tank roofs	FF	12
	excavate roadworks	FS	
Precast culverts	construct culvert	SS	70
	construct culvert	FF	40
	excavate roadworks	FS	
Excavate tanks	blinding and bases	SS	5
	blinding and bases	FF	20
Blinding and bases	tank linings	SS	10
	tank linings	FF	20
Tank lining	tank walls	SS	20
	tank walls	FF	25
Tank walls	backfill tanks	SS	25
	backfill tanks	FF	25
Backfill tanks	tank roofs	SS	5
	tank roofs	FF	12
	excavate pipe track	SS	30
	excavate tanks	SF	20
Excavate pipe track	construct culvert	SS	20
	construct culvert	FF	40
Construct culvert	culvert pipework	SS	40
	culvert pipework	FF	20
	construct prefabricated office	SS	40
Culvert pipework	pipe connections	FF	15
Pipe connections	tank roofs	FF	12
Tank roofs	finishes	FS	
Offices and roads topsoil strip	excavate office site	FS	
	excavate roadworks	FS	
Excavate roadworks	site drainage	FS	
	construct roads	FS	
Excavate office site	sewage works	FS	
	sewers	FS	
	construct prefabricated office	FS	
Site drainage	landscape	FS	
Construct roads	landscape	FS	
Construct prefabricated office	landscape	FS	
	culvert pipework	FS	
Sewers	landscape	FS	
Sewage works	blinding and bases	FS	
	landscape	FS	
Landscaping	finishes	FS	
	finishes		

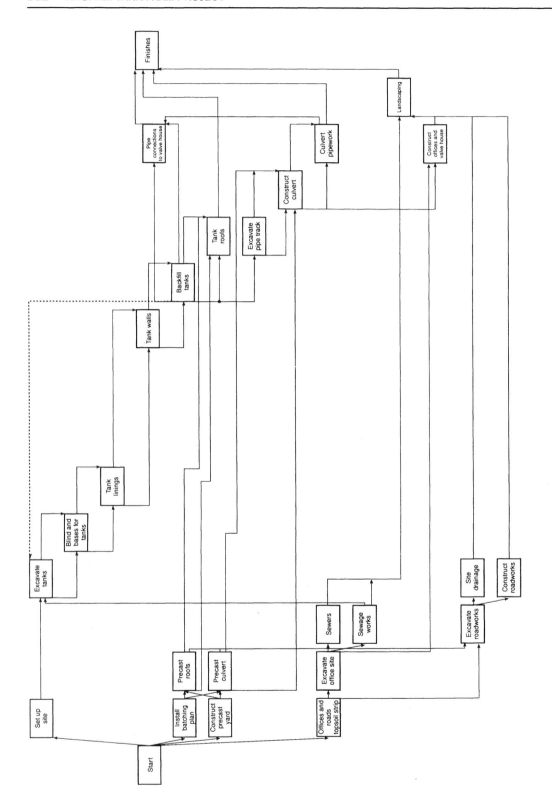

Figure 15.10 Precedence network for contractor's use.

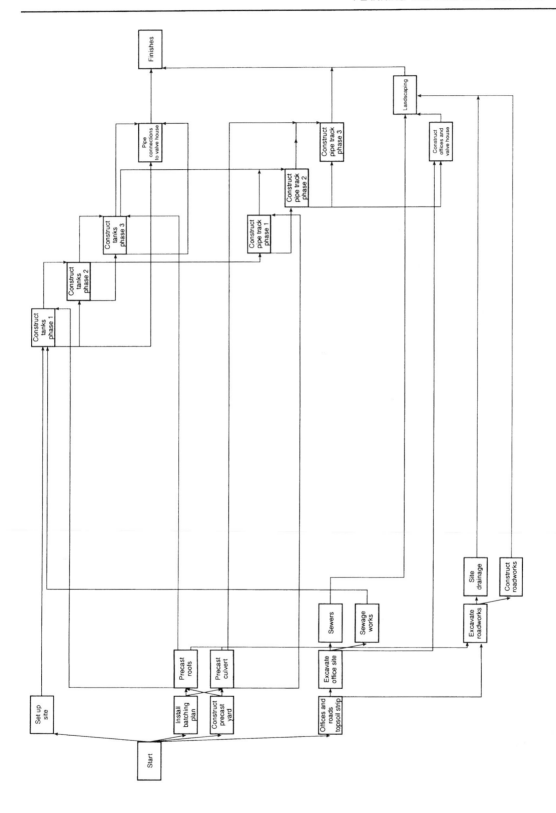

Figure 15.11 Precedence network for client's use.

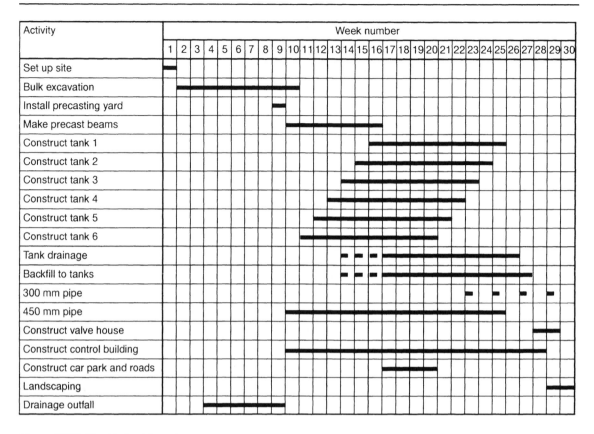

Figure 15.12 Gantt chart long-term programme.

ACTIVITIES

For the long-term plan for the project, it is important to be able to explain the construction method without including excessive detail. The planner has the choice of two approaches in deciding activities. First, the activities can be chosen to represent the work types, for example, the fixing of reinforcement for the tank walls. This choice would provide activities which are essentially single resource and constant resource use (Section 4.4). The resulting plan would be suitable for the contractor's use since it would enable planning and control of the resources involved in the construction. The networks shown in Figures 15.9 and 15.10 use activities largely selected by this mechanism.

The second approach is to consider activities in terms of the major structural elements. In this method, activities such as construct tanks 5 & 6 would be used. Such a choice would provide a plan which would be useful for the client or for the higher levels of management of the contractor. It would enable progress to be monitored and controlled but would not provide any information for resource control because the activities would be large and complex. The network shown in Figure 15.11 and the bar chart shown in Figure 15.12 use activities generated using this approach.

Although the plans produced using the two different approaches are obviously different (Figures 15.9 to 15.12), they should produce the same overall project parameters such as minimum project duration. It is important to realise that this may not be the case because of the various assumptions regarding the nature of the activities (splittable, non-splittable and stretchable) and the connections between the activities (work and delay lags, progress constraints, inclusive/exclusive lags and so on). The effects of these are fully discussed in Sections 5.4 to 5.6.

It should be noted that some networks imply certain assumptions about the activities. For example, the start–finish connection shown as a dotted line in Figure 15.10 between *Backfill tanks* and *Excavate tanks* means that the activities in the ladder between the two ends of the connection must be splittable or stretchable in order for the calculations to be carried out (Section 5.6 and Figure 5.20). The connection means that the last two tanks cannot be excavated until the first two tanks have been backfilled. Many computer packages would assume that there was an error in the logic of this network suggesting that the activities in the ladder and the connections as shown form a loop which cannot be resolved. This would be the case if the activities were non-splittable.

The various assumptions would probably not provide the same resource profiles if these were to be calculated. The discussion continues on the activities, concentrating on their durations and the resources they require.

DURATIONS AND RESOURCES

In order to decide on the optimum levels of resource to assign to the activities (and hence determine the activity durations) the planner must consider:

- the productivities of the resources available on each of the types of work involved;
- the way in which the resources work together and interact;
- the space available for the work;
- the method of working to be employed.

Neither the networks nor the bar chart provide an ideal base for the treatment of activity durations, resources and resource productivities but each is now considered in turn. The activity-on-node network (Figure 15.9) is perhaps the best to consider first. The activities are single work type and the quantities of work involved can be relatively easily determined.

This is what has been done here. Alternatively, the planner could have decided on the required durations and assigned the required resources to the activities in order to achieve these. A list of the activities and their durations is shown in Table 15.7. The key resources assigned to the activities are assumed to be constant throughout the duration of the activity. Note that not all resources are listed.

The contractor's precedence diagram (Figure 15.10) is the next best to consider from a resource and duration point of view. Although the activities are assumed to be splittable (see previous section) the resources are assigned and durations calculated assuming that the work will be carried out in a continuous manner on each activity. This model is identical to the activity-on node network shown in Figure 15.9 in all respects and the durations for the activities are as shown in Table 15.8. In this model, the durations of the lags are important and Table 15.9 shows these. With the values shown, the two networks shown in Figures 15.9 and 15.10 can be analysed to give the same results for the project duration and the earliest and latest start and finish times for the types of work. The assumptions required are that the lags are inclusive work lags and that the activities are splittable. To ease calculations it is sensible to assume that work is carried out 7 days per week.

The activity durations have, in many cases, been generated by adding together the durations of the three individual phase activities (for the three pairs of tanks) whenever this gives a duration which is less than or equal to the difference between the earliest finish time of the last (third) phase and the earliest start time of the first phase. This means that the phases will be constructed in sequence. A few of the activities cannot be carried out in sequence and still maintain the original project duration. In these cases, a maximum duration to maintain the project duration has been given, being the difference between the earliest finish time of the last phase and the earliest start time of the first phase. An example of such an activity is *Precast culvert* where the individual phases sum to 210 days but a duration of 180 days has been assigned to maintain the project duration. These reduced durations mean that the individual phases of the combined activity are carried out partially in parallel and the constant resource model is only an approximation. It in fact produces an infeasible lower bound to the resource requirements for the project.

The durations of the lags are given in Table 15.9. These have also been chosen to ensure the precedence network of Figure 15.10 models the activity-on-node network of Figure 15.9. The SS lags are equal to the duration of the first phase of the originating activity, whereas the FF lags are equal to the last phase of the end activity.

The client's precedence network shown in Figure 15.11 has multiple resource and variable resource demand activities. The durations here are targets rather than the results of calculations based on resource productivities. Because of this, they are difficult or impossible to check. These types of activities should be avoided whenever possible for resource and duration planning. Similar points apply to the bar chart shown in Figure 15.12. It is a good representation of a fictional view of the project which cannot really be justified or argued with.

In addition to the activity resources, it is important to determine the project resource demand and to ensure that this is realistic. Figures 15.13 to 15.15 show the project resource demand (labour, carpenter, steelfixer) assuming all the activities are carried out as early as possible. It can be seen that there is a peak requirement for each resource early in the project

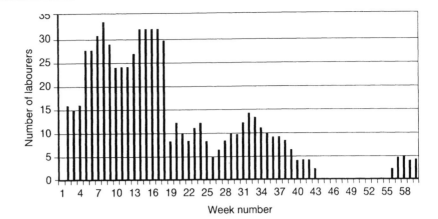

Figure 15.13 Earliest start time resource profile for labour.

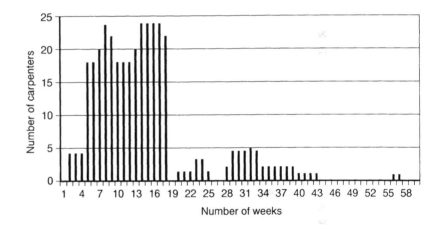

Figure 15.14 Earliest start time resource profile for carpenters.

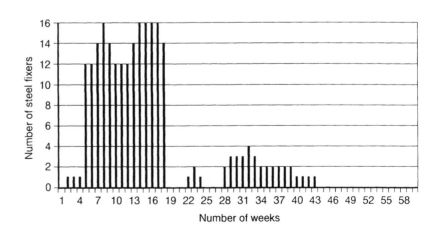

Figure 15.15 Earliest start time resource profile for steelfixers.

with a steady tailing off. This is typical of an earliest start schedule because of the large number of activities which could be done at the start of a project as modelled by a non-resource-constrained network. These profiles might be considered unrealistic and Figures 15.16 to 15.18 show the resource profile for a schedule produced by limiting the number of each resource to 20. It can be seen that this limitation only had an effect in the initial part of the project and a manager might wish to limit the resources to a smaller number later in the project.

During the determination of the durations and the resource demands, the planner should ensure that a safe system of work is being employed, that all the resources required to ensure safety are available and that the sequencing of the work is safe.

FINANCES

Project finances should be evaluated during the planning process. Figure 15.19 shows the cost of key labour (labourers, carpenters and steelfixers) for the project plotted at earliest, latest and scheduled times as calculated from the activity-on-node network shown in Figure 15.9 and using the resource demands shown in Table 15.7. Similar graphs should be produced for all financial parameters which are to be controlled. It is suggested that at least the graphs for earnings and liabilities be produced.

Figure 15.19 shows the 'S-curve' shape of a typical schedule and the way in which the earliest and latest time costs form an envelope into which the schedule curve fits. Although a manager will try to maintain the progress on the schedule line, deviation does not become too serious until the curve strays outside the earliest-latest envelope. When this happens, it becomes necessary to perform activities at durations other than their optimum duration (see previous section) or to develop a new construction method (not represented by the network) if the programmed finish date is to be achieved.

Managers should plot financial progress at regular intervals since the trends in the line may indicate potential problems before they actually occur.

INFORMATION TO BE GIVEN TO THE CLIENT

There is a considerable difference of opinion as to what information should be given to the client by the contractor. This ranges from the minimalist view under which the contractor passes the absolute minimum of information to the client to satisfy the contract, to the view that the contractor should pass over all the information available.

The minimalist information would normally be a simple bar chart of a few (20–50) activities. No indication of criticality or float would be provided and no resource or financial information would be given. This view tends to be taken by those who have not thought through the project thoroughly; it is often an indicator of problems in the future; and it provides no means for resolution of any disputes which arise.

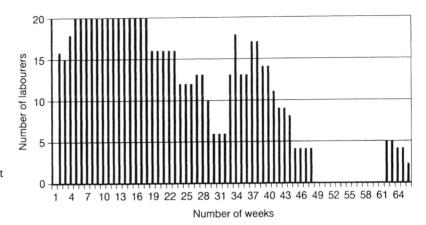

Figure 15.16 Scheduled start time resource profile for labour.

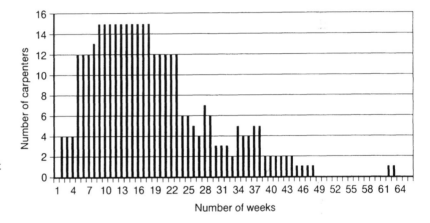

Figure 15.17 Scheduled start time resource profile for carpenters.

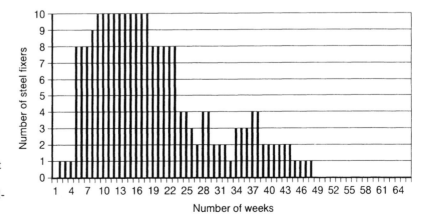

Figure 15.18 Scheduled start time resource profile for steel-fixers.

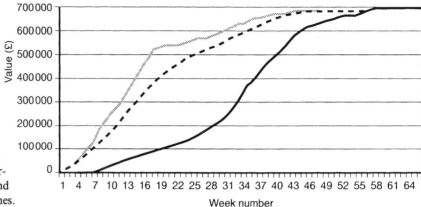

Figure 15.19 Labour costs plotted at earliest, latest and scheduled times.

The provision of all information would mean providing all the plans produced, the information used to produce them and the assumptions made. Typically it would consist of

- a network;
- the activity durations;
- the resource requirements of the activities;
- the resource productivities used to produce the activity durations;
- bar charts produced from the network showing float;
- the intended project resource levels;
- a resource schedule showing the project's use of resources;
- an earnings and liabilities against time curve;
- project phasing diagrams.

This information would be used to show the client exactly how the contractor intended to carry out the work, where potential problems were and what had to be done to minimise their effects. It would allow the client to plan their own work and would tell the client about the financial demands of the project. Should problems arise during the project, the information would provide a basis from which to calculate the entitlements of the parties in terms of money and time.

The approach (and attitude) of providing the client with all information is strongly recommended.

15.3.3 Short-term plans

For short-term planning, the resource bar chart (Section 2.2.4) has been selected. This is because it is good for ensuring that all the available resources are utilised fully and for communicating the work to be done to the resources that have to do it. It is a little more complex to draw than the equivalent Gantt chart but the advantages outweigh this slight disadvantage.

The short-term programme is drawn for one week with a day's lead in to the following week. The smallest time period considered is half a day. The plan should be drawn up on a Thursday, following consultation with all concerned. It should be agreed on a Thursday afternoon with both the senior managers and the foremen and communicated to the gangers and operatives on Friday in order that they can be prepared for a prompt start on Monday morning. In this project the short-term programme is to be produced for week 13.

In order to produce short-term planning, it is necessary to know what resources are on site for the period of the plan. These are shown in Table 15.10 for weeks 13 and 14 for the Tank Farm project. The resources considered for use at this stage may be different from those assumed in the long-term plan. This is because either more information becomes available or the work changes or the work method is changed. Whichever of these occurs, it is important to ensure that it is recorded and agreed by the parties concerned in order to avoid disputes later.

It is also necessary to know the current state of the project exactly. This obvious statement often causes problems. As has been discussed in Section 12.3, monitoring progress is more difficult than it first appears. The state of this project at the start of week 13 is described in Table 15.11. This table is open to interpretation and is given only for illustration purposes.

In addition to the resources predicted to be available, the planner should be aware of their expected costs and the allowances for the work. These are shown in Tables 15.12 and 15.6 respectively. It should be noted that the costs and allowances are both only for the controllable parts of the finance — labour and plant — as discussed above and in Section 10.3.

15.3.4 Evaluation of short-term plans

Any plan should be evaluated before acceptance and use. Managers should satisfy themselves of the goodness of the plan both qualitatively and quantitatively.

QUALITATIVE EVALUATION

The short-term programme should be evaluated in terms of its achievement of the objectives set out in the long-term or master programme, subject to the new constraints imposed upon it and the variations which occurred since the long-term programme was produced.

In this case, there must be differences between the long- and short-term plans because the work position described for the start of week 13 differs from the planned position, in so far as it can be determined from the long-term plan. These differences may well lead to inefficient use of the resources available on site. In addition there may not be the resources available to carry out some of the work scheduled on the long-term programme (as for the culvert construction in this project).

It is therefore impossible to meet the detailed objectives of the long-term plan. However, the short-term plan should be checked to ensure that it obeys the philosophy of the longer term plan in terms, for example,

Table 15.10 Resources predicted to be on site in weeks 13 and 14.

No. of gangs			
week 13	week 14	Type of gang	Make-up of gang
3	4	formwork gangs	1 foreman 2 joiners 1 labourer
3	3	steel-fixing gangs	1 ganger 4 steel fixers 1 labourer
1	1	concrete gang	1 ganger 2 skilled operatives 4 unskilled operatives
2	2	cranes	22RB or similar
1	1	tractor and trailer	
1	1	scaffolding gang	1 ganger 2 scaffolders
0	1	pipe-laying gang	1 ganger 1 banksman 2 pipe layers 1 Hymac 580D or similar 1 lorry
3	3	general labourers	

Table 15.11 Work position at start of week 13.

Work area	Current state
Set up site	completed
Bulk excavation	completed
Install precast yard	completed
Drainage outfall	completed
Tank 6	base poured at the end of week 12
Tank 5	base formwork in position, reinforcement ready to start
Control building	at foundation stage
Precast beams	6 moved at the end of week 12 to curing position — no other work on precast beams, awaiting possible redesign information

Table 15.12 Planned resource costs for week 13.

Resource	Labour cost (per hour)	Plant cost (per hour)
Formwork gang	£23.21	
Steelfixing gang	£36.87	
Concrete gang	£40.90	
Pipelaying gang labour	£28.80	£15.00
Scaffolding gang	£12.00	
Labourer	£4.00	
Tractor and trailer		£4.00
22RB crane		£30.00

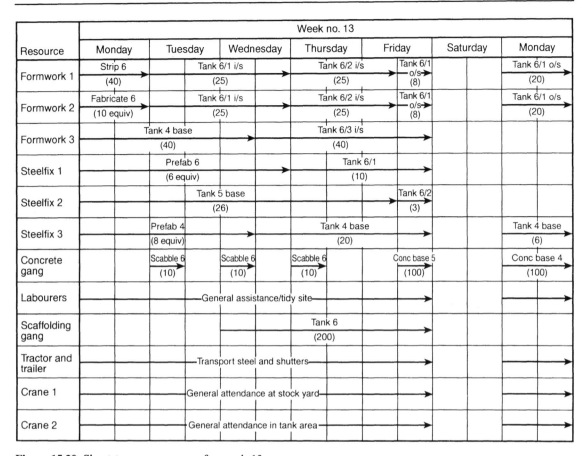

Figure 15.20 Short term programme for week 13.

of the sequencing of the major parts of the work, the spend and earn rates and the use of major resources.

The manager should check that all the resources predicted to be available on site are fully utilised. This is much simplified by the use of the resource bar chart for short-term planning since it relies on a list of the resources and shows what each is doing at all times.

All the work in the plan should be work for which allowances are available or for which payment will be made (Section 15.3.2).

The planner should check that the plan communicates to the resources exactly what they are supposed to be doing. They will then be able to prepare their own very detailed plans to ensure they achieve these objectives. The names of the work to be done should be informative and unambiguous.

On this point, it is important to note from the short-term programme (Figure 15.20) that it is impossible to tell what the crane and general labour are planned to do. Terms such as *general attendance* should be avoided wherever possible since they do not describe the work to be done and are uncontrollable. It is virtually impossible to know if there is sufficient cranage for the week because the possibly conflicting demands on the crane by the other resources cannot be seen on the programme as they have not been scheduled.

QUANTITATIVE EVALUATION

From the plan, it is a relatively simple matter to take off the items of work to be done. This is shown in tabular form in Table 15.13 which also contains the allowances for the work. A section engineer would have relatively few different types of work to consider and should need only 10–20 rates at any time to produce this. The task is made considerably easier by including the quantities of work on the programme as shown in Figure 15.20 where the quantities of work to be done are shown in brackets under the bars on the resource bar chart.

The ratios described in Section 10.4 can be calculated as shown in Table 15.14. Lack of entry in a particular cell indicates that no information exists or that no ratio can be calculated.

From these ratios, it can be seen that overall, the site is planning to work approximately 6% better than estimate but that this is made up from labour being 22% better and plant being less than half as good. Looking at the individual gangs the site is planning to lose money on formwork and to make considerable amounts of money on steelfixing.

These figures should enable the site manager to examine particular types of work to ensure the plan is reasonable and indicate whether or not pro-active control action is necessary either to alter the plan or the resources.

It can be seen that because the labourers, cranes and tractors have not been assigned to specific tasks there is no allowance for them. This makes it impossible to evaluate a planned performance and consequently makes it impossible to take any pro-active control action. As shown later, it also makes it impossible to take meaningful re-active control action.

15.3.5 Monitoring and control

Figure 15.21 shows the work done in week 13 in the same form as the weekly plan. This is by no means the only way of recording the information and, for example, a coloured-up general arrangement of the project or a shaded bar chart could be used. Often it is advantageous to keep all of the above since they show different things and no matter what records are kept, it is invariably the one that has not been kept that is required.

The resource availability during the week is also shown. Such information should be available in approximate form at the end of the week under consideration. Slight changes will not be important but speed

Table 15.13 Planned earning for week 13.

Item	Quan-tity	Labour rate	Plant rate	Labour total	Plant total
Strip base 6	40.00	2.00	0.20	80.00	8.00
Formwork tank 6/1 i/s (wall1 inside)	25.00	12.00	2.20	300.00	55.00
Formwork tank 6/2 i/s	25.00	12.00	2.20	300.00	55.00
Formwork 6/1 o/s (wall1 outside)	8.00	12.00	2.20	96.00	17.60
Total formwork gang 1				*776.00*	*135.60*
Fabricate formwork 6	10.00	12.00	2.20	120.00	22.00
Formwork 6/1 i/s	25.00	12.00	2.20	300.00	55.00
Formwork 6/2 i/s	25.00	12.00	2.20	300.00	55.00
Formwork 6/1 o/s	8.00	12.00	2.20	96.00	17.60
Total formwork gang 2				*816.00*	*149.60*
Formwork tank 4 base	40.00	12.00	2.00	480.00	80.00
Formwork tank 6/3 i/s	40.00	12.00	2.20	480.00	88.00
Total formwork gang 3				*960.00*	*168.00*
Steelfix prefabricate 6	6.00	165.00	10.00	990.00	60.00
Steelfix tank 6/1	10.00	165.00	10.00	1650.00	100.00
Total steelfix gang 1				*2640.00*	*160.00*
Steelfix tank base 5	26.00	160.00	10.00	4160.00	260.00
Steelfix tank 6/2	3.00	165.00	10.00	495.00	30.00
Total steelfix gang 2				*4665.00*	*290.00*
Steelfix prefabricate 4	8.00	160.00	10.00	480.00	80.00
Steelfix tank 4 base	20.00	165.00	10.00	3200.00	200.00
Total steelfix gang 3				*3680.00*	*280.00*
Scabble 6	10.00	8.50	0.50	85.00	5.00
Scabble 6	10.00	8.50	0.50	85.00	5.00
Scabble 6	10.00	8.50	0.50	85.00	5.00
Concrete base 5	100.00	7.00	2.10	700.00	210.00
Total concrete gang				*955.00*	*225.00*
Scaffold tank 6	200.00	2.00	0.20	400.00	40.00
Total scaffolding gang				*400.00*	*40.00*
Total labourers, cranes & tractor				*0.00*	*0.00*
Total				14 892.00	1448.20

in obtaining the results is essential if proper use is to be made of them. Tables 15.15 to 15.17 for resource costs, earnings and performances show, respectively, the actual situation for week 13 comparable with the planned situation presented in Tables 15.12 to 15.14

From these calculations it can be seen that the site has actually achieved overall 14% (that is 1/0.88) better than originally anticipated).

Table 15.14 Planned performances for week 13.

Resource	Labour liability	Plant liability	Labour allowance	Plant allowance	Labour perfor- mance	Plant perfor- mance
Formwork gang 1	1150.00		776.00	135.60	1.49	
Formwork gang 2	1150.00		816.00	149.60	1.41	
Formwork gang 3	1150.00		960.00	168.00	1.20	
Steelfix gang 1	1850.00		2640.00	160.00	0.70	
Steelfix gang 2	1850.00		4665.00	290.00	0.40	
steelfix gang 3	1850.00		3680.00	280.00	0.50	
Concrete gang	2000.00		955.00	225.00	2.08	
Pipelaying gang	0.00					
Scaffolding gang	600.00		400.00	40.00	1.49	
Labourers	600.00					
Cranes & tractor		3200.00				
Total	12200.00	3200.00	14892.00	1448.20	0.82	2.22

Total overall performance = ((12200.00 + 3200.00)/(14892.00 + 1448.20)) = 0.94

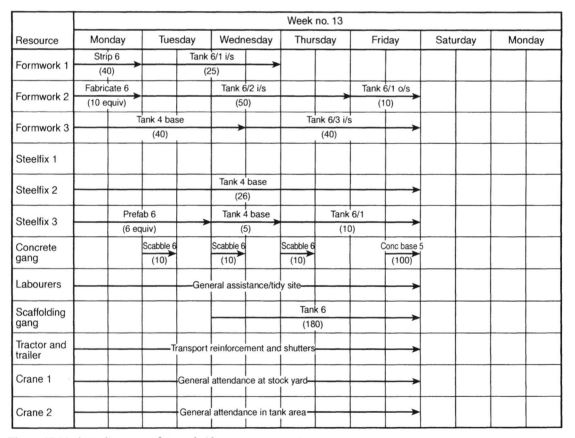

Figure 15.21 Actual progress for week 13.

Table 15.15 Actual resource costs for week 13.

Resource	Comments	Labour cost (£)	Plant cost (£)
Formwork gang 1	only available Monday–Wednesday	752.00	
Formwork gang 2	available all week	1172.50	
Formwork gang 3	1 joiner missing all week	862.50	
Steelfix gang 1	not on site at all	0.00	
Steelfix gang 2	available all week	1724.50	
Steelfix gang 3	available all week	1820.30	
Concrete gang	only available intermittently (working on a number of sites)	1050.20	
Pipelaying gang labour	not available	0.00	0.00
Scaffolding gang	only available Monday–Wednesday	360.00	
Labourer 1	not available	0.00	
Labourer 2	available all week	190.00	
Labourer 3	available all week	197.50	
Tractor and trailer	available all week		204.50
Crane 1	broken down Monday and Friday		962.30
Crane 2	available all week		1420.30

Table 15.17 Actual performances for week 13.

Resource	Labour liability	Plant liability	Labour allowance	Plant allowance	Labour performance	Plant performance
Formwork gang 1	752.00		380.00	63.00	1.96	
Formwork gang 2	1172.50		840.00	154.00	1.39	
Formwork gang 3	862.50		960.00	168.00	0.90	
Steelfix gang 1	0.00		0.00	0.00		
Steelfix gang 2	1724.50		4160.00	260.00	0.41	
Steelfix gang 3	1820.30		3410.00	210.00	0.53	
Concrete gang	1050.20		955.00	225.00	1.09	
Pipelaying gang	0.00					
scaffolding gang	360.00		360.00	36.00	1.00	
labourers	387.50					
cranes & tractor		2587.10				
Total	8129.50	2587.10	11065.00	1116.00	0.73	2.32

Actual overall performance = ((8129.50 + 2587.10) / (11065.00 + 1116.00)) = 0.88

Table 15.16 Actual earning for week 13.

Item	Quantity	Labour rate	Plant rate	Labour total	Plant total
Strip 6	40.00	2.00	0.20	80.00	8.00
Formwork tank 6/1 i/s	25.00	12.00	2.20	300.00	55.00
Total formwork gang 1				*380.00*	*63.00*
Fabricate formwork 6	10.00	12.00	2.20	120.00	22.00
Formwork 6/2 i/s	50.00	12.00	2.20	600.00	110.00
Formwork 6/1 i/s	10.00	12.00	2.20	120.00	22.00
Total formwork gang 2				*840.00*	*154.00*
Formwork 6/2 i/s	40.00	12.00	2.00	480.00	80.00
Formwork 6/1 i/s	40.00	12.00	2.20	480.00	88.00
Total formwork gang 3				*960.00*	*168.00*
Steelfix tank 5 base	26.00	160.00	10.00	4160.00	260.00
Total steelfix gang 2				*4160.00*	*260.00*
Steelfix prefabricate 6	6.00	160.00	10.00	960.00	60.00
Steelfix tank 4 base	5.00	160.00	10.00	800.00	50.00
Steelfix tank 6/1	10.00	165.00	10.00	1650.00	100.00
Total steelfix gang 3				*3410.00*	*210.00*
Scabble 6	10.00	8.50	0.50	85.00	5.00
Scabble 6	10.00	8.50	0.50	85.00	5.00
Scabble 6	10.00	8.50	0.50	85.00	5.00
Concrete base 5	100.00	7.00	2.10	700.00	210.00
Total concrete gang				*955.00*	*225.00*
Scaffold tank 6	180.00	2.00	0.20	360.00	36.00
Total scaffolding gang				*360.00*	*36.00*
Total labourers, cranes & tractor				*0.00*	*0.00*
Total				11065.00	1116.00

This is made up of a 36% better labour performance and a plant performance that is only half as good. It can also be seen that the site made a loss on formwork and a profit on steelfixing.

These actual performances do not tell the whole story since they do not indicate whether the plan was achieved. To determine this the efficiencies have been calculated as described in Section 10.4.2. They are shown in Table 15.18.

Table 15.18 Performance and efficiency figures for week 13.

Resource	Planned labour performance	Planned plant performance	Actual labour performance	Actual plant performance	Labour efficiency	Plant efficiency
Formwork gang 1	1.49		1.96		1.31	
Formwork gang 2	1.41		1.39		0.99	
Formwork gang 3	1.20		0.90		0.75	
Steelfix gang 1	0.70				-	
Steelfix gang 2	0.40		0.41		1.03	
Steelfix gang 3	0.50		0.53		1.06	
Concrete gang	2.08		1.09		0.52	
Pipelaying gang						
scaffolding gang	1.49		1.00		0.67	
labourers						
cranes & tractor						
Total	0.82	2.22	0.73	2.32	0.89	1.05

Overall efficiency = 0.88/0.94 = 0.94

From this it can be seen that, approximately:

- formwork gang 1 worked 31% worse than planned;
- formwork gang 2 worked almost exactly to plan;
- formwork gang 3 worked 25% better than planned;
- steelfixing gangs 2 and 3 worked approximately 5% worse than plan;
- the scaffolders worked 33% better than planned;
- overall the labour worked 11% better than planned;
- overall the plant worked 5% worse than planned;
- the site overall worked 6% better than planned.

A plan should be a target and therefore would be expected to be approximately 5% optimistic. It can be seen that only the 2 steelfixing gangs approached this. The planner should use this information to improve future planning.

15.4 CONCLUDING COMMENT

The chapter has used some of the techniques explained in this book to produce long- and short-term plans for the Tingham Tank Farm project. Good points and bad points have been included and discussed. The plans have been used to generate information for the control and future planning of the project. A system for planning and controlling the whole project has been introduced.

16

TINGHAM INTERNATIONAL HOTEL PROJECT

16.1 INTRODUCTION

This chapter discusses the planning of the Tingham International Hotel refurbishment project. Section 16.2 contains all the available information for the project including drawings, quantities and resource availability.

The project contains many of the elements of general building planning. There are many instances of tasks to be done in different places by similar or the same resources. These tasks can be done in almost any order and the planning process becomes one of sequencing the resources through the processes of the project rather than ordering the work correctly as in a civil engineering project like that described in Chapter 15.

Refurbishment is a particularly difficult area of construction to plan, monitor and control with any accuracy. It forms an increasing proportion of the construction work in many countries. With generally declining markets and decreasing margins, the need for good planning and control is obvious.

Several aspects of refurbishment projects are contained in this example. There are frequently problems encountered which, although having little to do with construction, have a significant effect on the programme. Contractors refurbishing multi-storey housing blocks have found the residents uncooperative and household electrical appliances being thrown from windows endangering the workers below; newly installed central heating equipment being removed and sold before inspection for verification of installation; and residents deliberately refusing entry to workers despite having been given all the necessary and agreed notice. Such problems must not be underestimated at the planning stage despite not being covered here.

16.2 TINGHAM HOTEL

The Tingham International Hotel was built in the late 1960s in an attractive part of the town and was then the best hotel in the area. It is now showing its age and the owners have decided on a complete renovation of all facilities to provide once again a first-class individual hotel.

16.2.1 Project description

The project consists of the refurbishment of all the accommodation and public and staff facilities of the 19 storey town centre hotel. The client has specified that the hotel must remain open and operational at all times.

Because of the complexity of the work, the required quality and the importance of customer relations, the client has requested that wherever possible, the main contractor's own labour should be used. The main contractor has agreed to this except in the kitchen area where the suppliers of the equipment are to install it, the air conditioning and the soft furnishings which are to be supplied and installed by specialist subcontractors.

The client's desire for the hotel to remain operational has been taken into account and it has been specified that:

- For the guest-room refurbishment, work will only take place on a floor which is completely empty of guests.
- When working in a guest-room, the whole of the wing of the floor below that room must be unoccupied.
- Access for all resources and materials must be through Lift 3.
- Work can take place only between 08.30 and 17.30 Monday to Friday and no weekend working is allowed.
- Vehicle access is available only at one entrance (as shown).
- No material or waste may be stored outside the working area.
- No more than one restaurant/public access area may be worked on at any one time.
- No more than one kitchen may be worked on at any one time.
- Services to occupied rooms must be maintained at all times.

The work will be carried out in two phases:

Phase 1 The renovation of floors 2 to 18 including all guest-rooms, cleaners' rooms and public access areas. This is a 30 week contract but it is unlikely that construction work can start before week 4 because of the difficulty of moving the guests from their rooms.

Phase 2 The renovation of the basement, and floors 1 and 19, the restaurants and kitchens, store areas, staff accommodation, lounges and bars and the construction of two new fire escape routes at the ends of the two wings.

All the facilities in the rooms are in need of removal and replacement.

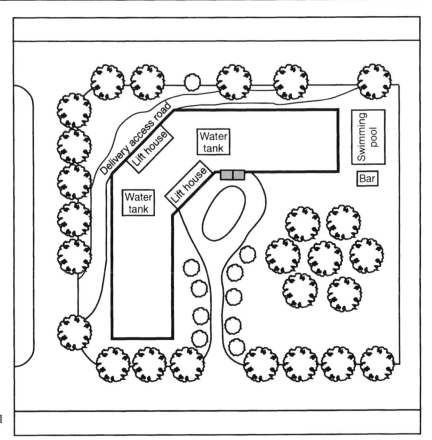

Figure 16.1 Tingham International Hotel – general site layout.

This project covers only Phase 1 but the contractor would hope to win the contract for Phase 2 on a negotiated basis during the work on the first phase.

16.2.2 Existing structure

The existing structure, which is structurally sound, is a reinforced concrete frame with internal walls made of 300 mm blockwork plastered internally. External walls are 300 mm blockwork with concrete cladding panels hung from the frame. Floors comprise prestressed concrete T-beams with 50 mm screed.

Ceilings are suspended from the T-beams with a 450 mm gap between the ceiling and the soffit of the beam. Services run through the ceilings.

Existing windows are double-glazed units which are not being replaced.

16.2.3 Existing services

The services provided at present are

- hot and cold water,
- electricity (220 v AC),
- cable television,
- telephone,
- air-conditioning and heating.

16.2.4 Drawings

The drawings in Figures 16.1 to 16.7 illustrate the general arrangement of the site and provide details of the major construction.

16.2.5 Quantities

The quantities for the major items of work are shown in Table 16.1.
Some items are to be supplied and installed by specialists. In particular:

- The air-conditioning, heating and ventilating (HVAC) will be installed by a specialist subcontractor who will require:
  - One gang for one working day per guest room; no other people can work in the room at the same time because of access problems and the work must be completed in a room before the suspended ceiling is installed.
  - One gang for one working week per corridor and must be completed before the suspended ceiling is installed.
- The carpets in the rooms are to be fitted by a specialist sub-contractor requiring 4 hours per guest room and 1 day per wing per floor for the public areas.

16.2.6 Allowances

The allowances that appear in the contractor's estimate are given in Table 16.2.

16.3 PLANNING THE HOTEL REFURBISHMENT

This project is complex. Planning is essential in order to ensure that the work proceeds in the most efficient, safe and cost-effective manner. From a contractor's point of view, no single technique will be able to communicate all the necessary information to all the parties involved and it is essential to have a method statement for the project. This should include:

- Sketches showing layout of the project during construction, verifying that the method adopted will adhere to the imposed restrictions. The access points to be used will be of particular importance in this project and they should be included in these sketches to demonstrate that their practicality has been considered.

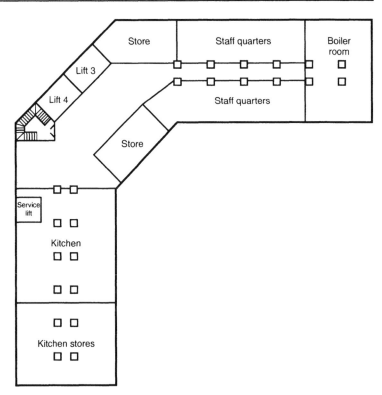

Figure 16.2 Tingham International Hotel – general arrangement of basement.

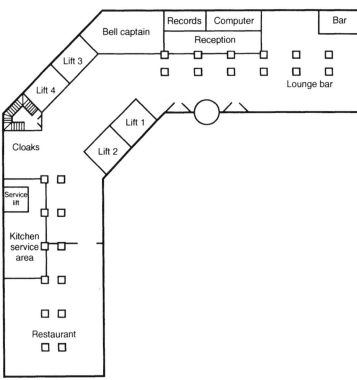

Figure 16.3 Tingham International Hotel – general arrangement of ground floor.

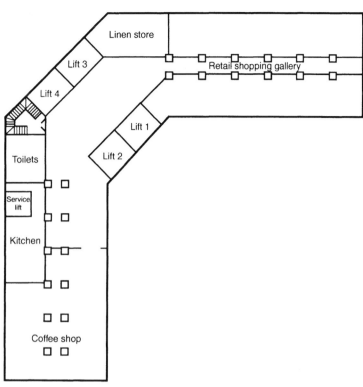

Figure 16.4 Tingham International Hotel – general arrangement of floor 1.

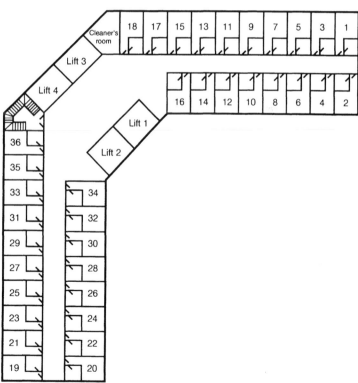

Figure 16.5 Tingham International Hotel – general arrangement of floors 2–18.

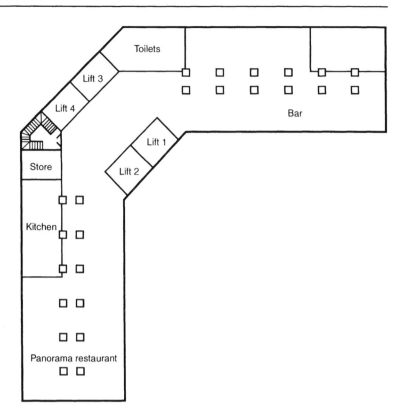

Figure 16.6 Tingham International Hotel – general arrangement of floor 19.

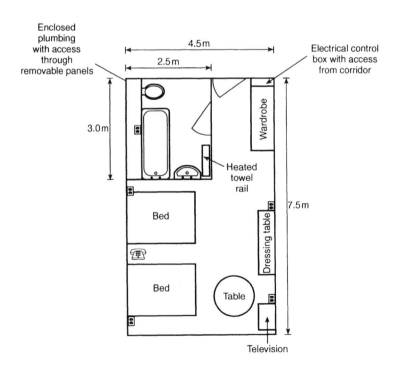

Figure 16.7 Tingham International Hotel – general arrangement of guest-room.

Table 16.1 Estimated quantities.

Item	Quantity	Unit
Bathroom		
Floor tiling	5	m^2
Wall tiling	9.5	m^2
Plumbing	1	item
low level, WC		
wash hand basin		
bath with shower		
Electrical	1	item
extractor fan		
light fittings		
Bedroom		
Decoration	1	item
wall paper		
painting		
Carpet	26	m^2
Electrical fittings	1	item
Light fittings		
Telephone fittings	1	item
Furniture	1	item
wardrobe		
beds		
table		
dressing table		
Air conditioning	1	item
Ceiling	26	m^2
Finishings	1	item
Quantities for floor		
Decoration	1	item
Paintwork		
Wall paper		
Light fittings	28	nr.
Fire doors	4	nr.
Carpeting	460	m^2
Sprinkler system	1	item
Ceiling	460	m^2

- A programme of work for the whole project.
- A list of the resources to be used and their allocation to the project programme, showing that the rates required are achievable.
- A financial appraisal of the project.

These items are discussed in the following sections.

With a significant amount of the work being subcontracted, there is a great temptation to ignore the detail of subcontracted work when

Table 16.2 Allowances and quantities per room in the contractor's estimate.

Item	Quantity	Unit	Allowance (£/unit) Labour only
Bathroom			
Floor tiling	5	m²	30
Wall tiling	9.5	m²	114
Plumbing	1	item	162
low level, WC			
wash hand basin			
bath with shower			
Electrical	1	item	45
extractor fan			
light fittings			
Bedroom			
Decoration	1	item	121
wall paper			
painting			
Carpet	26	m²	0.80
Electrical fittings	1	item	45
light fittings			
Telephone fittings	1	item	10
Furniture	1	item	45
wardrobe			
beds			
table			
dressing table			
Air conditioning	1	item	450
Ceiling	26	m²	378
Finishings	1	item	50

planning. In general, this leads to problems later in the project. Not only will each subcontractor not know how their work impinges on that of others but also good control will be impossible, extra work will tend not to be recognised and the likely financial performance of the project will not be known. All this leads to more expensive and less efficient construction. If subcontractors are to be used, all the work should be viewed as the main contractor's own and treated in enough detail in time, resources and money to allow full control to be exercised.

16.3.1 Choice of activities and planning technique

The hotel refurbishment project consists mainly of repetitive work but with important non-repetitive elements. The repetitive parts are the rooms on Floors 2 to 18 and the public access areas on the same floors. The activities in the plan depend on the techniques selected and their selection is discussed with the choice of planning technique.

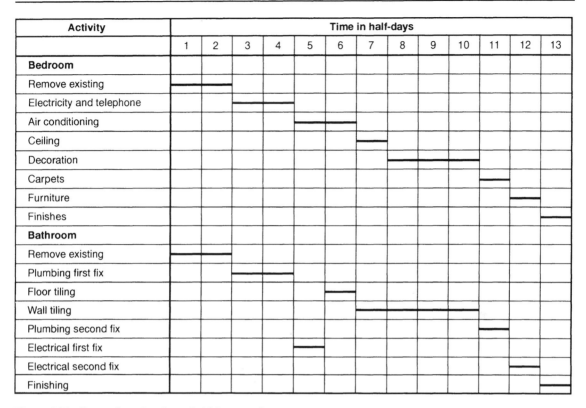

Figure 16.8 Gantt chart for the refurbishment of one guest-room.

PLANNING FOR THE REPETITIVE WORK

The repetitive work in this project is repetitive in two ways: each of the 36 rooms on a floor has the same work to be done and each of 17 floors has the same work. Although there may be slight differences between the work in each room and on each floor, they can be considered to be identical for planning purposes. The planning techniques used for this project must be able to take advantage of the repetition.

For the planning of the rooms on a floor, two techniques are suggested. First, it is necessary to plan the work in a single room. A Gantt chart is ideal for this and one possible one is shown in Figure 16.8

The guest-room and bathroom are combined into one chart. Another equally good approach would be to have one chart for each since they contain significantly different work and are independent except for access through the bedroom doorway.

The activities have been chosen to be single-trade ones and their durations have been developed from the outputs of the trades on the various tasks. These outputs should come from company records, trade literature or subcontractors' estimates. The latter should not be used without verification since they could have significant adverse effects on

Table 16.3   Resource outputs used for planning of room.

Resource	Unit	Planned output (per gang per day)
Plumbing first fix	item	1 room
Plumbing second fix	item	2 rooms
Floor tiling	m²	3 m²
Wall tiling	m²	3 m²
Electrical	item	1 room
Ceiling installation	m²	25 m²
Carpeting	m²	40 m²
Decoration	item	0.5 room

the planning and therefore the outcome of the project. The outputs used here are shown in Table 16.3. In this project, as with much building work, the outputs are difficult to define in detail because of the complexity of the task involved (for example, *plumbing first fix* involves such things as pipe cutting, bending, fixing and joining and the installation of various fittings). For such tasks the output is in terms of *items* in the table.

To plan the work for the whole floor, it is useful to maintain the level of detail present in the plan for a single room. The use of these individual activities gives an unmanageable number of activities in total on a Gantt chart. It is therefore suggested that a line-of-balance approach is adopted. For this, it is necessary to have a target duration for the completion of all the rooms on a floor. Figure 16.9 shows the production charts (Section 2.4) for the whole of one floor based on a target completion time of 20 days.

Twenty days has been chosen as the target completion time by considering an overall plan (Figure 16.10). This itself was drawn up obeying the contract and constraints and such that the work on the floors could continue requiring only three floors to be worked on at any time. This has the effect of keeping each floor closed for $5\frac{1}{3}$ weeks and allows some slack into the work on each floor. This would allow for slight alterations to take account of unforeseen problems without totally disrupting the cycles. It is therefore a low-risk plan and should generally be considered to be better than a high-risk plan in which a slight change would affect all the cycles.

At this point, the planning process itself becomes obviously iterative. The plan obtained, based on the assumption of the target completion time, determines the resources to be used, the overall project duration and the number of operations going on together. If any of these is not satisfactory, the process can be gone through over and over again until satisfactory answers are obtained.

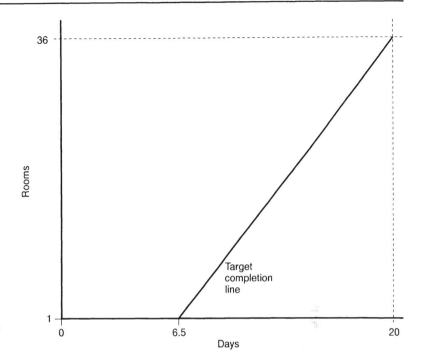

Figure 16.9 Production targets for a floor.

Activity	Week number																													
	1	2	3	4	5	6	7	8	9	10	11	12	13	14	15	16	17	18	19	20	21	22	23	24	25	26	27	28	29	30
Finishes																														
Floor 18																														
Floor 17																														
Floor 16																														
Floor 15																														
Floor 14																														
Floor 13																														
Floor 12																														
Floor 11																														
Floor 10																														
Floor 9																														
Floor 8																														
Floor 7																														
Floor 6																														
Floor 5																														
Floor 4																														
Floor 3																														
Floor 2																														
Initial set up																														

Figure 16.10 Long-term bar chart programme.

PLANNING FOR THE NON-REPETITIVE WORK

The non-repetitive work in this project is concerned with the 'enabling' work. Specifically this includes the initial organisation of the site; the development of a stores area; the setting up of the offices; the ordering of the various materials; the protection of the lifts from damage by the transport of the materials and workers; and the temporary access facilities if allowed (an external material hoist for example). The finishes at the end of the project are concerned with the removal of the facilities and the final tidying up at the end of the project.

The obvious way to do this planning is with a Gantt chart (Section 2.2).

PRESENTATION OF THE PLAN

For this type of project, there is no single technique which allows representation of all the work in satisfactory detail. It is suggested that the following be used:

- an outline Gantt chart showing all the work on the project (Figure 16.10);
- a detailed Gantt chart for the work in a single room (Figure 16.8);
- a detailed Gantt chart for the non-repetitive work;
- a target-completion-time graph for the completion of a typical floor (Figure 16.9).

The combination of the repetitive and non-repetitive work in the planning phase is not difficult in this example since in reality they do not interfere with each other. Separate plans could be used but it is usually convenient to show all the work on the same programme. A Gantt chart is used for this and is shown in Figure 16.10.

From the programmes presented, it can be seen that there is an almost infinite order in which the detailed activities can be done. The rooms on each floor can be done in any order (there are 36! ways of doing this) and even the work within each room can be done in a large variety of sequences. There are therefore something like 4×10^{42} feasible orders for carrying out the activities. With such freedom, most computer planning software is of little use since it is almost invariably based on networks. Computers can assist in presentation and record keeping and the more general computer packages such as spreadsheets and databases can be effectively employed for resource and financial planning.

Obviously with such a large number of elements making up the overall plan, monitoring becomes rather more complex than on a standard project. This is discussed in Chapter 12.

Table 16.4 Resources predicted to be on site in week 13.

	No. of gangs	No. in gang	No. of workers	Hourly cost (£) per gang (estimated)
Bathroom				
First and second fix plumbing	3	2	6	12.00
Bathroom flooring	1	1	1	6.00
Bathroom wall tiling	4	1	4	6.00
Electrical	1	1	1	7.50
Bedroom				
Electrical and telephone	1	2	2	7.50
Air conditioning	2	2	4	12.00
Ceiling installation	1	2	2	11.64
Decorating	3	2	6	9.23
Carpet laying	1	1	1	6.00
Furniture installation	1	2	2	10.00
General labour gang (remove existing, transport, attendance)	3	2	6	10.00

16.3.2 Short-term plans

As with all short-term planning, the objectives in this project are to ensure that all the resources are employed as effectively as possible to realise the objectives of the longer term plans. They should also be used as major means of communication with the various operatives.

To ensure that the first objective is achieved, the resources that are predicted to be available for the period are first listed. In this case, the resources predicted to be available for week 13 of the project are shown in Table 16.4. To help achieve the second objective (communication), it has been decided to use resource bar charts showing each gang which areas/rooms they are to work on in half-day periods. The whole programme should be issued to all the gangs/subcontractors with their own work highlighted.

An equally important consideration when drawing up any plans, but particularly short-term ones, is that the plan should start from the current situation. It therefore relies on the monitoring and record keeping carried out for the previous periods on the project. The current state of the project is described and discussed in various ways in Section 16.3.3.

The short-term plan should in general be drawn up weekly for a period of 10 days. Any less frequently and it would fail to recognise the inevitable changes on the project; any more frequently and it would occupy too much management time without significantly increasing its effectiveness. Obviously, if a major change occurs, such as one of the floors not becoming available when required, management should replan as soon as the necessary information becomes available. Figure 16.11 shows the short-term programme for week 13 (and start of week 14) for the example based on the predicted resource availability.

Week number	Monday		Tuesday		Wednesday		Thursday		Friday		Monday		Tuesday	
Resource gang	am	pm	am	pm	am	pm	am	pm	am	pm	am	pm	am	pm
Plumbing 1	8/13/1		8/15/1		8/17/1		8/19/1		8/21/1		8/23/1		8/25/1	
Plumbing 2	8/16/1		8/18/1		8/20/1		8/22/1		8/24/1		8/26/1		8/28/1	
Plumbing 3	7/26/2	7/25/2	7/24/2	7/23/2	7/22/2	7/21/2	7/20/2	7/19/2	8/1	8/2	8/3	8/4	8/5	8/6
Bathroom flooring	7/22	7/20	7/19	8/13	8/16	8/15	8/18	8/17	8/20	8/19	8/22	8/21	8/24	8/23
Bathroom walling 1			7/22				8/13				8/17		8/21	
Bathroom walling 2	7/25		7/21				8/16				8/20			8/24
Bathroom walling 3	7/24				7/20		8/15				8/19			
Bathroom walling 4			7/23		7/19				8/18				8/22	
Bathroom electrical	8/11/1	8/14/1	8/13/1	8/16/1	8/15/1	8/18/1	8/17/1	8/20/1	8/19/1	8/22/1	8/21/1	8/24/1	8/23/1	8/26/1
Electrical and telephone	7/27		7/30		7/25		7/28		7/23		7/26		7/21	
Air conditioning 1	7/32		7/27		7/30		7/25		7/28		7/23		7/26	
Air conditioning 2														
Ceiling installation					7/34	7/29	7/32	7/27	7/30	7/25			7/28	7/23
Decorating 1			7/15		7/17		7/33				7/29			
Decorating 2			7/36		7/18		7/31				7/30			
Decorating 3			7/35		7/34		7/32				7/27			
Carpeting	7/12	7/14	7/16	7/15	7/36	7/35	7/17	7/18	7/34	7/33	7/31	7/32	7/29	7/30
Furniture installation	7/10	7/11	7/12	7/13	7/14	7/15	7/16	7/17	7/18	7/36	7/35	7/34	7/33	7/32
Labour 1 – material supply														
Labour 2 – bathroom finishes	7/29	7/27	7/26	7/25	7/24	7/23	7/22	7/21	7/20	7/19	8/1	8/2	8/3	8/4
Labour 3 – bedroom finishes		7/10	7/11	7/12	7/13	7/14	7/15	7/16	7/17	7/18	7/36	7/35	7/34	7/33

Figure 16.11 Short-term programme for week 13.

The programme should be evaluated using the financial ratios (Section 12.5.3.2) but for this project there are only labour ratios since the only plant used is small tools and is deemed to be included in the labour rates.

For the programme shown in Figure 16.11, the ratios are shown in Table 16.5.

Computer based planning packages in general are of little help in the preparation of short-term programmes. However, a spreadsheet can be very useful especially in the evaluation process.

Another way of presenting this information would be in the form of a list such as is shown in Table 16.6. This has the advantage of being simpler to follow and hence rather more useful for communication when used in a large building project. It is of little use in the formulation of the plan and the recognition of the interactions of the gangs and the tasks on which they are working. For this reason, the resource bar chart is preferred as a working document but a list may be abstracted for passing to the resources concerned.

Table 16.5 Table of planned performances for week 13.

Resource	Planned liability (£)	Planned earning (£)	Planned performance
Plumbing 1	648.00	728.00	0.89
Plumbing 2	648.00	728.00	0.89
Plumbing 3	648.00	728.00	0.89
Bathroom flooring	324.00	420.00	0.77
Bathroom walling 1	324.00	399.00	0.81
Bathroom walling 2	324.00	399.00	0.81
Bathroom walling 3	324.00	399.00	0.81
Bathroom walling 4	324.00	399.00	0.81
Bathroom electrical	405.00	630.00	0.64
Bedroom electrical and telephone	405.00	385.00	1.05
Air conditioning 1	648.00	450.00	1.44
Air conditioning 2	0.00	0.00	—
Ceiling installation	628.56	630.00	1.00
Decoration 1	443.00	484.00	0.92
Decoration 2	443.00	484.00	0.92
Decoration 3	443.00	484.00	0.92
Carpet laying	324.00	291.20	1.11
Furniture	540.00	630.00	0.86
Finishings	1440.00	260.00	5.54
Overall	9283.56	8928.20	1.04

16.3.3 Monitoring and record keeping

With all large building projects, monitoring and record keeping are important and difficult functions to carry out. They are important to ensure that progress is maintained, that the project is run efficiently, that the project is profitable and to provide information for future projects. They are difficult because of the large amount of information and the need to be able to see cause and effect throughout perhaps long periods of time.

Many techniques have been tried, most with the utter conviction of the management team that they are exactly what is required and that they will provide all the necessary information. Unfortunately, all techniques have been shown to be individually lacking in some aspects, especially when detailed information is required after the project for claims' information. The methods range from a basic coloured bar chart to an immensely complicated chart relying on the use of different symbols and colours on a time–location grid to represent the status of the different activities in that place at different times. The simple techniques have the advantage of being simple but the disadvantage of not showing all the information. The complicated techniques have the disadvantage of being complicated to the point of being impossible to

Table 16.6 Short-term plan for week 13 produced as a list.

Resource	Tasks
Plumbing gang 1	8/13/1, 8/15/1, 8/17/1, 8/19/1, 8/21/1 (floor/room/1st or 2nd fix)
Plumbing gang 2	8/16/1, 8/18/1, 8/20/1, 8/22/1, 8/24/1
Plumbing gang 3	7/26/2, 7/25/2, 7/24/2, 7/23/2, 7/22/2, 7/21/2, 7/20/2, 7/19/2, 8/1/2, 8/2/2
Bathroom flooring	7/22, 7/20, 7/19, 8/13, 8/16, 8/15, 8/18, 8/17, 8/20, 8/19 (floor/room)
Bathroom walling 1	7/22, 8/13, 8/17
Bathroom walling 2	7/25, 7/21, 8/20
Bathroom walling 3	7/24, 7/20, 8/16
Bathroom walling 4	7/23, 7/19, 8/18
Bathroom electrical	8/11/1, 8/14/1, 8/13/1, 8/16/1, 8/15/1, 8/18/1, 8/17/1, 8/20/1, 8/19/1, 8/22/1, (floor/room/fix)
Electrical and telephone	7/27, 7/30, 7/25, 7/28, 7/23 (floor/room)
Air conditioning 1	7/32, 7/27, 7/30, 7/25, 7/28
Ceiling installation	7/34, 7/29, 7/32, 7/27, 7/30, 7/25
Decorating 1	7/15, 7/17, 7/33, 7/29
Decorating 2	7/36, 7/18, 7/31, 7/30
Decorating 3	7/35, 7/34, 7/32, 7/27
Carpeting	7/12, 7/14, 7/16, 7/15, 7/36, 7/35, 7/17, 7/18, 7/34, 7/33
Furniture installation	7/10, 7/11, 7/12, 7/13, 7/14, 7/15, 7/16, 7/17, 7/18, 7/36
Labour	
1 attending on trades	
2 bedroom finishes	7/29, 7/27, 7/26, 7/25, 7/24, 7/23, 7/22, 7/21, 7/20, 7/19
3 bathroom finishes	7/10, 7/11, 7/12, 7/13, 7/14, 7/15, 7/16, 7/17, 7/18

interpret, while still being incomplete. Chapter 12 covers monitoring and record keeping in more detail.

Recognising that no single technique is ideal, several methods are included here, in particular:

- A listing of the work situation at the end of a period (week 12) shown in Tables 16.7 and 16.8.
- A listing of the actual resources available throughout the period, their actual costs and comments on their availability where appropriate (Table 16.9).
- Floor diagrams 'coloured' (here, shaded) to show progress to date (Figures 16.12 and 16.13). These include a date and show cumulative progress only.
- An overall bar chart 'coloured' (here, shaded) at the end of each period (week) to show progress in that week and cumulative progress to date (Figure 16.14).
- Completion graphs for the various tasks (Figure 16.15).

Table 16.7 Progress on floors 6 and 7 to the end of week 12.

At the end of week 12, the following situation existed:

Floor 6 completed and handed back to hotel but closed due to work on floors above.

Floor 7 – work in progress – the following have been completed:

Bathrooms

Remove existing	36 rooms
Plumbing first and second fix	36 rooms
Floor tiling	33 rooms
Wall tiling	27 rooms
Electrical installation	33 rooms
Finishing	25 rooms

Bedrooms

Remove existing	36 rooms
Electrical and telephone installation	25 rooms
Ceiling installation inc. HVAC	21 rooms
Decorating	15 rooms
Carpet laying	12 rooms
Finishing and furniture installation	9 rooms

Table 16.8 Progress on Floor 8 to the end of week 12.

Floor 8 – work in progress – the following have been completed:

Bathrooms

Remove existing	36 rooms
Plumbing first and second fix	13 rooms
Floor tiling	10 rooms
Wall tiling	2 rooms
Electrical installation	10 rooms
Finishing	0 rooms

Bedrooms

Remove existing	36 rooms
Electrical and telephone installation	0 rooms
Ceiling installation inc. HVAC	0 rooms
Decorating	0 rooms
Carpet laying	0 rooms
Finishing and furniture installation	0 rooms

Table 16.9 Actual resource costs for week 13.

Resource	Comments	Actual cost (£)
Plumbing 1	480.00	680.00
Plumbing 2	480.00	650.00
Plumbing 3	480.00	640.00
Bathroom flooring	only 3 days work	240.00
Bathroom walling 1	240.00	350.00
Bathroom walling 2	240.00	350.00
Bathroom walling 3	only 4 days work	300.00
Bathroom walling 4	only 4 days work	300.00
Bathroom electrical		300.00
Bedroom electrical and telephone		300.00
Air conditioning 1	no work done	
Air conditioning 2	no work done	
Ceiling installation		465.60
Decoration 1	paid on measured work	369.20
Decoration 2	paid on measured work	369.20
Decoration 3	paid on measured work	369.20
Carpet laying	paid on measured work	300.00
Furniture	paid on measured work	400.00
Labour	two labourers available all week	600.00
Overall		6983.20

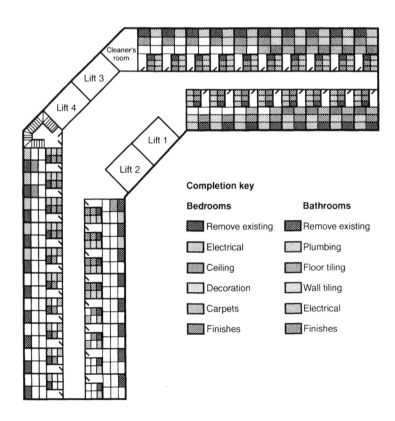

Figure 16.12 Floor 7 progress to week 13.

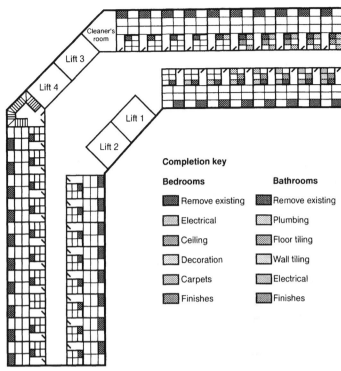

Figure 16.13 Floor 8
progress to
week 13.

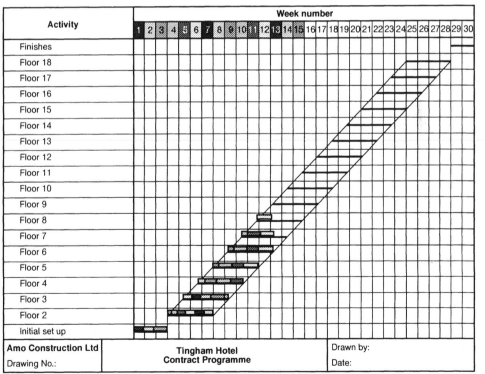

Figure 16.14 Updated bar chart to week 13.

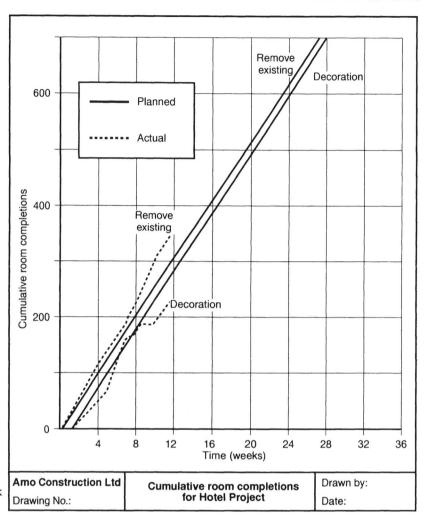

Figure 16.15 Target completion chart for work in rooms.

Such records should be kept at each period but it is not sufficient just to keep them. They must be used to monitor and control progress. Managers must determine whether the programme has been achieved, recognise deviations, understand why deviations happened and take corrective action to ensure either that they do not happen again, that the correct party pays for them, and that future plans take them into account (or a mixture of all of these).

This can be done to some extent by inspection. Unfortunately this is usually the only method used and much information is missed. It is suggested that the records of performance are translated into actual performance and thence efficiency figures as described in Chapter 12 and worked through in Chapter 15 for the Tingham Tank Farm.

16.4 CONCLUDING COMMENTS

This chapter has shown the different use of planning techniques for a hotel refurbishment project compared with the Tank Farm project (Chapter 15). Where the techniques used here converge (in for example the control phase) with those used in the Tank Farm project the analysis is not continued but the reader may wish to do so. In particular it is hoped that the diversity of available planning techniques and the need to evaluate them carefully before planning the project have been emphasised in this chapter.

Appendix A

TINGHAM DEVELOPMENT SCHEME

A.1 INTRODUCTION

Planning is both an art and a science. No matter how much theory is learned, skills can only be developed with practice. Books cannot really provide all that is necessary in a practical manner. In an attempt to overcome such problems, several projects are introduced in this appendix to help illustrate some of the points discussed in this book. All these projects are based on developments in the fictitious town of Tingham.

The projects considered are:

- Tingham Bridge
- Tingham Bypass
- Tingham Offices
- Tingham Hotel refurbishment
- Tingham Tank Farm.

The range of development projects at Tingham offers two new civil engineering works, a linear civil engineering project, a repetitive building project, and a building refurbishment project. This section contains a brief description of the town and all the projects.

The Tingham Bridge project is used throughout the book; the Bypass and Offices are used where examples of linear or repetitive projects are required. Chapter 2 includes examples of planning outputs for these projects.

Chapters 15 and 16 contain fuller treatment of two of the projects, the new tank farm and the hotel refurbishment projects. They include:

- a discussion on the choice of planning technique;
- examples of some of the planning, monitoring and control techniques discussed in the body of the book.

A.2 TINGHAM

Tingham is a medium-sized metropolis set in the English heartland with easy access to many tourist attractions. It is served by road and rail transport although the roads in the area are very congested and there is a long-term plan to provide a ring road. There is also a small airport nearby which may be developed in the future. A sketch showing main roads and general layout of the town is shown in Figure A.1.

The area has mixed industry and commerce and because of this does not suffer from the highs and lows of the economic cycle as some

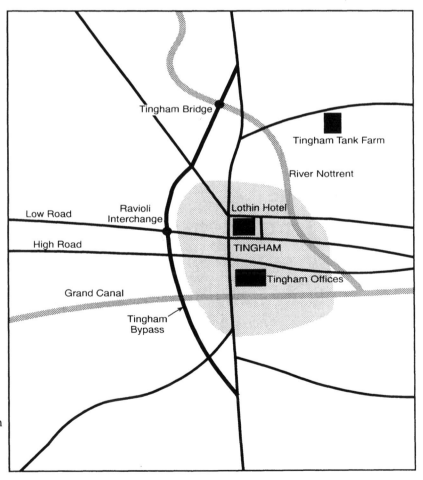

Figure A.1 Plan of Tingham town showing main roads and major projects.

other towns do. It has a steady stream of construction projects providing continuous employment for the skilled construction workers of the area.

A.3 TINGHAM BRIDGE

This project considers the construction of a small two-span bridge. One span is over land and is of *in situ* reinforced concrete, while the other span crosses the River Nottrent and is constructed using precast concrete beams. All precast beams are imported ready made. The substructure comprises *in situ* reinforced concrete and the foundations are driven steel piles. Access is available to both sides of the river at all times. Sketches of the construction are shown in Figure A.2.

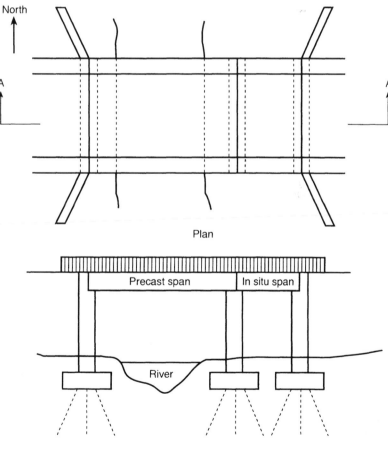

Figure A.2 Sketch of Tingham bridge project.

The project is to be constructed for Tingham Town Council, a local authority client, by Amo Construction plc. The design has been produced for Tingham Town Council by Diligo Consultants. (Readers with a classical education may see hidden meanings to the names of the construction company and the consultant. In fact the contractor's name simply reflects the names of the book's authors; the consultant's name reflects a Latin equivalent to *amo* to show the total impartiality of the authors.)

Two scenarios are considered for the completion of this bridge.

Scenario 1   The bridge is to be considered as a complete project in its own right.

Scenario 2   The bridge is to be considered as part of the Tingham Bypass project which is introduced below and contains structures which are similar to the Tingham Bridge.

A.4   TINGHAM BYPASS

This project is for the construction of 8 km of dual carriageway road with 12 structures. It is illustrated on the plan, Figure A.1. The project was designed to be let in three separate contracts and the one illustrated in detail in the text is the centre section. This is 1.5 km long with bridges at the Grand Canal, the High Road and the complex Ravioli interchange. The project is essentially linear in nature. Amo is again the contractor.

A.5   TINGHAM OFFICES

This project is for the construction of a 10 storey block of offices. The building is an *in situ* reinforced concrete frame. It is a relatively simple project as the foundations have already been constructed under a separate contract. This project has extensive repetition of structural work (columns, walls and slabs), building services and finishings on every floor. Amo has secured this contract as well.

A.6   TINGHAM TANK FARM

This project comprises the construction of a number of underground reinforced concrete tanks, a valve house, a control and administration building, roadways, car park and pipework. An existing pumping station is located approximately 1150 m from the valve house. It is yet another Amo project. Further details appear in Chapter 15.

A.7   TINGHAM HOTEL

The Tingham Hotel was built in the late 1960s in an attractive part of Tingham and was then the best hotel in the area. It is now showing its age and the owners have decided on a complete renovation of all facilities to provide once again a first class hotel. Amo has amazingly secured another project. Further details appear in Chapter 16.

Appendix B

UNCERTAINTY IN ACTIVITY SEQUENCE

B.1 REVIEW OF TECHNIQUES

When the uncertainty problem in planning was first tackled by Eisner it was suggested that there was a need to include a decision capability within networks. Working with activity-on-node networks, a simple decision box could be included with little or no problem. The network in Figure 8.2 illustrates the use of such a decision box to produce a probabilistic network for Tingham Offices (Appendix A, Section 2.4 and Section 8.1).

Although allowing the network to be drawn for an uncertain process, the technique has problems in use. Consider, for example, the situation in the example network. In the normal planning process, following the production of the network and the estimation of the activity durations, the earliest and latest start and finish times of the activities would be calculated using the forward and backward pass and the critical path identified (Section 5.2). The decision box in Figure 8.2 means that the earliest start time of the *construct superstructure* activity cannot be calculated and therefore such networks cannot be used in the same way and for the same purposes as conventional ones.

The solution, which is basically an extension of network-based planning techniques, was originally suggested in 1963 by Elmaghraby. He built on previous work by attempting to overcome some of the problems inherent in planning when decisions are required. The method has since been extended and now appears in essence in some national standard documents. The approach of generalised networks offers help in decision making at an early stage of a project and might thus be considered to be a planning technique. Its uses are, however, potentially greater in decision making and risk management than in traditional project planning.

B.2   STRUCTURE OF GENERALISED NETWORKS

Generalised networks are of the activity-on-arrow type with arrows representing the jobs running into and out of nodes or events. These nodes represent decisions and are made up of two distinct halves: the *emitter* out of which come all activities from the node and the *receiver* into which go all activities entering the node.

There are three types of receivers. The *and*, the *inclusive or* and the *exclusive or*. These are illustrated in Figure B.1.

The AND receiver   This is represented in networks as a semicircle with the arrows (or activities) coming into the curved part. Figure B.1(a) means that event 4 cannot happen until *activity a*, *activity b* and *activity c* have all been completed. This is the normal job-on-arrow network or arrow diagram event receiver.

The INCLUSIVE OR receiver   This is represented as a triangle pointing back towards the start of the network with all the activities which enter the receiver entering the point of the triangle, as shown in Figure B.1(b), which means that event 4 can happen if *activity a*, or *activity b* or *activity c* or any combination of *activities a, b* and *c* have been completed.

The EXCLUSIVE OR receiver   This is represented as a triangle pointing back towards the start of the network with a vertical line at that point. All the activities which enter the receiver enter the point of the triangle. Figure B.1(c) indicates that event 4 can happen if *activity a* alone, or *activity b* alone or *activity c* alone has been completed. Combinations of *activities a, b* and *c* are not allowed.

Theoretically there are corresponding types of emitters for each of the three receivers (Figure B.2).

The three types of emitter are therefore:

The AND emitter   This is represented in networks as a semicircle with the arrows (or activities) coming out of the curved part. The diagram in Figure B.2(a) means that after event 1, *activity a*, *activity b* and *activity c* will occur. This is the normal job-on-arrow event emitter.

The INCLUSIVE OR emitter   This is represented in networks as a triangle pointing forwards towards the end of the network with all the activities which leave the emitter leaving the point of the triangle. The diagram of Figure B.2(b) means that after event 1, *activity a*, or *activity b* or *activity c* or any combination of *activities a, b* and *c* will occur. An INCLUSIVE OR emitter can cause problems in network diagramming and considerable care should be taken if it is required.

The EXCLUSIVE OR emitter   This is represented in networks as a triangle pointing forwards towards the end of the network with a vertical line at that point. All the activities which leave the emitter leave the point of the triangle. Figure B.2(c) indicates that after event 1, *activity a* alone,

(a)

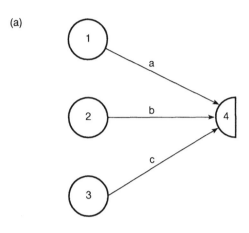

(b)

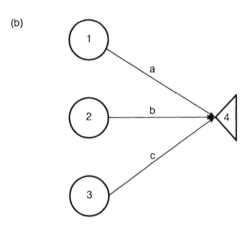

(c)

Figure B.1 Types of receiver:
(a) AND;
(b) INCLUSIVE
OR;
(c) EXCLUSIVE
OR.

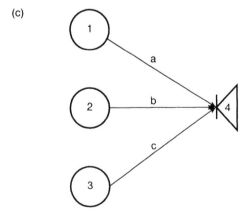

(a)

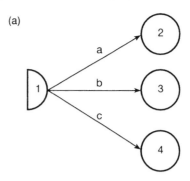

(b)

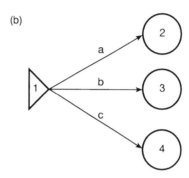

(c)

Figure B.2 Types of receiver:
(a) AND;
(b) INCLUSIVE
OR;
(c) EXCLUSIVE
OR.

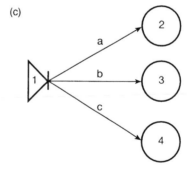

or *activity b* alone, or *activity c* alone will occur. Combinations of
activities a, b and *c* are not allowed.

Using these ideas, it is possible to draw networks to represent an
uncertain process. Figure B.3 shows such a network for a simplified
construction project.

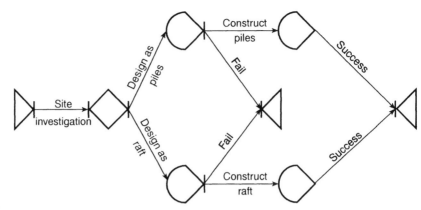

Figure B.3 Generalised
network for
a small project.

B.3 CALCULATIONS USING GENERALISED NETWORKS

Because of their uncertain nature, it is not possible to perform
conventional network analysis (Section 5.2) even when durations are
assigned to each of the activities. However, on assigning a probability of
occurrence to each activity (in addition to a duration), it is possible to
evaluate the likelihood of each possible outcome of the project and the
expected time to achieve the outcome.

The calculations of likelihood of different outcomes and their
expected durations can be achieved in a number of ways. The simplest is
described here. It is the one which is most likely to be employed by a
computer program and the most straightforward to explain. In this
method, the required results are achieved by reducing the networks to
their simplest form (Section B.4). This would be where all the paths from
the start to a particular end node are combined and the resulting path
reduced to a single arc with the properties of the combined paths.

For example, a three-ended network is reduced to the network
shown in Figure B.4, where arcs a, b, and c have probabilities of
occurrence p_a, p_b, p_c and durations (if they do occur) of t_a, t_b and t_c
respectively.

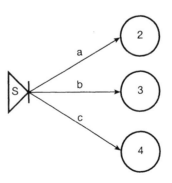

Figure B.4 The reduction
of a three-ended
network.

Obviously if node S is not itself to be an end to the project, then

$$\sum p_i = 1$$

where the sum operates over all the possible outcomes. In the network in Figure B.4, this would be

$$p_a + p_b + p_c = 1$$

Similarly at any intermediate node in a network, the sum of the probabilities out of that node must also be unity if it is not to be an end. As mentioned earlier, there are potential problems with INCLUSIVE OR emitters and this may not always be possible. Care should be taken when defining the activity probabilities so that the person doing it also defines the exact meaning of the probability. For example, is it the probability of the activity occurring at all after the event or the probability of it occurring by itself after the event. The definitions will affect the method of calculation employed. In all the examples given here, the probabilities are the probabilities of the activities occurring at all after the relevant event; this is the usual definition. Using this definition following an INCLUSIVE OR emitter implies that the emitter can always be an end to the network, a situation which is potentially confusing and should be avoided whenever possible.

In practice, when drawing generalised networks, end nodes should be included explicitly and the probabilities out of nodes checked for their summation to unity.

B.4 REDUCTION OF NETWORKS

Reduction of the networks to their simplest forms can be done by many methods. For example, in simple networks the obvious method is to evaluate the probability and duration of each and every path through the network from the start to the individual ends. These are then combined using statistical expectation.

In realistic project networks, however, this is a very difficult technique to employ because of the difficulty in recognising all the individual paths. Reduction is therefore done by a process of successive simplification in which several arcs of the network are reduced to one. One of the ways to do this is to consider the several simple elements of which all networks must be composed. Throughout, assume the probability of *activity a* occurring at all from a node is p_a and its duration, should it happen, t_a.

It is not practical to cover here all the situations which could arise in practical networks (and even if it were, it would not be sensible to do so). Each combination or reduction should, wherever possible, be evaluated

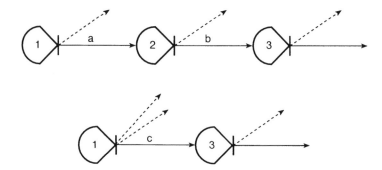

Figure B.5 Arcs in series and their reduced representation.

from basic probability theory and equations and formulae relied on as little as possible.

In these examples, special attention should be paid to the forms of the emitters and receivers.

B.4.1 Arcs in series

Arcs in series are shown in Figure B.5 together with the reduced representation. Essentially arcs a and b can be reduced to arc c as shown, where the probability of c occurring is

$$p_c = p_a + p_b$$

and the expected lag of node 3 after node 1 is

$$t_c = t_a + t_b$$

B.4.2 Arcs in parallel

Here there are three different possible situations and trying to apply formulae can cause trouble if the model used is not correct. First the difficult situation referred to above (Section B.3) is presented, namely the use of INCLUSIVE or emitters. The sum of the probabilities on the arrows leaving these may not sum to unity because of the probability of them occurring together or not at all. Figure B.6 shows arcs a and b in parallel and their reduced representation for EXCLUSIVE OR receivers, where:

$$p_c = p_a + p_b - 2p_a p_b$$

This is made up from the probability of a occurring by itself

$$= p_a - p_a p_b$$

plus the probability of b occurring by itself

$$= p_b - p_a p_b$$

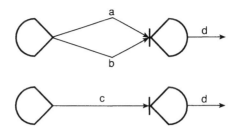

Figure B.6 Arcs in parallel and their reduced representation (EXCLUSIVE OR receiver).

The expected lag of node 2 after node 1 is given by

$$t_c = p_a t_a + p_b t_b - p_a p_b (t_a + t_b)$$

Arcs in parallel with INCLUSIVE OR receivers are shown in Figure B.7 together with their reduced representation, where p_c, using the same reasoning as above, is given by

$$p_c = p_a + p_b - p_a p_b$$

Here the expected lag of node 2 after node 1 is given by

$$t_c = \max(t_a, t_b) p_a p_b + t_a(p_a - p_a p_b) + t_b(p_b - p_a p_b)$$

$$t_c = p_a t_a + p_b t_b - p_a p_b \min(t_a, t_b)$$

Figure B.8 shows arcs in parallel and their reduced representation for the case of AND receivers. Both a and b must occur for node 2 to happen and

$$p_c = p_a p_b$$

The expected lag of node 2 after node 1 is given by

$$t_c = \max(t_a, t_b)$$

If the emitter in these three cases (Figures B.6–8) is changed to an EXCLUSIVE OR emitter, then $p_a p_b = 0$. This alters the solution for the INCLUSIVE OR receiver, and the AND case is not possible.

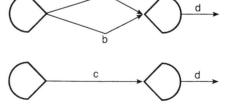

Figure B.7 Arcs in parallel and their reduced representation (INCLUSIVE OR receiver).

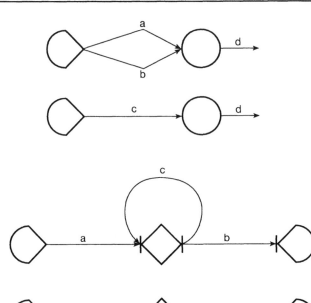

Figure B.8 Arcs in parallel and their reduction (AND receiver).

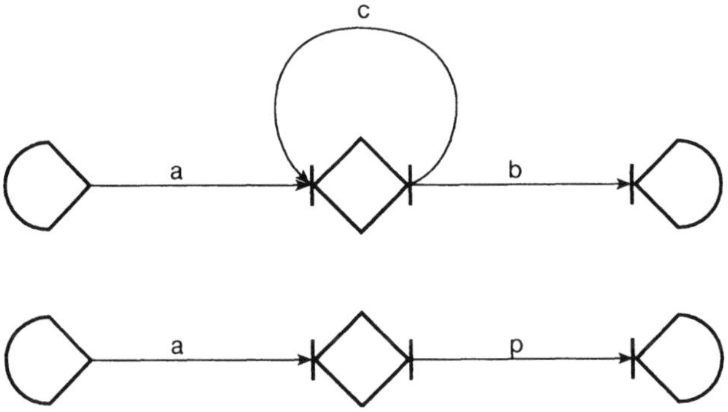

Figure B.9 Reduction of un-limited self-loop.

Should the emitter be an AND type, then the probability of each occurring on its own is zero, and the solutions hold for the EXCLUSIVE OR receiver which is impossible.

These basic elements are complemented by the addition of two further situations which are used to model a repetitive scenario (procedure followed by inspection).

B.4.3 Unlimited self-loop

Figure B.9 shows the reduction of an unlimited self-loop, a situation which can arise if a procedure allows any number of inspections of a process.

The probability of n loops occurring and then b is

$$p_b p_a^n$$

and the lag will be

$$n t_a + t_b$$

The probability of node 3 occurring at all in n loops or less (p_e) is:

$$p_e = \sum_{i=0}^{i=n} p_b p_c^i$$

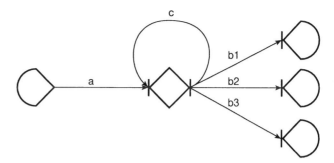

Figure B.10 Several arcs following a self-loop.

that is,

$$p_e = p_b\left(\frac{1 - p_c^{n+1}}{1 - p_c}\right)$$

Since $p_c < 1$, as n tends towards infinity, the probability becomes

$$p_e = \frac{p_b}{1 - p_c}$$

If $p_b = 1 - p_c$ then $p_e = 1$ which means that node 3 must occur eventually. Similarly, the expected lag (t_e) is given by:

$$t_e = \sum_{i=0}^{i=n}(it_c + t_b)p_b\, p_c^i$$

$$= p_b t_c \sum_{i=0}^{i=n} i p_c^i + p_b t_b \sum_{i=0}^{i=n} p_c^i$$

$$= \frac{p_b t_c p_c}{(1 - p_c)^2} + \frac{p_b t_b}{1 - p_c} \quad \text{for } p_c < 1$$

$$t_e = \frac{p_b}{1 - p_c}\left\{\frac{p_c t_c}{1 - p_c} + t_b\right\}$$

When $p_b = 1 - p_c$ this reduces to

$$t_e = \frac{p_c t_c}{1 - p_c} + t_b$$

In the situation where more than one arc follows a self loop, the loop acts equally on each and the results apply as shown in Figure B.10.

$$P_{e_i} = \frac{p_{b_i}}{1 - p_c}$$

$$t_{e_i} = \frac{p_{b_i}}{1 - p_c}\left\{\frac{p_c t_c}{1 - p_c} + t_{b_i}\right\}$$

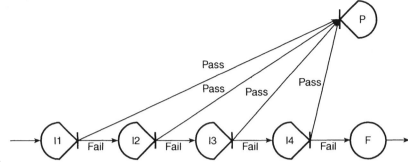

Figure B.11 Limited self-loop.

B.4.4   Limited self-loop

Should the number of inspections be limited, the element is called a limited self-loop (Figure B.11). This can be drawn using the elements described earlier, which can be reduced as described in Sections B.4.1 and B.4.2.

   If the number of inspections is limited to n, then the probability of reaching event 2 is

$$\sum_{i=0}^{i=n} p_{\text{fail}}^{i} \, p_{\text{pass}}$$

with lag

$$= p_{\text{p}} t_{\text{p}} \frac{1 - p_{\text{f}}^{n+1}}{1 - p_{\text{f}}} + p_{\text{p}} t_{\text{f}} \frac{p_{\text{f}}}{(1 - p_{\text{f}})^{2}} [1 + n p_{\text{f}}^{n+1} - (n + 1) p_{\text{f}}^{n}]$$

Similarly, the probability of event $3 = p_{\text{p}}^{n}$ with lag

$$= n t_{\text{f}}$$

As n tends to infinity, this tends towards an unlimited self loop.

B.5   COMMENTS ON GENERALISED NETWORKS

Generalised networks provide a means of modelling uncertain processes and evaluating them in terms of duration, cost, value or any other linear measure. As such they are excellent decision aids at the initial phase of a project to determine whether or not to sanction the work. They do not provide a way of planning a project in detail and planners should not attempt to use them as such. They do not provide a schedule of dates for the activities and cannot easily be used to give resource or financial loadings throughout the period of the work.

Several techniques exist for the reduction of generalised networks. The basic principle which it is advised to follow is that each combination should, wherever possible, be evaluated from basic probability theory and equations and formulae relied on as little as possible.

BIBLIOGRAPHY

This bibliography includes a selection of books, papers and articles which have been chosen for inclusion because they give a broader insight into certain aspects of planning and control. They are grouped into general categories which reflect the layout of the book. Some of them focus on a very narrow area; some of them could be in more than one category. They have all been useful at times to the authors of this book. The list is by no means exhaustive and many good works have undoubtedly been omitted; some of the works are now out of print and can be traced only through reference libraries. The last category covers a rapidly changing field and can only give a brief view of an exciting research area; works relating to all the other categories are included in it.

PLANNING TECHNIQUES

Arditi D. and Albulak M. Z. (1979). Comparison of network analysis with line of balance in a linear repetitive construction project. In *Proceedings of the 6th INTERNET congress*, Garmisch Parten Kirchen, Germany

Askew W. H., Mawdesley M. J. and Patterson D. E. (1995). Computer Aided Generation of Earthwork Activities in Project Planning, Civil-Comp 1995. In *Proceedings of the 6th International Conference on Civil and Structural Engineering Computing, Developments in Computational Techniques for Civil Engineering*, Cambridge, September, pp 17–23

Battersby A. J. (1970) *Network analysis for planning and scheduling* (3rd edn). Basingstoke: Macmillan

Birrell G. S. (1980). Construction planning – beyond the critical path. *ASCE Journal of the Construction Division*, **106**, 389–497

Booth J. A. N. (1993). The evaluation of project models in construction. *PhD Thesis*, University of Nottingham

Booth J. A. N., Askew W. H. and Mawdesley M. J. (1990). Evaluating the completeness of project plans – an automated approach. In *Proceedings of the 7th International Symposium on Automation and Robotics in Construction*, Bristol, 455–462

British Standards Institution (1987). *Glossary of terms used in project planning and control* (BS 4335). London: BSI

Burman P. J. (1972). *Precedence networks for project planning and control.* McGraw Hill

Clark-Hughes J., Mawdesley M. J. and Askew W. H. (1995). Construction Project Planning – Strategies For Clients And Their Representatives. In *Proceedings of the 1st International Conference on Construction Project Management*, Singapore, January, 365–372

Elmaghraby S. E. (1977). *Activity networks: project planning and control by network models.* Chichester: Wiley

Fischer W. A. (1975). Line of balance obsolete after MRP?, *Production inventory management*, Part 4, October

Halpin D. W. and Riggs L. S. (1992). *Planning and analysis of construction operations.* Chichester: Wiley

Handa V. K. and Barcia R. M. (1986). Linear scheduling using optimal control theory. *ASCE Journal of the Construction Division*, **112**

Jaafari A. (1984). Criticism of CPM for project planning analysis. *ASCE Journal of the Construction Division*, **110**

Lees J. L. (1991). Planning and control of linear construction projects. *PhD Thesis*, University of Nottingham

Lockyer K. G. and Gordon J. H. (1996). *Project management and project network techniques*, 6th edn, London: Pitman Publishing

Lumsden P. (1968). *The line of balance method.* Pergamon, Oxford

Mansour M. W. (1982). Full precedence networks for construction planning, *PhD Thesis*, University of Nottingham

Mawdesley M. J., Askew W. H., Lees J. L., Stevens C. S. and Taylor J. (1989). Computerization of time-chainage charts for the planning and control of linear construction projects. In *Proceedings 6th Conference on Computing in Civil Engineering*, ASCE, Atlanta

Morgan D. (1986). Site layout and construction planning. *PhD Thesis*, University of Nottingham

O'Brien J. J. (1975). VPM – scheduling for high-rise buildings. *ASCE Proceedings of the Construction Division*, **101**, 895–905

Patterson D. E., Mawdesley M. J. and Askew W. H. (1995). Site Layout for Road Construction Projects. In *Proceedings of the Annual Conference of the Canadian Society for Civil Engineering*, Ottawa, June, **III**, 349–357

Stradal O. and Cacha J. (1982). Time space scheduling method. *ASCE Journal of the Construction Division*, **108**

Suhail S. A. and Neale R. H. (1994). CPM/LOB: new methodology to integrate CPM and line of balance. *ASCE Journal of Construction Engineering and Management*, **120**(3)

Waldron A. J. (1976). Precedence diagramming – the great time robbery! In *Proceedings 3rd INTERNET Conference*, Stockholm

Willis E. M. (1986). *Scheduling construction projects*. Chichester: Wiley

RESOURCES AND PRODUCTIVITY

Abdul Ahafur N. H. (1984). Multi-project scheduling. *PhD Thesis*, University of Nottingham

AbouRizk S. M. and Halpin D. W. (1990). Probabilistic simulation studies for repetitive construction processes. *ASCE Journal of Construction Engineering and Management*, **116**(4)

Arditi D. (1985). Construction productivity improvement. *ASCE Journal of Construction Engineering and Management*, **111**(1), 1–14

Baxendale A. T. (1985). *Measuring site productivity by work sampling* (Technical Information Service Paper no. 55). Chartered Institute of Building, Ascot

Baxendale A. T. (1985). Site production efficiency. *Building technology and management*, October, 19

Burton F. M. (1986). Methodology for measuring construction productivity. *Transactions of AACE*, June, L3.1–L3.4

Carr R. I. (1979). Simulation of construction project duration. *ASCE Journal of the Construction Division*, **105**

Chang L. M. (1991). Measuring construction productivity. *Cost Engineering*, **33**(10), 19–25

Clapp M. A. (1980). Productivity on building sites. *Building Research Establishment News*, **51**, 17–18

Clark F. D. (1985). Labour productivity and manpower forecasting. *Transactions of AACE*, June, A1.1–A1.7

European Construction Institute (ECI) – Productivity taskforce (1994). *Total productivity management – guidelines for the construction phase.* ECI, Loughborough University of Technology

Hussain A. (1979). Construction productivity factors. *Engineering (USA)*, **105**(E14), 189–195

Ireland J. (1994). Improving construction site productivity. *Construction Industry Computing Association (UK) Bulletin*, September, 7

Jarle P. (1987). Gauging productivity in construction. *Building Research and Practice*, Feb/Mar, 43–49

Kohen E. and John M. (1989). Subcontractors – concern regarding productivity. In *Proceedings of Construction Congress 1*, ASCE, 278–283

Kohen E. and Brown G. (1985). International labour productivity factors. *Journal of Construction Engineering and Management*, **112**(2), 129–137

Mandel T. (1991). Determining labour productivity. *Cost engineering*, **33**(5), 21–23

National Economic Development Office (1989). *Promoting productivity in the construction industry.* London: NEDO, December

O'Neal R. P. (1989). Productivity – guess it or measure it. *Transactions of AACE*, June, R11.1–R11.3

Pascoe T. L. (1965). An Experimental Comparison of Heuristic Methods for Allocating Resources. *PhD Thesis*, Cambridge University

Rau N. A. (1988). Management of productivity in construction. *Journal of the Institution of Engineers (India), Part CI*, **69**(2), 113–116

Skoyles E. R. and Skoyles J. R. (1987). *Waste prevention on site.* London: Mitchell

Thomas H. R., Mathews C. T. and Ward J. G. (1986). Learning curve models of construction productivity. *ASCE Journal of the Construction Division*, **112**

Urie R. (1983). Learning and forgetting in construction. *BSc dissertation*, University of Nottingham

Vanegas J. A., Bravo E. B. and Halpin D. W. (1993). Simulation technologies for planning heavy construction processes. *ASCE Journal of Construction Engineering and Management*, **119**(2)

UNCERTAINTY AND RISK

Al-Bahar J. and Crandall K. (1990). Systematic risk management approach for construction projects. *ASCE Journal of Construction Engineering and Management*, **116**(3), 533–546

Chapman C. B. (1990). A risk engineering approach to project risk management. *International Journal of Project Management*, **8**(1)

Cooper D. F. and Chapman C. B. (1987). *Risk analysis for large projects: models, methods and cases.* Chichester: Wiley

Eisner H. (1962). A generalized approach to the planning and scheduling of a research program. *Journal of the Operations Research Society*, **10**(1), 115

Elmaghraby S. E. (1964). An algebra for the analysis of generalized activity networks. *Management Science*, **10**(3), 494–514

Malcolm D. G., Roseboom J. H., Clark C. E. and Fazar W. (1959). Application of a technique for research and development program evaluation. *Journal of Operations Research Society*, **7**(5)

Perry J. G. and Hayes R. W. (1985). Risk and its management in construction projects. *Proceedings of the Institution of Civil Engineers*, **78**, 499–521

Thompson P. A. and Norris C. (1993). The perception, analysis and management of financial risk in engineering projects. *Proceedings of the Institution of Civil Engineers*, **97**(1)

Van Slyke R. M. (1963). Monte Carlo methods and the PERT problem. *Operational Research*, **11**, 839–860

MONITORING AND CONTROL

Al Jibouri S. H. (1985). Cost models for construction. *PhD Thesis*, University of Nottingham

Association of Project Managers (1993). *Project management techniques*, APM Specific Interest Group on Integrated Cost and Schedule Control, Project, **5**(7), 10–11

Barnes M., ed. (1990). *Financial control*. London: Thomas Telford

Beaumont E. (1970). Construction monitoring techniques. *Architects Journal*, **152**, 1379–1382

Berny J. and Howes R. (1987). Project management control using growth curve models applied to budgeting, monitoring and forecasting within the construction industry. In *Managing Construction Worldwide, Vol 1, Systems for Managing Construction* (Langley P. P. and Harlow P. A., eds), 304–313, E & FN Spon

Lowe J. G. (1987). Cash flow prediction and the construction client – a theoretical analysis. In *Managing Construction Worldwide, Vol 1, Systems for Managing Construction* (Langley P. P. and Harlow P. A., eds), 327–335, E & FN Spon

Neal J. M. (1982). *Construction cost estimating and project control*. Prentice-Hall

Ninos G. E. and Wearne S. H. (1986). Control of projects during construction. *Proceedings of the Institution of Civil Engineers*, Part 1, August

Pearson N. (1975). *The control of subcontractors*. (Occasional Paper no 7). Chartered Institute of Building, Ascot

Price A. D. (1985). *Methods of measuring production times for construction work* (Technical Information Service Paper no 49). Chartered Institute of Building, Ascot

Rasdorf W. J. and Abudayyeh O. Y. (1991). Cost and schedule control integration: issues and needs. *ASCE Journal of Construction Engineering and Management*, **117**(3), 486–502

Teicholz P. M. (1987). Current needs for cost control systems. In *Project Controls: Needs and Solutions*, (Ibbs C. W. and Ashley D. B., eds), ASCE, 57–67

Thompson P. A. (1981). *Organisation and economics of construction*. McGraw-Hill

Wearne S. H. (1989). *Control of engineering projects*. London: Thomas Telford

Williams L. J. (1969). Techniques for the evaluation of a contractor's progress. In *20th International Conference and Convention*. Texas, May

Wyatt D. P. (1975). Monitoring materials on a housing site. *Building Trades Journal*, May 21–31

EVALUATING CHANGES

Bowers J. A. (1995). Criticality in resource constrained networks. *Journal of the Operational Research Society*, **46**(1)

Cree C. and Barnes M. (1989). Quantifying Disruption Costs. *Construction Law Journal*, **5**(4), 258–271

Diekmann J. E. and Nelson M. C. (1985). Construction claims: frequencies and severity. *ASCE Journal of Construction Engineering and Management*, **111**(1), 74–81

Dieterle R. and De Staphanis A. (1992). Use of productivity factors in construction claims. *Transactions of AACE*, June, C1.1–C1.7

Galloway P. D. and Nielson K. R. (1983). Schedule delay: a productivity analysis. In *Proceedings of the Project Management Institute, Annual Seminar Symposium*, PMI, Houston, USA, October, IJ.1–IJ.18

Halligan D. W., Demsetz L. A., Brown J. D. and Pace C. B. (1994). Action – response model and loss of productivity in construction. *Journal of Construction Engineering and Management*, **120**(1)

Householder J. L. and Rutland H. E. (1990). Who owns float. *ASCE Journal of Construction Engineering and Management*, **116**(1), 130–133

Jaafari A. (1984). Criticism of CPM for project planning analysis. *ASCE Journal of the Construction Division*, **110**, 222–233

Leonard C. A., Fazio P. and Moselhi O. (1988). Construction productivity: major causes of impact. *AACE transactions*, D.10.1–D.10.7

McLeish D. C. A. (1981). Manhours and interuptions in traditional house building. *Building and Environment*, **16**(1), 59–67

Mawdesley M. J. and O'Reilly M. P. (1994). The management of disputes: a risk approach. *Arbitration and Dispute Resolution Law Journal*, December, 72–79

Moselhi O., Leonard C. and Fazio P. (1991). Impact of change orders on construction productivity. *Canadian Journal of Civil Engineering*, **18**

O'Brien J. J. and Kreitzberg F. C. (1986). Network scheduling variations for repetitive work. *ASCE Journal of the Construction Division*, **112**

Reams J. S. (1989) Delay analysis: A systematic approach. *Cost Engineering*, **31**(2), February

Reams J. S. (1990). Substantiation and use of planned schedule in delay analysis. *Cost Engineering*, **32**(2), February

Revay S. (1984) Time extensions in construction contracts. *Construction Law Reports*, **6**, 253–260

Revay S. (1987) Calculating impact costs. *International Business Lawyer*, 400–409

Waldron A. J. (1976). Precedence diagramming – the great time robbery! In *INTERNET* (International journal of project management), September

ARTIFICIAL INTELLIGENCE

Askew W. H. and Mawdesley M. J. (1988). Site layout and an A.I. approach to planning. Updating the state of the art of civil engineering computing tools. In *Proceedings of the 3rd International Conference on Computing in Civil Engineering*, Vancouver, Vol 2, 413–420

Askew W. H., Mawdesley M. J. and Booth J. A. N. (1989). An A.I. approach to automating the recognition and evaluation of logic classes in project scheduling. In *Proceedings of the 6th International Symposium on Automation and Robotics in Construction*, San Francisco, 197–203

Askew W. H., Mawdesley M. J. and Booth J. A. N. (1992). Automating and integrating the management function. In *Proceedings of 9th International Symposium on Automation and Robotics in Construction*, Tokyo

Booth J. A. N., Askew W. H. and Mawdesley M. J. (1991). Automated budgeting for construction. In *Proceedings of the 8th International Symposium on Automation and Robotics in Construction*, IPA Stuttgart, 529–540

Carr V., Mawdesley M. J. and Askew W. H. (1995). The automated production of project networks from activity lists. In *Proceedings of the 1st International Conference on Construction Project Management*, Singapore, January, 391–398

Flood I. (1989). A neural network approach to the sequencing of construction tasks. In *6th International Symposium on Automation and Robotics in Construction*, San Francisco, June

Flood I. (1990). Simulating the construction process using neural networks. In *7th International Symposium on Automation and Robotics in Construction*, Bristol, June

Hendrickson C., Zozaya-Gorostiza C., Rehak D., Baracco-Miller E., and Lim P. (1991). Expert system for construction planning. *ASCE Journal of Computing in Civil Engineering* 5(4)

Ibbs C. W. and De La Garza J. M. (1988). Knowledge engineering for a construction scheduling analysis system, *Expert Systems in Construction and Structural Engineering* (Ed. Adeli H), London: Chapman & Hall

Kangari R. (1986). Application of expert systems to construction management decision making and risk analysis. In *Expert Systems in Civil Engineering* (Kostem C. L. and Maher M. L., eds), ASCE, New York, 78–86

Kartam N. A. and Levitt R. E. (1990). Intelligent planning of construction projects. *ASCE Journal of Computing in Civil Engineering*, 4(2)

Kartam N. A., Levitt R. E. and Wilkins D. E. (1991). Extending artificial intelligence techniques for hierarchical planning. *ASCE Journal of Computing in Civil Engineering*, 5(4)

Levitt R. E. and Kunz J. C. (1985). Using knowledge of construction and project management for automated planning and scheduling. *Project Management Journal*, **16**(5)

Mawdesley M. J. and Askew W. H. (1991). Automating project scheduling – a case study. Preparing for construction in the 21st century. In *Proceedings of Construction Congress '91*, American Society of Civil Engineers, Cambridge, USA, 360–365

Mawdesley M. J. and Carr V. (1993). Artificial neural networks for construction project planning. In *Neural Networks and Combinatorial Optimization in Civil and Structural Engineering* (Topping B. H. V. and Khan A. I., eds), Edinburgh: Civil-Comp Press, 39–47

Mawdesley M. J., Cullingford G. and Haddadi A. (1988). An approach to modelling movement around sites. In *Proceedings of the 5th International Symposium on Robotics in Construction*, Tokyo

Mawdesley M. J., Cullingford G. and Morgan D. C. (1987). An approach to automated project planning. In *Proceedings of the 4th International Symposium on Automation and Robotics in Building Construction*. Haifa

Moselhi O., Hegazy T. and Fazio P. (1991). Neural networks as tools in construction. *ASCE Journal of Construction Engineering and Management*, **117**(4)

Moselhi O. and Nicholas M. J. (1990). Hybrid expert system for construction planning and scheduling. *ASCE Journal of Construction Engineering and Management*, **16**(2)

Sirajuddin A. M. (1991). An automated project planner. *PhD Thesis*, University of Nottingham

Sirajuddin A. and Mawdesley M. J. (1989). A knowledge-based system for recognition of work required for construction. In *Proceedings of the 4th International Conference on Civil and Structural Engineering Computing*. London: Civil-Comp Press, 83–92

Sirajuddin A. and Mawdesley M. J. (1990). The automation of project resource scheduling using site layout mapping. In *Proceedings of the 7th International Symposium on Automation and Robotics in Construction*. Bristol Polytechnic, 535–542

Sirajuddin A. and Mawdesley M. J. (1991). A knowledge-based system for recognition of work required for construction. *Computing & Structures*, **40**(1), 37–44

Soh C. K., Phang K. W. and Wong W. P. (1993). An integrated intelligent system for scheduling construction projects. *Proceedings of Institution of Civil Engineers*, **93**, 156–162

INDEX